Proceedings of the 4th International Conference on

Robot Vision
and Sensory Controls

9-11 October 1984, London, U.K.   Edited by: Prof. A. Pugh

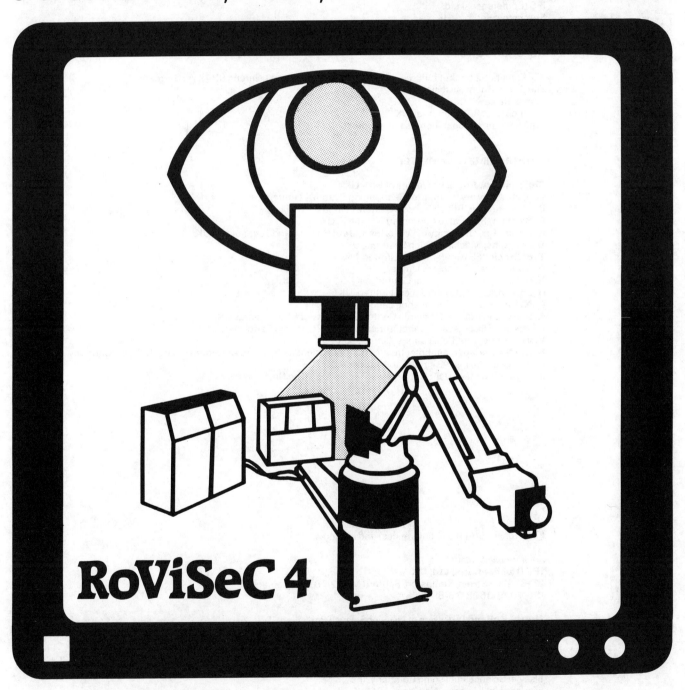

RoViSeC 4

Organised by: IFS (Conferences) Ltd, UK

Sponsored by: The Fraunhofer Institut fur Produktionstechnik und Automatisierung (IPA), W. Germany
The British Robot Association, UK
Sensor Review, UK
The Industrial Robot Journal, UK
British Pattern Recognition Association, UK

Co-published by: IFS (Publications) Ltd and North-Holland (a division of Elsevier Science Publishers BV)

Proceedings of the 4th International Conference on Robot Vision and Sensory Controls

An international event organised by:

IFS (Conferences) Ltd,
Kempston, Bedford, UK

CO-SPONSORS

The Fraunhofer Institut fur Produktionstechnik und Automatisierung (IPA), W. Germany
The British Robot Association, UK
Sensor Review, UK
The Industrial Robot Journal, UK
British Pattern Recognition Association, UK

PROGRAMME COMMITTEE

Chairman: A. Pugh, University of Hull, UK
W. G. L. Adaway, Computer Recognition Systems Limited, UK
B. M. Atkinson, British Robotic Systems Limited, UK
P. Bartlam, Integrated Photomatrix Limited, UK
B. Batchelor, University of Wales Institute of Science and Technology, UK
D. Braggins, Machine Vision Systems, UK
T. E. Brock, IFS (Conferences) Limited, UK
P. Davey, Meta Machines Limited, UK
K. Gibbs, IFS (Conferences) Limited, UK
R. Jennings, IFS (Conferences) Limited, UK
R. N. Kay, National Engineering Laboratory, UK
R. B. Kelley, Robotic Research Center, University of Rhode Island, USA
I. Powell, GEC Research Laboratories, Marconi Research Centre, UK
B. W. Rooks, IFS (Publications) Limited, UK
H.-J. Warnecke, Fraunhofer Institut fur Produktionstechnik und Automatisierung (IPA), W. Germany
A. Weaver, Philips Research Laboratories, UK
N. J. Zimmerman, Delft University of Technology, The Netherlands

Jointly published by
IFS (Publications) Ltd, UK
35-39 High Street, Kempston, Bedford MK42 7BT, England
ISBN 0-903608-75-8

North-Holland (a division of Elsevier Science Publishers BV)
PO Box 1991, 1000 BZ Amsterdam, The Netherlands
ISBN 0-444-87626-X

In the USA and Canada:
Elsevier Science Publishing Company, Inc.,
52 Vanderbilt Avenue, New York, NY 10017

Printed by: Cotswold Press Ltd, Oxford, UK

II

CONTENTS

Sensor Based Manufacturing

Vision Systems

Sensor Guided Welding

Three-Dimensional Sensing

Robot Guidance and Sensory Control

Non-Vision Sensing

Knowledge-Based Sensory Systems

Advanced Vision Techniques

Sensor Based Manufacturing II

Late Paper

Supplementary Papers

AUTHOR INDEX

V

D. McKeag. 55
Ulster Polytechnic, UK

C. Meier. 231
Siemens AG, W. Germany

Y. Mikami. 155
Tokyo Metropolitan Technical College, Japan

F. Monteil. 115
Ecole Normale Superieure de l'Enseignement
Technique, France

H. Mori. 477
Yamanashi University, Japan

D. H. Mott. 241
British Robotic Systems Ltd, UK

H. R. Nicholls. 241
University College of Wales, UK

L. Norton-Wayne. 65
Leicester Polytechnic, UK

C. J. Page. 123
Coventry Polytechnic, UK

G. J. Page. 1
British Robotic Systems Ltd, UK

K. Paler. 351
SERC Rutherford Appleton Laboratory, UK

D. Placko. 115
Ecole Normale Superieure de l'Enseignement
Technique, France

R. J. Popplestone. 371
University of Edinburgh, UK

J. M. Richardson. 261, 395
Rocwell International Science Center, USA

S. Sakane. 155
Ministry of International Trade and Industry, Japan

G. Sandini. 173
DIST-University, Italy

P. Saraga. 65
Philips Research Laboratory, UK

E.-J. Schmidberger. 27
IPA, W. Germany

J. S. Schoenwald. 261
Rockwell International Science Center, USA

P. Sholl. 209
Unimation (Europe), UK

Z. Smati. 91
Central Electricity Generating Board, UK

C. J. Smith. 91
Central Electricity Generating Board, UK

M. Straforini. 173
University of Genoa, Italy

S. Tachi. 251
Ministry of International Trade and Industry, Japan

K. Tanie. 251
Ministry of International Trade and Industry, Japan

C. J. Taylor. 325
Wolfson Image Analysis Unit (University of
Manchester), UK

V. Torre. 173
University of Genoa, Italy

A. N. Trounov. 285
Nikolaev Institute of Shipbuilding, USSR

W.-H. Tsai. 183, 379
National Chiao Tung University, Taiwan

Y. N. Tsveraidze. 487
Tbilisi State Medical Institute, USSR

A. Verri. 173
University of Genoa, Italy

J. M. Vranish. 269
National Bureau of Standards, USA

R. P. Walduck. 123
Coventry Polytechnic, UK

N. Wittels. 103
Automatix Inc., USA

D. Yapp. 91
Central Electricity Generating Board, UK

B. Yin. 371
Beijing Institute of Aeronautics, Peoples Republic of
China

V. D. Zotov. 487
Institute of Control Sciences, USSR

VI

FOREWORD

Professor Alan Pugh
Robotics Research Unit
University of Hull

SINCE its inception in 1981 the RoViSeC conference series has attracted over 1,000 delegates establishing it as the leading international conference devoted to intelligent manufacturing. The previous three events held in the UK, Germany and the USA have been noted for their ability to integrate the needs of the modern factory with the availability of advanced sensing techniques. This formula has attracted manufacturing managers and practising production engineers to discover for themselves the techniques likely to assist in the survival and prosperity of their companies. At the same time engineers and scientists involved in advanced manufacturing R&D have benefitted from the exchange of information with their fellow researchers as well as with those at the hard end of applying the advanced methods.

This year's conference continued the general theme of educating and introducing to industry advanced manufacturing methods based on vision and other sensing techniques. Now, industry is itself contributing to the application of these techniques and delegates were able to hear from several companies with experience of applying sensor based methods that were R&D topics when the first RoViSeC conference was held in 1981.

Whilst demonstrating the practical applicability of many sensing methods further advances in the subject will not be neglected. Software is a key feature in these developments which will be introduced to industry over the next five years. Any executive whose responsibility encompasses manufacturing should be making him or herself aware of these developments if his/her company is to first survive and then be profitable in the second half of the decade. The RoViSeC conference provided them with that opportunity.

Complementing the conference was the Machine Intelligence Show, the first exhibition devoted to the application of advanced computer and sensor based systems to manufacturing industry. Delegates had the opportunity to see and discuss with vendors some of the equipment and services being presented in the conference. They were able to compare the what's-available-now with the what's-to-come-in-future, thus providing a unique opportunity to come to terms with 'manufacturing of the future'.

SENSOR BASED MANUFACTURING

Vision driven stack picking in an FMS cell

G. J. Page
British Robotic Systems Limited, UK

ABSTRACT

The paper describes an implementation of a vision processing algorithm to perform robot stack picking in an FMS cell. The paper describes the stages of the algorithm from image collection to coordinate selection in detail.

THE PROBLEM/INTRODUCTION

The 'bin picking' problem as it is known has been around for some time as one of the classic problems in vision processing. The problem is for a mechanical device, such as a robot, to pick the top or most easily accessible component from a completely disordered pile within a bin, and to present the component to whatever tool or process requires it. The sensing involved can use visible light, or other forms like range finders. The reason this problem has received so much attention is that there are many practical applications in industry where bin picking and allied techniques could be applied to improve automation. It has been suggested that in the fully automated world of the future, robots will have packed the bin in the first place, hence there will be no need for any 'visual' sensing. This is an unrealistic view. Firstly, parts are placed in bins to allow them to be transported. If the parts are not held in place by jigs then they will become disorientated fairly easily. Therefore to preserve position the parts have to be transported in special containers or containers with special jigs, this raises the cost of transport and also makes the packing and unpacking more complex as both part and jig have to be unpacked. Secondly, there are examples where the manufacturing process itself produces unordered bins of parts.

Thirdly, without visual inspection the system is very inflexible and highly prone to silly errors of mechanical placing and alignment

Thus the 'bin picking' problem will have a relevance in many places and for a significant time to come.

This paper is not about the solution to the general bin picking problem. It describes some of the problems encountered, and the eventual solutions to those problems, when BRSL implemented a stack picking system.

1.1 Stack Picking

The generalised bin picking problem has far too many degrees of freedom to allow it to be solved with the hardware currently available within a time that is acceptable in a flexible manufacturing cell. The obvious approach to take from an applied scientific point of view is to reduce the number of degrees of freedom by constraining the problem.

The most constrained that the problem can become is to have the components in precisely defined positions and orientations. This completely constrained solution is being used in a number of cases where parts are automatically loaded and unloaded from pallets. However, as mentioned in the introduction it offers no flexibility whatsoever, and may have a cost penalty.

If we are to relax the stringent constraints, to produce a more flexible system, then it is necessary to choose a system that is sensible both from the vision processing and the robot handling points of view. Fortunately these two factors work with each other, and the constraints which were applied to simplify the vision processing also simplified the robot handling problem.

The first constraints to be applied were those relating to orientation. These were removed by stating that the parts were to be in layers, that is parallel to the bottom of the bin. It was also stated that the parts were to be circular, which implied that they have no specific orientation. The circular parts also help with other properties as we shall see later.

Thus the problem was reduced to finding the centres of stacks of circular parts which will be touching and may be packed very closely. The height of all these stacks should be uniform to assist with the robot unloading sequence.

1.2 The Hardware

This stack picking system was part of the vision sensing installed by BRSL for an FMS cell. As well as stack picking the vision system was responsible for checking that the parts were in the correct position prior to it being loaded into a CNC lathe and also to orientate semi-machined parts.

The main computer was a DEC PDP 11/23 with Winchester disk as backing store. Communication to the robot was via a serial link using the DDCMP protocol. Digital I/O lines were used to synchronise the vision system to the cell controller. The operator communicated with the vision system via a keypad and monitor. A hardware diagram of the whole system is shown in Figure 1. The camera was a standard 128 x 128 solid state array. The initial investigation into the problem using AUTOVIEW VIKING (BRSL's image prototyping system) demonstrated that using a 128 x 128 area produced sufficient resolution to obtain the desired accuracy.

The advantages of using a small array are:

o the camera technology is tried, proven and available

o the same processing over a 128 x 128 area is 4 times faster than a
   256 x 256 image.

o the image area does not have to be memory mapped, thus reducing the
   cost and complexity of the surrounding system.

o pixel speeds allow direct digitisation into memory without intermediate buffering.

The disadvantages are:

o pixel sizes are large (typically 7mm) this affects the minimum
   achievable resolution.

o the processing has to be more complex due to the coarser image.

As with all vision processing problems, the lighting played a large
part in the eventual success of the system. This system was no exception and several different lighting rigs were tried before selecting
the configuration actually used. The intention was to produce a
uniform illumination over the whole bin area from the top to the bottom
of the bin. The final arrangement was 4 wide angle spot lights
surrounding the camera which was at a height of 2 metres above the bin
floor. The lights were mounted in housings which were designed to
allow fine adjustments to be made. Lighting arrangements that were
rejected included various arrangements of fluorescent strip lighting
and reflectors.

2. VISION PROCESSING

The vision processing algorithm was broken down into several stages,
each of which will be described. These stages were:

o preprocessing
o feature location
o back-end processing
o adding intelligence
o final checking

These stages of processing and the intermediate results are shown
in Figure 2 and Photos 1-6.

2.1 Preprocessing

The aim of this stage of the processing was to reduce the grey level
image to a binary image suitable for the next stage. This was the most
difficult stage to develop and the algorithm may not be of general
applicability but local to these particular sets of components where all
the components were brighter than the bin bottom. The algorithm had
to give a single pixel wide line that ran around the edges of all the
parts. Various gradient operators were tried such as Sobel Edge and
Roberts Gradient, but none of these were found suitable over the
wide range of circular parts under consideration.

The final algorithm was arrived at empirically rather than by any
theoretical consideration.

A high pass filter was applied to the original image of the parts.

Then all the negative parts of this image were thresholded out, that is, all the negative areas were set to black, all the positive areas to white. This left lines around the inside edge of the parts, as the parts are all significantly brighter than the bin bottom. Although this highlighted the edges it also left a very noisy image. The noise was reduced by only keeping points with three or more neighbours and also removing connected regions that were small.

This had the effect of removing a lot of the noise whilst retaining the essential information.

The completion of the preprocessing left a binary image of the edges of the parts. Typically this would form a long web as opposed to discrete part edges.

2.2 Feature Extraction

Having obtained a binary edge image the next stage was to reduce this to a single feature for each part. This was where the constraint of circular parts was used. The centre of a circle can be found if three points on the circumference are known. The technique that was used started three "bugs" off around following the binary edge. These bugs were a fixed distance apart and reported their position every move. From these three positions the centre of the circle and the radius, on which these three points must lie, could be found. This is shown in Figure 3. If the radius was found to be within predetermined limits of the expected radius of the image of the part, then a marker was incremented at the circle centre. When the whole edge had been followed the resulting image consisted of a map of possible centre 'hits'. The higher the number of hits, the greater the possibility that there was a part centre in that position.

2.3 Back-End Processing

We were now left with an image containing clusters of points with values ranging from 1 to about 10 which represented the 'hits' found by the centre finder. The algorithm now had to reduce these clusters to a single point that represented the best estimate of the centre of the original circle. Various techniques were tried. The one that was finally selected coagulated all the hits that were adjacent to each other and then found the medium of these groupings. The total number of hits associated with this cluster was also stored.

2.4 Adding Intelligence

We now had a fairly robust algorithm of preprocessing, circle finder and coagulation. However it still gave some inconsistent results. In particular there were centres that were close together with the coagulation did not work properly. This was solved by adding a small degree of intelligence. The original problem constraints specified circular parts that were not overlapping. For uniform circles this implies that adjacent centres are at least one diameter apart, because the centres of two touching circles are separated by the sum of their radii. Thus we took the image from the coagulation stage, found the maximum intensity, that is the centre with the most 'hits', marked this as a definite centre and masked out a circle with the diameter of the parts as its radius. This had the effect of removing any misclassification of clusters.

2.5 Final Checking

The above algorithm gave results that were accurate to ± 17mm for a lm square bin (this represents an uncertainty of ± 2 pixels). To increase the accuracy a further stage that would give sub-pixel accuracy was required. This extra module was supplied with an estimate of the centre of a part that was correct to ± 2 pixels from the above 4 stages. The module had to return a corrected coordinate that was accurate to within ± ½ a pixel, thus giving an accuracy of ± 4mm for the centre of the part.

The main outline of the algorithm, for each part within the pallet, was as follows:

A circular window was set up around the part of interest to reduce the amount of processing required. A radial gradient operation was then performed, centred on the estimated centre of the part. This highlighted the edge of the circular part very well. This image was then used for subsequent processing. The next objective was to find the best approximation to the edge of the part. The properties of the circles again come to the rescue as we were able to predict the expected area and perimeter. When the best edge had been found the centre of the area enclosed by it was determined by standard centre of gravity techniques. Theoretical considerations showed that the centre of gravity of a circle of radius 5 pixels can be determined to better than 0.2 pixels, which is well within the required accuracy to ± 0.5 pixels.

3. USING THE VISION DATA

The algorithm that has been described above is only one part of guiding the robot to pick up the parts from the stacks. Other factors that led to the eventual solution include:

o calibration between the robot and vision
o correction for height of stacks

3.1 Calibration

The processing that has been defined above returned a result that was in pixel coordinates. To be of any use these coordinates had to be translated into robot coordinates. The procedure adopted was a simple one. A large white disk with a known area was presented to the vision system. The vision system then chose a threshold value that gave a binary image in which the largest blob had the pre-taught area. The centre of gravity of this blob was then found. At the same time the robot coordinates were determined by fixing a pointer to the robot arm in place of the gripper and positioning it above the centre of the disk. In this way four sets of coordinates were found in both the vision and the robot world. It was then a matter of solving a number of sets of simultaneous equations to determine a homogenous transform that could be applied to the vision coordinates to produce robot coordinates.

3.2 Correction

All measurements at this stage (and coordinates) had been made with respect to the bottom of the bin. The final correction was for parallax, necessary as the stacks of parts were of different heights. This was simply corrected by the ratio of the distance of the bin bottom to the camera and the camera and the top of the current stacks. As the vision system had no depth information available the number of parts in the stacks had to be communicated to the vision system prior to the bin picking operation being started.

4. THE WIDER SCENE

I have described in some detail the approach that was taken in solving one particular subset of the bin picking problem. What we have solved is for a simple system where the mechanical constraints imposed by the gripper and the robot were coincident with those that were required by the vision system, that is random stacks. The arm and gripper would only be capable of picking a core of parts in the centre if total randomness were allowed. What this system has given us is a thorough grounding in all the fundamental problems of illumination, communication, calibration etc that have to be solved before more complex vision processing is necessary or sensible.

We will use this experience to expand from this very constrained case to the more general. This migration will be performed in stages going from the current situation of touching layers of identical parts to total randomness with different parts. It is envisaged that the early stages of development will build on what we have already. The first area to attack will be the preprocessing and the edge following will probably be performed directly in the grey level image rather than transforming it to a binary image.[2] Another simple area will be the objects that can be recognised by generalising the shapes from circles to other shapes. At all stages of development the lighting and gripper technology will have to be developed to be able to make use of the increased flexibility of the vision processing.

At some stage in the near future we will come to a limit of processing capability using a conventional serial processor. Hardware modules like the Linear Array Processor will then be necessary to perform the base level processing similar to that described above in far shorter times.

These hardware modules will be driven by higher level routines that will have a software model of the parts that are being picked from the bin. Work is already being done that would allow these techniques to be developed.[3]

REFERENCES

1.   Batchelor, B G, Mott, D H, Page G J, Upcott, D N
     'The AUTOVIEW Interactive Image Processing Facility'
     Peter Peregrinus, Sept 1982

2.   Ballard and Brown, 'Computer Vision' Chapter 11
     Published by Prentice-Hall Inc ISBN O-13-165316-4

3.   David Hogg, 'Model-Based Vision: a program to see a walking
     person", Image and Vision Computing, Vol 1, Number 1, pp5-20
     (Feb 1983)

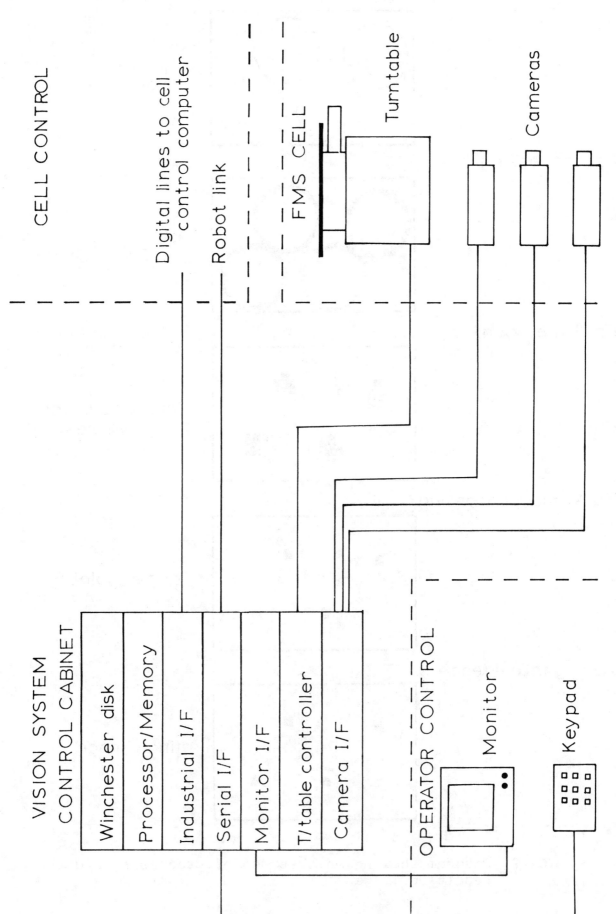

FIG 1. Hardware Components of Vision System in the FMS Cell

7

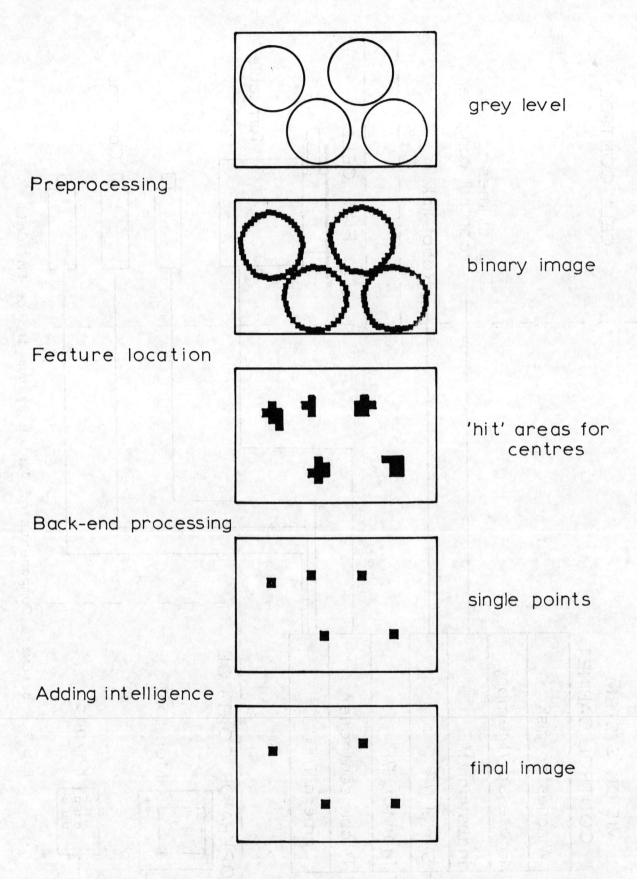

FIG. 2 Diagram showing main stages of processing and the resulting images

8

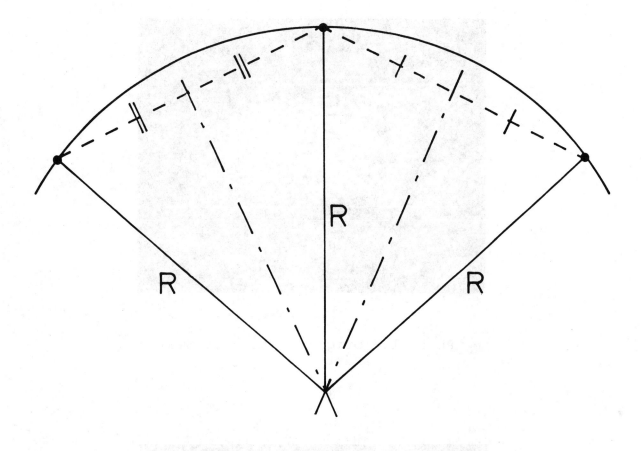

FIG. 3 Diagram showing the principles used to estimate the centre of circle given three random points on the circumference.

These show a sequence as diagrammed in Figure 2 for the various stages of processing.

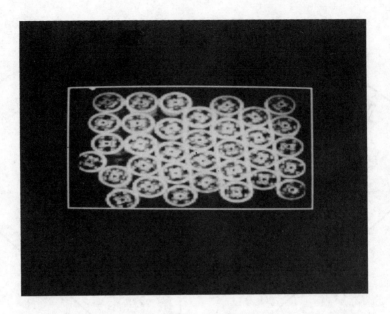

Photo 1 - Original Image of Small
Parts

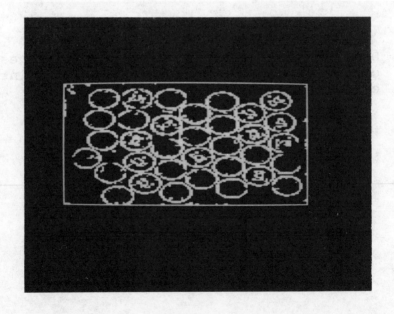

Photo 2 - Binary Image

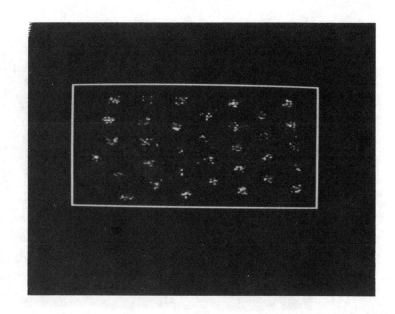

Photo 3 - 'Hit' areas for centre, the
brighter the dot the greater
the number of 'hits'.

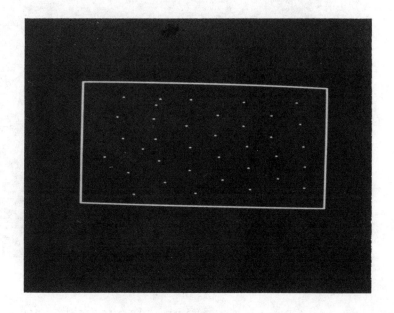

Photo 4 - All possible centres, note there
are two false indications.

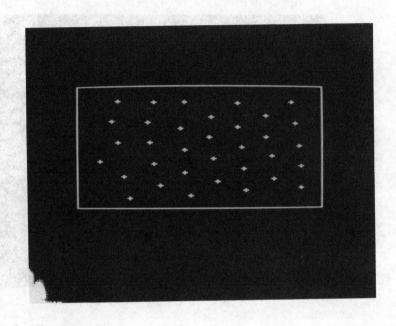

Photo 5 - False indications removed - these
are shown with reduced intensity.

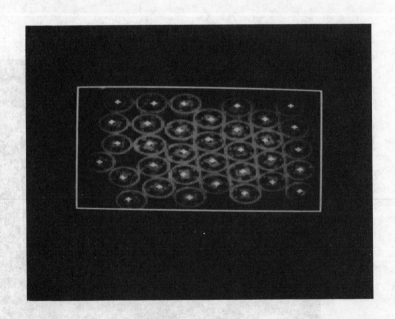

Photo 6 - Final image with found centres
marked on original image.

A New Approach to Programming Vision Systems

H. Jochnick

Asea Robotics, Sweden

Abstract

The design philosophy behind an alternate way to program a vision system is described. The realization of this philosophy is demonstrated with the aid of an example.

Introduction

Traditionally vision systems have been a product of the academic world. The reasons for this are obvious. Vision systems have always been on the leading edge of technology, requiring deep theoretical knowledge paired with vast amounts of computing power. However the recent years advances in technology have made it possible to realize the hardware needed for the vision systems to step out of the laboratories and into the industrial world. The first systems developed at the universities were primarily concerned with vision algorithms and also to some extent trying to establish how far one can go with vision. This led to systems that were general in their nature but lacked all but the most rudimentary operator interface. Although this was satisfactory in the academic environment these systems were used in, the user unfriendliness carried over into many of the commercial systems now on the market. Most systems require a good knowledge of Pascal or Basic in order to program even a simple object into the system. Once this is achieved, the user is confronted with coordinate transformations and communication problems to make his mechanical part obey the vision system. All this leads to an amount of engineering that is, if not impossible to overcome, at least exhausting and expensive. One leading vision manufacturer stated at a recent conference that his company has to program 80% of the systems for the customer. It must of course be a matter of great concern to the end user, that he himself should be able to make alterations and additions to his own system without having to call the manufacturer. With all this in mind ASEA has developed a new approach to programming the vision system for its second generation industrial robots.

Design philosophy

ASEA's vision system was developed in parallel with the second generation robot controller. Since the vision system is designed to be a subsystem to the robot controller, the same design criteria were applied to the vision system. Some of them are:

- Shop floor personnel should be able to program the system.
- Programming method should be adapted to the industrial environment, e.g it should be possible to program the system with gloves on.
- The system should not allow the operator to make mistakes.

In addition several new criteria were taken into the design:

- The operator shuld only be required to have a minimal knowledge of image processing.
- There should be facilities for holding and manipulating multiple models in its database.
- Maximum programming time for a model should be less than fifteen minutes.
- The operator shouldn't need to worry about the math involved.

Hardware

The above mentioned criteria clearly make the use of a conventional CRT terminal impractical. Instead the robot's programming unit is used. Fig 1.

The programming unit consists of a two row aplhanumerical display and a number of buttons. The upper row of characters are used to display prompting, informative or error messages, while the lower line of the display is used to label five keys to form a programmable menu. Only the five programmable keys together with the numerical keypad are used for vision programming. The button with the camera symbol is used to enter the vision subsystem. Any other button will return to robot mode. The operator is given the impression that he is dealing only with one homogenous system, when in fact he is talking to two different computers. A CRT monitor is added to the controller's front door. Fig 2. The monitor is used for displaying either camera video, edge data or the two superimposed on each other. When the operator during programming is asked to point out a significant part of the object, as displayed on the CRT, he uses the programming unit's joystick.

Programming dialogue

By depressing the "camera" button on the programming unit, vision mode is entered. As stated earlier the dialogue should be such that the operator avoids making mistakes. One way towards this goal is to present to the operator only such choices and information that are relevant at any given moment. Another way of helping the operator is made by placement of the programmable labels on the programming unit. The labels are placed so that the order in which they usually are depressed go from left to right. E.g. in the top menu, Fig 3, the labels CALIB, PROG, DISK and TESTMOD are found. This also reflects the order in which they are used. CALIB is used when initially setting up the system, defining windows and so on. PROG is used when creating a new model. DISK is used for storing or retrieving model data. TESTMOD performs vision processing without running the robot. This is useful when programming models before an actual robot program is ready and tested. Since this paper is supposed to deal with the programing of vision systems, emphasis will be placed on the hierarchy under the PROG label. Fig. 4. Again the flow of programming goes form left to right. Under TPICT the operator is given the option of adjusting the analog input circuitry with a contrast and brightness control, much the same way one does with a TV set at home. It is also possible to set a sensitivity control for the classification of gradients. If the system is correctly set up at installation, the default values supplied by the system could be used in 90% of the cases. Returning to the PROG sub menu the next step will be to place the object to be programmed into the system under the camera and depressing the DEFINE button. The vision system will now automatically extract contours of the object under the camera. Since the system is a gray scale system contours inside the perimeter of the object are also displayed and can be used. The operator is prompted to move a cursor to a surface that he regards as significant for the object. Fig. 5. He can select this surface alone by depressing SURFACE or he can include all surfaces inside the surface he points to by depressing COMSURF. Once a surface is selected the system regards this as a significant feature of the object. Up to sixteen different features are allowed. Every time a selection is made the system will present data about the feature, Fig. 6. The operator is given the option of either surpressing the feature by depressing WRONG, selecting a new feature with the N.FEA button, which will return him to the contour display with the cursor, or he can finish the programming by depressing VIEWRDY. The last choice will cause this view of the model to be examined by the system, several parameters will be extracated based on the features and an orientation algorithm will be selected before the view is linked into the database. The next step is to collect statistical data about the object by showing it to the system in various positions. This is performed under the TRAIN label. To conclude the programming, the gripping point of the object is taught to the system. This is done by depressing ROBOTP. Fig. 7. With the aid of the robot the object is placed under the camera and the position is recorded by depressing ENDPOS. The operator opens the robot's grippers and removes the robot from the field of view. By depressing READY the sytem processes the object and creates a transformation from its internal representation of the object to the gripping point. This is done automatically without any operator intervention. The transformation is stored in the database.

These easy and straightforward steps form the entire process of progamming an object into the system. The remaining two labels in the menu are for modifying and reviewing the contents of the model database.

Conclusion

The programming method as described in this paper constitutes a great simplification of the task of programming vision systems as compared to many other systems. When a vision systems is used as part of flexible production environment, ease and speed of programming are of vital importance. This is demonstrated by the system described in this paper.

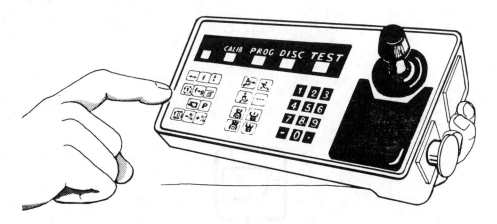

Fig. 1.

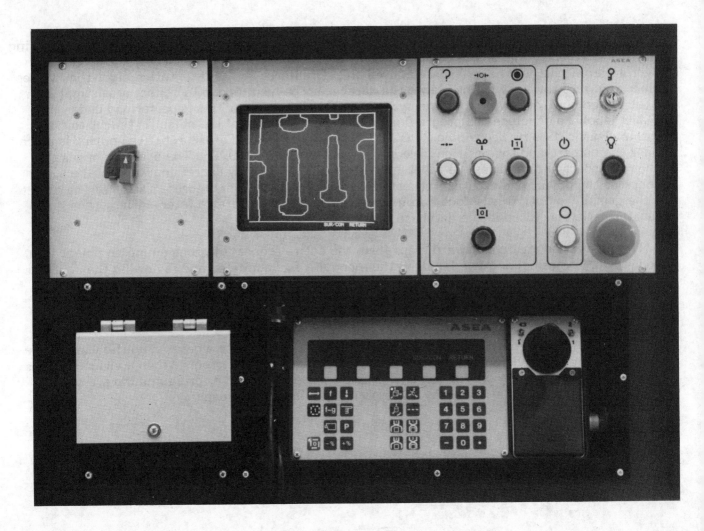

Fig. 2.

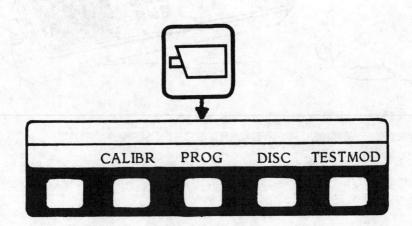

Fig. 3.

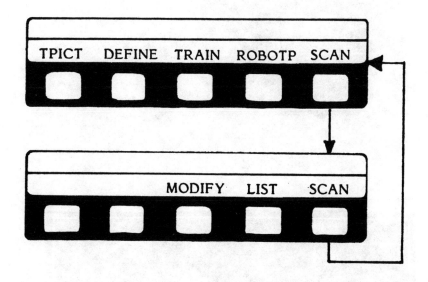

Fig. 4.

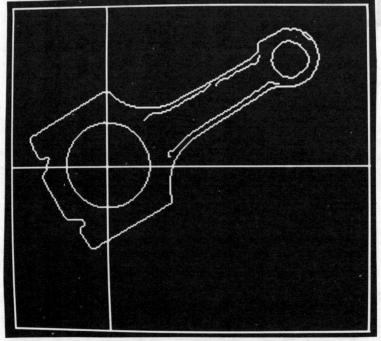

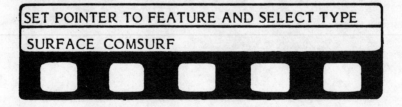
SET POINTER TO FEATURE AND SELECT TYPE

SURFACE COMSURF

Fig. 5.

18

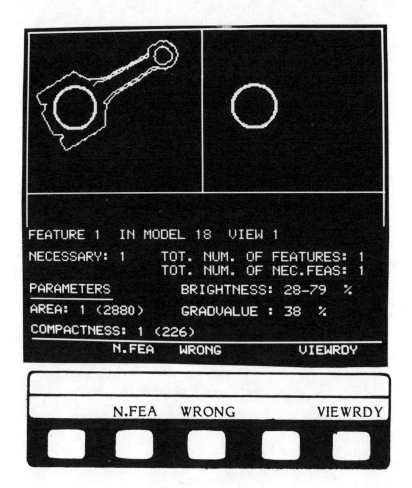

Fig. 6.

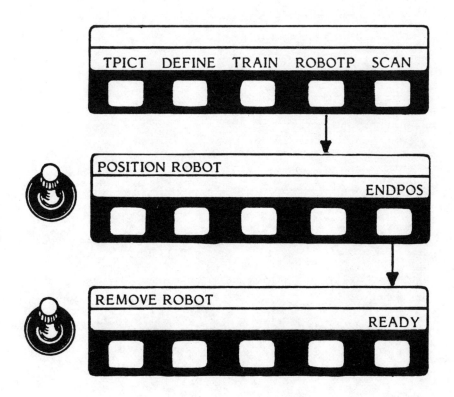

Fig. 7.

Machine Vision – The Link Between Fixed and Flexible Automation

D. Banks

Object Recognition System Inc., USA

Machine Vision is in its infancy both technically and as a manufacturing tool. Even at this early stage, machine vision can play a vital role in evolving manufacturing from today's fixed automation to flexible future automation. This paper describes the elements of productivity improvement currently achieved with machine vision and predicts how machine vision can be a bridge technology. Sensory automation is vital if computer integrated manufacturing is to become a reality. Use of machine vision now expedites practical development of the technology as well as user expertise.

INTRODUCTION

Manufacturing is currently undergoing widespread and drastic changes. One Senior GE executive has coined the phrase "automate, emmigrate or evaporate". Others call the current period the second industrial revolution as computers penetrate manufacturing. John Nesbitt author of Megatrends indicates we're approaching a global economy where information processing, handling and control via computers will revolutionize the entire process of making, distributing and even using goods and services. How does the manufacturing company get from where we are now to the factory of the future; meaning a competitive, efficient enterprise optimizing use of plant, equipment and labor through computer control and communication?

In order to start from a common point we should define terms. Websters New Collegiate Dictionary defines automation as "automatically controlled operation of an apparatus process or system by mechanical or electronic devices that take the place of human organs of observation effort and decision". It is significant that observation is mentioned first - to control one must observe. I describe fixed or hard automation as automation which must be changed by physical (human) or mechanical intervention. Flexible automation is automation that can be changed through program control via computer. The most typical description of a flexible automation system is an industrial robot. We will define the term quality as conformance to specification. The term CAD is Computer Aided Design. The term CAE is Computer Aided Manufacturing. CIM is Computer Integrated Manufacturing.

INDUSTRIAL MACHINE VISION TODAY

The availability of micro-processors coupled with advanced pattern recognition software makes computer vision an industrial reality. Vision systems, most simply put, are the first available element of sensory artificial intelligence enabling machines to monitor, modify or be guided by visible manufacturing elements on the factory floor. Vision is not a leading automation technology. It fits into processes already in place where some level of automation is utilized.

Early applications of machine vision have been associated with fixed automation systems set up to perform large batch runs. Machine tools, printing and labeling equipment, painting or finishing equipment and assembly systems are automated machines whose performance and output quality vision is improving. Most of these applications are currently turnkey in nature - the vendor supplying the total solution including all mechanical, electronic, optical, and interface components required to meet a fixed performance requirement.

Examples of production use of machine vision in Figure 1 indicate the value of in-process monitoring as pay back from cost reduction.

APPLICATION	SPEED	SYSTEM PAYBACK PERIOD
○ Credit card inspection for magnetic signature strip quality	450/min.	2 months
○ Wheel bearing inspection (auto assembly) 3 features	1/second	1 year
○ Flash bulb orientation (auto assembly) in camera	2/second 2 lines	4 months
○ Tape cassette assembly	2/second	6 months

Fig. 1. Typical Machine Vision Application.

All these applications were performed by the same model vision system with r reprogramming or software modification There were material handling requirements for application one which were provided as part of the inspection sys tem. In the other applications the er users supplied the material handling and were able to implement the applica tion successfully with applications engineering assistance from ORS.

Parameters that need to be considered in implementing a vision oriented solution are shown in Figure 2.

° Throughput Rate
    continuous motion
    indexed motion

° Resolution/detail required to detect defect/condition/orientaion

° Field of view

° Normal - acceptance - variations
    position
    illumination
    appearance

° Machine errors/false rejects

° Gentleness

° Reliability

Fig. 2. Vision Applications Specifications.

Successful addition of machine vision to manufacturing processes using fixed or hard tooled automation results in real-time process control. Better product quality, increased tool lifetime and real-time yield information are the results obtained. Unfortunately the predominant approach to use of vision is currently as a retrofit technology rather than integral to the control process. As such, a turnkey - retrofit vision system places the learning burden on the vendor rather than the user. Efficient use of vision-based process control will become common only when the end user understands and can implement the technology with minimum help from the vision vendor.

THE VISION LINK

Now that CAD and Computer Aided Engineering can do product and manufacturing process design at the same time, the need exists to implement the link from these computerized technologies into production. Both fixed and flexible automation in the form of robotic work cells must tie back to the design and

process planning function to enable process simulation, quality monitoring, statistical quality control, yield data, trend analysis for efficient maintenance of the manufacturing process and equipment.[1] Manufacturing companies share a number of common elements; engineering or design, process planning, material procurement and control, manufacturing, quality assurance, and the administrative functions. To some extent these functions are automated and interact but not nearly to the extent that utilizes todays technology efficiently.[2] Computer vision is the key technology that links design, production and quality assurance by providing the contributions listed in Figure 3.

FIXED AUTOMATION	FLEXIBLE AUTOMATION (ROBOTIC)
° Product yield (output)	° Adaptive guidance
° Defect rejection	° Object differentiation/identification
° Defect classification	° Safety
° Trend analysis/prediction	° On-line gauging
° System characterization	° Unoriented part acquisition

Fig. 3. Machine Vision Contributions

Experience with use of vision in fixed automation is a prerequisite to intelligent design of flexible work cells. Thus machine vision links together existing manufacturing functions for immediate feedback and control as well as acting as a bridge technology leading to flexible work cells.

FLEXIBLE AUTOMATION

Programmable flexible automation (robotics) is not truly flexible if part orientation or presentation must be precisely defined for a given operation. Part feeders, molded dunnage and conveyors add to work cell cost and usually are specific to a given manufacturing process thus, robot systems cannot realize their flexibility potential in small batch applications without some level of intelligence which is either visual or tactile in nature.

Vision provides guidance as well as quality measurement or inspection to robotic manufacturing operations such as material handling, welding, spray painting and assembly.[3] Vision with integrated tactile (eye-hand) coordination is a very powerful part of a robot system. A simple application example is a vision controlled work cell to acquire, inspect and place large bolts in an assembly.

This complete process would be costed in Figure 4.

Vision Guided		Hard Tooled
Robot	$55K	$55K
Vision Module	$25K	$ 0
Gripper	$2.5K	$1.0K
Bowl feeder and Chutes	$0	$28K
	$82.5K	$84K

New part/
process change
costs

Replace Bowl feeder and Chutes	$0	$28K

Fig. 4. Bolt Acquisition, Inspection Assembly Cell Costs.

If the cell process changes if another bolt size is needed or, the cell is set to do a different manufacturing operation the hardware used to orient this bolt and present it is lost, the tooling must be totally replaced at significant tooling cost. Vision also enables the use of a less precise less expensive robot because the vision system can guide the robot to slightly misplaced target parts or target locations.

The evolution of the modern factory - true computer integrated manufacturing - will be just that, an evolution. Few companies can start with a blank slab and build a flexible production facility from the ground up. Even if this were possible, automated manufacturing processes need to be designed based upon experience with sensory technology. This experience can be gained now by the manufacturing engineers using a vision as a routine tool. Machine vision provides the best mechanism to integrate current fixed automation into processing planning, process monitoring and process control systems via the CAD data base. Vision enables flexible automation (or robotic systems) to be flexible, to change through software, to adapt to small batch needs. Vision provides benefits far beyond the replacement of the traditional "inspector" at the end of the line by a tireless unblinking machine. It reduces scrap, operates at high speed with very high accuracy, provides feedback to people when a process begins to fall out of spec, gives yield information and generally supplies computers with the inputs necessary to integrate discreet manufacturing operations with the other elements in a business. The pay back is improved productivity through better machine and people utilization.

SUMMARY OR HOW DO WE GET FROM HERE TO THERE?

Use of machine vision can provide a range of benefits for manufacturing. Recognition of this is leading to use of on-line vision in dedicated or turnkey systems designed for a specific production task, requiring heavy vendor applications engineering involvement.

To evolve more flexible automation vision will be a prerequisite. Understanding vision based sensory systems, their applications and their potential is a burden that must be shared by the vision system vendor and end user. This understanding at the user level will expedite efficient, flexible application of vision in appropriate production processes as well as providing the feedback to vision vendors required to advance the technology.

Experience with machine vision must be gained by manufacturing engineering people on real problems. End user companies reliance on vendor applications engineering is dangerous in the long run because no outside resource understands manufacturing processes as well as those who design and maintain them. Foreign attitudes regarding experimentation with and implementation of new manufacturing technologies are still more aggressive than those the majority of U.S. manufacturers display. The opportunities for vision are great, the technology is available and the competitive motivation is clear. As the link between fixed and flexible automation vision is helpful in the factory of the present, but is _essential_ in the evolving CIM facility of tomorrow.

BIBLIOGRAPHY - Books

(1) Krause John K. Computer Aided Design and Computer Aided Manufacturing the CAD/CAM Revolution. Marcel Dekker, Inc.

(2) IEEE Spectrum, May 1983; Data Driven Automation: Toward a Smarter Enterprise.

(3) Machine Vision Systems - A Summary and Forecast - Tech Tran Corporation.

Quality Control with a Robot-guided Electro-optical Sensor

E.-J. Schmidberger

and

R.-J. Ahlers

Fraunhofer-Institut fur Produktionstechnik und Automatisierung (IPA), W. Germany

Abstract

Quality control for automobile chassis parts includes recognizing deformations such as: cracks, overlaps, folds, missing punched holes, etc.
Up to now, this was primarily accomplished by means of stationary sensors. The system described in this paper has to perform quality control
for arbitrarily large chassis parts.

Based on experiences in the manufacturing process, it is assumed that
the deformations occur at certain typcial locations on the part's surface. The system includes an industrial robot which guides a camera to
these locations.

1 Introduction

Today, production systems are largely characterized by reusable, flexible equipment and subassemblies. This calls for measuring and test equipment with the appropriate performance characteristics. The economical use of even complex test systems with powerful sensors for quality inspection is enhanced by the great progress that has been made in the field of information technology. Optical sensors together with a wide variety of systems for information processing have provided a basis today, to efficiently automate visual and dimensional test operations on a broad scale.

Among the things contributing to this are improvements realized in the hardware structures, which permitted the development of inexpensive and powerful systems for the rapid processing of the video information, even of complex inspection tasks. Furthermore, results have been achieved in the field of software development which, in the ineraction with the hardware mentioned, permit the realization of extremely flexible and problem-orientated systems /1/.

Stationary cameras /2,3,4/ have mainly been used so far, so that the test objects have had to be moved into the confined field of view of the associated sensor target. Dependent on the workpiece size and on the structures to be covered and characteristics, it has been necessary to break down the total inspection area into separated frames. If the sensor is fixed in its position the sectioning of the whole scene into frames is achieved by the defined positioning of the test object under the camera. Such test systems have predominantly been used for testing flat workpieces /5,6/.

In large objects of any volume, it is expedient to move the camera towards the inspection location. This can be achieved by various translatory and/or rotatory positioning units /7,8/, but it always involes specialized and mostly expensive special-purpose solutions. The probably most flexible system arrangement is based on the robot-guided imaging sensor; in this arrangement, an industrial robot is used to guide the sensor to the inspection location /9/. This method of camera positioning is promising, especially so since the use of industrial robots falls back on flexible standard products.

2 Image processing systems for quality inspection

The information flow and the processing of information of inspection scenes are governed by the type and size of the products to be tested. The process, ranging from the optical imaging of the test object to the test result, requires various hardware and software components for performing the computer-controlled image processing. Dependent on the characteristics to be evaluated, it is necessary to establish task-specific test systems, as described by the following example.

2.1 Image processing hardware

In the order of the information flow, the hardware can be classified as follows:

- Illumination
- Optical imaging system
- Imaging sensor
- Picture digitization and processing
- Peripherals (printer, monitor, etc.)

The optimized object illumination plays an important role both in visual and in automated inspection. Basically, efforts are made to obtain a high-contrast representation of the characteristics to be evaluated out of the inspection scene through specially adapted illumination equipment. For this purpose, various illumination methods are available, as e.g. transmitted light illumination, vertical illumination, flashing light, etc. These permit directed or diffuse object illumination.

To unambiguously recognize the most varied defects in their entirety, it will be necessary to integrate the illumination equipment into the system in such a way that no concealed areas will develop and that the shadow structures will not produce any pseudo-defects.

The quality of the image and its magnifiation sensor are dependent on the optical system used. It is basically possible to use any commercially available high-quality camera lens.

Semiconductor and tube cameras are normally available as imaging sensors. Used in our set-up is a Fairchild CCD matrix camera with a resolution of 488 x 380 picture elements. The advantage of this sensor in comparison to the tube type is that the former features smaller mounting dimensions, lower weight, freedom from distortion, long service life, and less susceptibility to mechanical and eletronic faults. It is particularly the insensitvity to dynamic stress which makes it appear reasonable to use semiconductor sensors in connection with industrial robots.

It is principally possible to directly process the digital information as obtained from the matrix sensor. In the inspection tasks discussed in the following, however, the video signal (composite video signal) has been used for further digital processing.

In the system shown in Figure 1, the video signal present in the analog-digital converter (ADC) is divided into a raster of 784 x 512 picture elements with 256 grey-levels. The digitized image intensity can be modified in different ways through the look-up tables shown (Figure 1), which perform a selective conversion of the grey levels. It is thus possible to intensify or weaken the grey-level picture in selected contrast zones (e.g. to fade out blooming), to reduce the variety of information by selectively varying the intensities - to two levels in the simplest case - and a lot more. The actual black-and-white representation, the so-called binary image, is obtained from appropriate threshold values in the intensity profile of the look-up tables: the intensity "white" may for example be set above a threshold and "black" below. Both contrast modfication and intensity reduction had to be made in the inspection tasks described in the following. The digital image information thus modified is now stored in the image memory (Maximum storage volume: 784 x 512 picture elements and 256 grey levels per memory page). The image transfer from the memory output to the input (Figure 1) permits high-speed image operations such as image addition, subtraction, integration /10/. During this process, the image manipulations performed can be monitored on a television monitor linked to the memory output. Dependent on the processing algorithms, the computer reads the data stored in the image memory via direct memory access (DMA). The storage capacity is 64 Kbytes. The peripherals used are a terminal, a printer, an external storage unit with a capacity of 10 Mbytes, and the interface to the industrial robot.

2.2 Image processing software

The software for image processing can be divided into three areas:

- Dialog software
- Control software, and
- Image processing software.

The digital software ranks higher than the control and image processing software. Its task is to enable optimum man-machine communication. Basically, there are structures which are based on the dialog software, but in general it is largely problem-orientated.

The control software predominantly consists of program modules programmed in assembler language. Various system directivies can be linked into the image processing and dialog software in the form of subroutines. These include routines for initializing, sychronizing of the memory hardware to the video signal, organizing of the image transfer from the camera via the memory to the monitor or between image memory and computer, loading and activating the various lookp-up tables, superimposing of auxiliary functions (cursor, windows, contour lines) and a variety of others.

The third partial area is the actual image processing software (Figure2). Its task is the automatic recognition and evaluation of objects with respect to their condition and their interrelation. Up to now there is nothing like a standard in this kind of software, since the science of applied image processing is still rather young. So new methods and techniques are being devised worldwide every day, and even completely new research programmes are being defined. The individual tasks of the image processing software comprise image preprocessing, finding of characteristics, classification and class grouping. Various methods and techniques of applied mathematics are necessary to cope with these tasks /1/. These are applied to both grey-level pictures and binary images; the following implementation is also based on them.

3   Solution to special inspection tasks

As mentioned earlier, there are two possible ways to produce a relative movement in the three-dimensional space between the imaging sensor and the test object. Depending on the task involved, it will be meaningful to either move the test object or the sensor. A distinction can be made in this respect by using the terms "robot-guided test object" /11/ and "robot-guided sensor" /9/.

The test with robot-guided camera described in the following was the outcome of work based on the experience available at the IPA in the field of computer-controlled image processing.

During the inspection of deep drawing parts it is essential to detect and evaluate reforming defects of many different kinds. These are:

- cracks
- overlaps
- folds and waviness
- reductions in area, and
- missing punched holes.

These defects may occur in any spatial position on a workpiece. The production process involved, however, permits the conclusion that these defects will only occur in defined, workpiece-specific locations. If these defects go unnoticed, this will result in considerable consequential costs. To avoid these consequential costs, it is endeavoured today to perform a 100% visual inspection. To cut the inspection costs and to achieve objective inspection, the automation of this process will be necessary in this field. The image processing system mentioned earlier is particulary suitable for application to visual inspection tasks. The requirements made - large inspection area, high resolution, flexibility regarding the spatial arrangement of the defects - make it necessary to move the sensor of the test system to the test location. The robot (ASEA IRb 6/2) carries the camera (Figure 3) on its tool flange and it positions the camera at the critical control locations. The camera is mounted in such a way that it can be easily positioned according to the robot's hand coordinate system. The robot control recognizes the centre point of the image section as "work point" (tool centre point) and, therefore, it can selectively change the incidence angle of the camera, without changing the distance to the object. In this manner, the motion program can be entered within the shortest time and changes necessitated by changing geometries of the parts under test will be possible without great difficulty.

The communication between the image processing system and the industrial robot is carried on via the digital input/output lines of the robot control. Information obtained from image processing can be fed back via interrupt inputs to the robot and its motion process, thus e.g. causing the refocusing by changing the distance of the camera or the finding of an additional test position.

To adequately illuminate the inspection scenes, a ring-shaped light fitting (Figure 4) has been integrated, which supplies adequate lighting for all types of defects.

The total test operation goes on as follows: the camera is moved to the roughly preprogrammed test position, it covers the scene, e.g. a punched cut-out in a surface section (Figure 5). The image processing system stores the converted optical information in the image memory and starts selective image enhancement. The interfering structures resulting from blooming have to be filtered out first, the contrast of the scene must be increased, and a conversion into binary form (conversion into a pure black-and-white image) must be made from the resulting grey-level picture by the computed threshold values. The contours produced in the binary image will be covered automatically, and then evaluated and identified according to a parameter list (Figure 6).

The decision as to whether the object contour is defective is particularly critical, if cracks are present which may occur in the marginal zones (Figure 7). In their apperance, these cracks are similar to properly produced contour patterns (Figure 8), so that a real distinction can only be made with the aid of specially developed software algorithms. In this case, that is, if the structure is defective, this will be communicated to the operator by identifying the defective zones with a rectangular window (Figure 7).

Various techniques have been investigated for evaluation defined workpiece zones regarding folds and waviness. One of these possibilities is based on the so-called light sectioning method /12/. A coded illumination source, e.g. a light bar, is projected on to the surface to be inspected. Depending on the surface shape these coded structures are distorted, thus leading to information about the surface of the inspection object. Another approach to cover the type of defects described above is the evaluation of the intensity profile of the grey-level picture in the respective test zone. Generally speaking this method resembles a frequency selective evaluation of the intensity-distribution included in the whole scene (impermissible waviness or folding will be displayed on the monitor by marking, see Figure 9).

The latter method gives the advantage that the second (coded) light source can be omitted so that all defects - folds and waviness included - can be detected with one light source.

Upon completion of one test process, the test system passes the priority to the robot control, so that the camera can be moved to the next inspection position. On arrival at the next test location, priority passes back to the image processing.

The test results will be documented on both the display screen and on the connected printer. The principle of integrating the described test system into a production line can be seen from Figure 10.

4   Summary and prospects

Wherever requirements such as large inspection area, high resolution and flexibility regarding the spatial position are made in the field of quality inspection, it will be necessary to achieve a relative movement between sensor and test object. A combination between an image processing system and an industrial robot is an obvious solution to this kind of tasks. Depending on the task involved, it may be suitable to have the test object moved to the camera or to move the camera to the test location by a robot.

This combination results in an optimum utilization of the capabilities and characteristics of the individual systems participating in the communication.

"Image processing" as partial system permits the automated evaluation of two-dimensional image scenes in a confined image field, as covered by the matrix sensor. An extension of the work area to several metres in any spatial position is achieved by the mobility of the industrial robot. The test process goes on automatically, due to the interactive communication between the robot control and the test system. The analysis of the image scene which then follows permits an evaluation which - dependent on the test result - causes the robot to move to further positions or to discontinue testing on this workpiece (reject).

The image processing system principally permits changed positions and orientations of the test objects to be covered and to reproduce the change by computing the coordinates accordingly. Therefore, one of the objectives of future work will be an improvement of the communication between the individual systems. For this purpose, it will be endeavoured to communicate the object coordinates determined anew by the image processing system to the robot control, with the robot then being moved to the new position of the object after appropriate coordinate transformation. This approach permits the total system to respond flexibly, and even "intelligently" to the changing test conditions. Any displacement or rotation of the test objects will thus no longer constitute problems that cannot be overcome.

References

1. Rueff,M., Melchior,K., Schmidberger,E.; "Structure of a software-library for tasks in visual inspection problems". 3rd International Conference on Robot Vision and Sensory Controls, p. 85-91(November 1983), Cambridge, USA.
2. Jentner,W., Schmidberger,E.; "Solving problems of industrial quality control by means of image processing systems". 2nd International Conference on Robot Vision and Sensory Controls, p. 367-378, (November 1982), Stuttgart, Germany.
3. Schmidberger,E., Melchior,K.,Rueff,M.; "Electro-optical In-line Quality Control in Industrial Production". 3rd International Conference on Robot Vision and Sensory Controls, p. 573-578, (November 1983), Cambridge, USA.
4. Warnecke,H.-J., Melchior,K.; "Pattern recognition: an element of automation". 2nd International Conference on Robot Vision and Sensory Controls, p. 1-8, (November 1982), Stuttgart, Germany
5. Doemens,G.; "Application of Sensors in the Production and Quality Control of Microelectronic Devices and electromechanical components". 5th International Conference on Automated Inspection and Product Control, p. 187-197, (June 1980), Stuttgart, Germany.
6. Loughlin,C., Taylor,G., Pugh,A., Taylor,P.; "Visual Inspection Package for Automated Maschinery". 5th International Conference on Automated Inspection and Product Control, p. 77-85, (June 1980) Stuttgart, Germany.
7. Jäger,H.; "Qualitätskontrolle feinmechanischer Präzisionsteile mit Fernsehbildanalyse am Beispiel einer Schiltzmaskenmessmaschine". 12. IPA-Arbeitstagung, p. 57-65, (June 1980), Stuttgart, Germany.
8. Kashioka,S.; "An approach to the integrated intelligent robot with multiplel snesory feed back, visual recognition techniques". 7th International Symposium on industrial Robots, p. 531-538, (1977), Tokyo.
9. Brunk,W.; "Geometric control by industrial robots". 2nd International Conference on Robot Vision and Sensory Controls, p. 223-231, (November 1980), Stuttgart, Germany.
10. N.N.; "User Manual of MVD (modular video digitizer)". VTE GmbH, Waller Weg 22, Braunschweig, Germany.
11. Schmid,D.; "Industrial robot with video camera for detection of material defects". 2nd International Conference on Robot Vision and Sensory Controls, p. 19-25, (November 1980), Stuttgart, Germany.
12. Warnecke,H.-J.,et.al.; "Applied Pattern Recognition in Sensor Systems for Quality Control and Automated Production Systems". 16th CIRP International Seminar on Manufacturing Systems, (July 1984), Tokyo, Japan.

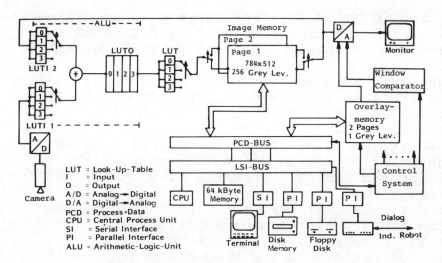

Fig.1. Schematic representation of the computer-controlled image processing system

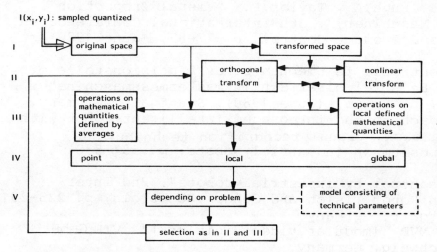

Fig.2. Software for industrial image processing

Fig.3. System layout of the robot-guided camera

Fig.4. Control system (ring-shaped light fitting mounted on camera lens)

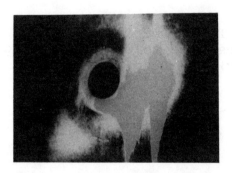

Fig.5. Test characteristic: punched hole (grey-level picture with blooming)

Fig.6. Test characteristic: punched hole (binary image after contrast improvement)

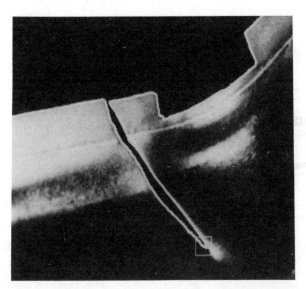

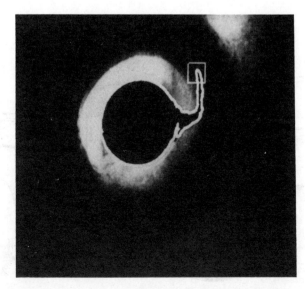

Fig.7. Result of the automatic contour assessment (identification of the contour with crack pattern and origin of crack)

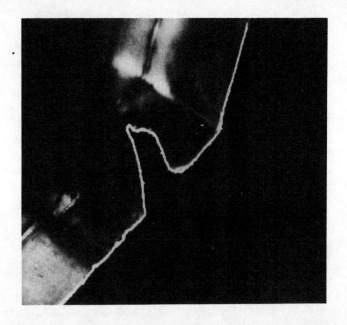

Fig.8. Tested part with proper, but partially crack-like contour pattern (see bright contour)

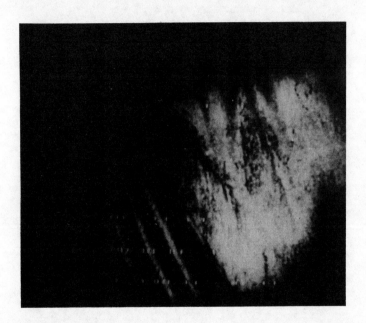

Fig.9. Tested part showing impermissible folding (see bright marking)

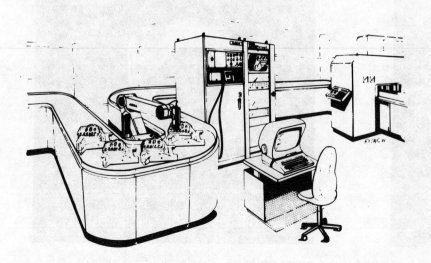

Fig.10. Flexible test system in the future production environment

VISION SYSTEMS

Solving Illumination Problems

D. A. Hill

C.U.E.L., UK

ABSTRACT

Illumination is an essential part of machine vision and related systems. There is now sufficient application experience available for many problems to be anticipated and solutions proffered. The pre-requisite is a thorough understanding of lamps and optical components, not only in broad terms, but more particularly as they relate to this small, highly specialised field of machine vision.

INTRODUCTION

Lighting and illumination are directly relevant to machine and computer vision, image processing, except that incorporating x-ray and ultra-sound inputs, pattern recognition, and also to optical non-contact sensing and gauging. Although the illumination is predominantly at visible wavelengths, that is not always the case.

Its role is one of acquiring and transferring information about an object to a sensor.

ILLUMINATION PROBLEMS

Three groups of problems are encountered in connection with this role. The first is more attitudinal than technical. Lamps and lighting are so much part of everyday life that they receive the contemptuous treatment which is often associated with familiarity. Illumination is all too frequently relegated to a position where it receives residual attention after the specifications of other components of a system have been finalised. The solution to that problem is not a technical one.

The second group of problems, to which reference will be made from time to time, are those resulting from the incompatibility of the illumination equipment and the sensor which it serves.

The third group is that of which the object being illuminated is the pivotal point. The salient features of the many different objects, the constraints of the working environment and the characteristics of off-the-shelf illumination equipment are rarely compatible without optical engineering. Since this is the largest group it is considered first, under two headings - optical and environmental problems.

Optical problems

The two most common optical problems are directly opposite to one another. On the one hand there is the often disconcertingly high reflectivity of shiny objects, even translucent and non-metallic ones. The reflectivity may be so high as to saturate certain types of camera completely. As much as 96% of the light received may be, in the worst condition, reflected by an object into a sensor. The resulting saturation, if the initial light level is too high, is known generally as either blooming or, if more localised in its distribution, a flare spot.

On the other hand there are dark coloured objects, especially matt ones, which absorb as much as 98% of the light which falls on them. If the illumination system has to provide sufficient illumination to enable beam splitters or polarisers to be used in association with the sensor, the illumination intensity required is multiplied by a factor of approximately 2.5 in practice.

Further problems of reflectivity are caused by the shape and micro surface geometry of industrial objects. They are not optically flat. Surface irregularities generate localised spurious reflections as if they were micro mirrors set at many different angles.

However it must be noted, in parenthesis, that there are applications in which these spurious reflections can provide information for non-contact surface gauging and measurement.

The shape of an object is potentially another source of optical problems. The radius of curvature may act in the one direction like a cylindrical or spherical convex mirror. If it is curved in the other direction it behaves like a concave mirror. Unfortunately none of these quasi-mirrors have the high grade ray patterns of real mirrors and optical components.

Not only is the object itself a source of optical problems but also its background. With the wrong background a sharp edge deteriorates into an indistinct smudge.

Such problems are not uniform throughout the spectrum. It is dangerous to assume that the machine sees what the eye sees. A distinction has to be drawn between human and electronic vision. Some apparently opaque materials are infra-red transparent. Other materials, on account of their chemical composition, selectively reflect some wavelengths and absorb others. Unfortunately data on the optical properties of only a relatively small number of the materials used in industrial applications is available.

Matching the shape of the illumination to that of the object is often a problem. If the emission angle of the illumination does not match the object, light is not used effectively.

It is as well to remember that lamps are mass produced and, like other industrial objects, are subject to tolerances. They also have minor defects which, although acceptable for 99% of their vast range of diverse applications, cause optical problems when they are used in conjunction with machine vision systems. There are occasions when lamps have to be hand selected by the user for the more demanding applications.

The need to reduce cycle times also adds to these problems another one. Keeping industrial objects moving has many advantages but it usually requires the illumination to be more intense than would be the case if they were stationary. This requirement is most noticeable when the objects are moving quickly and only a few milliseconds are available for sensing.

Environmental problems

Lamps which give off excessive heat cannot be sited near personnel. Considerable discomfort would be caused, especially when ambient temperatures are high, as they may be in summer.

Although they are always protected by a suitable housing, lamps which shatter disruptively on rare occasions, e.g., xenon ones, are best avoided, if only for psychological reasons.

Physiological reasons require that human eyes should not be exposed to very bright lights. Fortunately harmful uv-c rays are not used for illumination

Machines are a source of more difficult problems. The "geography" of the workplace, for various engineering and economic reasons, restricts the space available for machines and machine vision systems. This affects the design, installation and maintenance of any illumination system. It is by no means unusual for parts of machines to protrude and obstruct the optical axis. This was a more obvious problem before the advent of fibre optics.

Vibration caused by machinery is another cause of fluctuation in the level and direction of illumination. In fast scanning applications tiny luminous irregularities wobbling over an object make the data acquired by the system suspect. Vibration also affects the life of some lamps.

Sensor problems

Since sensors and objects are next to one another in a machine vision system, the characteristics of the one sometimes interfere with those of the other. Saturation, blooming glare and "starvation" have already been mentioned.

The problems of interference are most acute in fast scanning applications. Lamps connected to an alternating current supply, even after allowing for the smoothing effect of thermal inertia, have a cyclical variation which may not match the scanning rate. Direct current or rectified alternating current is the only solution.

Gradual ageing of the lamp and variations in the electrical supply voltage assume great importance when a consistent optical datum is needed against which the sensor can detect momentary, minute deviations.

Incompatibility between the bandwidth and peak sensitivity of a

sensor and its associated illumination is often a problem. Although
data on the spectral distribution of sensors and lamps is available,
similar data on the total illumination system is not so readily
available. For example few manufacturers of fibre optic light
sources publish such data. Many sensors peak in the infra-red but
fibre optic light sources, of necessity, incorporate infra-red
filters.

The sensitivity of some sensors is such that they sense the
illumination as well as the object. Since this problem usually
manifests itself at the commissioning stage, further consideration is
deferred until the final section of this paper - "Systematic
Solutions".

ILLUMINATION HARDWARE

A pre-requisite for overcoming the problems of illumination is a
knowledge of lamps, accessories and ancillary equipment.

Types of lamps

Not all lamps from the enormous range available are suitable for
machine vision systems. Those which merit consideration are listed
below.

Tungsten filament These lamps range from miniature, through low
voltage car lamps, to the ubiquitous domestic clear and pearl ones.
Ratings are from 0.25A to 150 watt for machine vision purposes.

There are several variations. Miniature ones often incorporate
a lens at the tip of the envelope. Domestic lamps have integral
reflective coatings so that they may be used as narrow beam angle
spotlights.

Tungsten halogen These are also known as quartz halogen and
quartz iodine, or simply abbreviated to QI. A halogen, iodine
is added to the filling gas. They are bright, much smaller and more
efficacious than tungsten filament types but their operating
temperature is higher. Although there are both smaller and larger
ones, those between 20 and 250 watts are preferred for machine vision
illumination. Their burning position is sometimes restricted and
lamps in excess of 50 watts need forced air cooling. Fibre optic
light sources use quartz halogen lamps. (The 1,000 watt version is
very large because it needs so much cooling.)

Fluorescent All wattages of these lamps are ideal for providing
low intensity, even illumination over a relatively large area. Some
types of fluorescent lamps are used for ultra violet light.

Compact source iodide These are cinema projector lamps available
in various versions up to 4,500 watts. They are small in size and
very bright. Modern ones are more efficacious than xenon lamps and
they require less control gear.

Mercury arc These are used as a spot of ultra violet rays, e.g.
in fibre optic uv light sources.

Particular features

Each and every lamp has a number of features which might either
contribute to the solution of illumination problems or at worst they
may even acerbate them if the wrong choice is made.

All the following factors may have to be considered when choosing lamps.

1. Physical size.
2. Shape - cylindrical, bulbous.
3. Rating - normally in watts, miniature ones are rated in milliamps.
4. Luminous efficacy - lumens per watt.
5. Operating voltage.
6. Burning position - horizontal, vertical or any
7. Beam angle.
8. Cooling requirements.
9. Environmental hazards.
10. Compatibility with ancillary equipment.
11. Costs - initial, running and replacement.
12. Susceptibility to damage by vibration and movement, e.g., on a robot arm.
13. Envelope material - clear or pearl.
14. Evenness of distribution.
15. Stability, e.g., arcs in high pressure mercury lamps.
16. Integral components - lenses, reflectors.

Accessories and ancillary equipment

The following summary shows how many of the above features may be harnessed by the incorporation of suitable accessories.

Shape changing   Masks and fibre optic components change the shape of light reaching an object.   Solid circles may become hollow annuli or rectangular slits.

Deflection   Prisms and mirrors deflect light at fixed geometric angles.

Distribution and spreading   Diffusers, reflectors and condensers spread the light from a lamp over an object. Several of the readily available plastic translucent materials in sheet form, e.g., opal acrylic of various densities and thicknesses, and polypropylene, are useful diffusers. Materials such as PTFE have a surface which, when a sheet is used as a reflector it dissipates shadow from filament lamps.

Splitting   This takes two forms. Plate, cube and pellicle beam splitters sub-divide into two beams at rightangles to one another. Multi-branch fibre optic light guides, which may have arms of different diameter and end section, are used for multiple splitting from a single source and aligning each arm as required for optical illumination of the object. Filters and lenses are often attached to the emission end.

Intensity regulation   Diaphragms - crescent and iris, discs and neutral density filters, regulate intensity by optical means. Rheostats are used for altering the voltage reaching the lamp. Although they suffer from the disadvantage of effecting the colour temperature they are popular because they enable lamps to be under-run and so conserve lamp life.

For more sophisticated, fine regulation, a photo-electric cell, with compensatory electronic controls, is incorporated in the circuit.

Wavelength selection   Absorption and interference filters are used to select particular wavelengths and bandwidths of the light emitted by a lamp. Although unsuitable for permanent use, gelatine based photographic filters are a low cost aid to problem diagnosis.

   Polarisation   Those machine vision systems which need polarisation
for improving edge definition use two crossed pole filters. One
filter is fitted to each of two sources of light or alternatively one
filter is fitted to a light source while the other is placed in
front of the image acquisition lens.

   Protection   In harsh environments, windows of various materials,
glass, quartz and sapphire, protect illumination equipment.
Operating temperatures in the object zone dictate the choice of
window material.

   Redirection   Fibre optic light guides "pipe" light from lamps
round obstacles and enable it to be directed more precisely than
would be the case if the lamp were closer to the object.

   Synchronisation   Electronic shutters and rotating discs
synchronise the emission of illumination equipment with the movement
of objects and the sampling speed of sensors. Exposure times as
small as 1.8 milliseconds are possible.

   Optical referencing   Graticules facilitate machine vision by
superimposing a wide variety of optical datum lines on industrial
objects when they are in the sensing position.

SYSTEMATIC SOLUTIONS

   The state of the art is such that even the most careful
consideration, at the design stage, of typical problems and illumination hardware, may fail to anticipate all the problems which may
manifest themselves once machine vision systems are installed and
commissioned.

   The commissioning process of any major piece of equipment, is one
of problem diagnosis followed by experimental adjustment and change
until all the problems which might impede its performance have been
solved.

   As a preparation for this final section, much of the material for
the diagnosis of illumination problems and their solution has already
been introduced. However no attempt was made in the earlier sections
to fit it into a diagnostic-experimentation sequence. The advance
of such a sequence is that when the first suggested solution does
not solve the problem a second suggestion may be tried. If the basic
feasibility of a system has been established at the design stage the
possibility of a particular problem still persisting after all the
suggestions have been tried is unlikely although not impossible. The
following are common problems which can usually be overcome by a
systematic empirical approach.

Excess illumination

   The excess has to be related to the peak sensitivity and bandwidth
of the sensor. For that reason reducing the intensity of the lamp
by optical means is the first step.

   Introducing filters, which unlike diaphragms and neutral density
filters are selective of wavelengths, wavelengths potentially
troublesome may then be removed.

   Changing the direction of illumination is another means of

overcoming the problems of excess illumination. The angle at which the light falls on an object is often crucial.

When the direction of illumination is from the back of an object a diffuser may be needed in front of the light source in order to distribute the light more evenly.

If, on the other hand, the direction of illumination is from the front, there may be some advantage in replacing a large point source with several small ones.

Insufficient illumination

The problem of insufficient illumination is not automatically solved by using a larger lamp. Before changing the lamp certain tests should be made. Is the light properly directed? It is by no means unknown to find the surrounding area better illuminated than the object!! Secondly, are the accessories compatible with the lamp? Under-utilisation of lamps is common. When fibre optic light sources are involved there is often a mistaken belief that the amount of light energy falling on a target is the same irrespective of the light guide area. This is not the case.

Another example of incompatibility is that of fluid light guides. They are much more efficient than optical glass and acrylic fibre ones but only at shorter wavelengths.

If there is still insufficient illumination, two medium sized lamps may be substituted for one larger one. The light from the one may be focussed on top of that from the other.

Lack of uniformity

This problem is sometimes loosely known as "optical noise". There are several causes. One cause is the superimposition on the object of images, or shadows, of components, which are part of the lamp providing the illumination.

Another cause which effects back-lit objects is the sensitivity of cameras. They are capable of detecting individual 50um fibres, including broken ones, in cross section converters as well as small variations in linearity.

If some loss of intensity can be tolerated, these problems are not insurmountable. The effectiveness of condensing lenses can be further improved by etching. The most suitable diffusing material, if the solution lies in that direction, may only be chosen by experimentation.

Front illumination problems may benefit from the use of quartz halogen lamps with integral multi-mirror (or multi-facet) dichroic reflectors. They bring considerable improvement although there is a noticeable fading away in intensity towards the outer of the circle of light.

Another possible solution is that of moving the lamp to a position where the troublesome components are no longer on the optical axis.

Lack of contrast

The techniques of microscopists and photographers are a source of ideas for overcoming this problem. Before resorting to polarisation, different, contrasting backgrounds should be tried behind the object. Matt backgrounds are first choice. In many applications a matt, black, background introduces the right amount of contrast which enables shiny objects to be seen.

Offsetting one point of light or lamp against another so that they "interfere" with one another often reveals salient features. Fibre optic swan necks, positioned at different angles, are often used to heighten contrast. Sometimes both arms have a lens fitted. At other times only one lens is fitted. The use of lenses is not obligatory but if they are used, focussing the honeycomb pattern of the fibres on the object should be avoided.

Other problems

In order to overcome those problems to which attention was drawn in the section at the beginning on "Typical problems", the illumination specialist needs help from other engineering disciplines.

There are no short answers to problems such as vibration and "dirty" electricity supplies. Fibre optics has done much with respect to overcoming problems associated with access.

The solution to the problem of the gradual build-up of atmospheric dirt is usually routine cleaning. Air purges can be fitted to some illumination components for the purposes of continuous cleaning. However routine checks have still to be made so as to ensure that one form of dirt is not being replaced by another borne along in the air supply.

GVS: A Grayscale Vision System for Iconic Image Processing

P. Levi

University of Karlsruhe, W. Germany

ABSTRACT

This paper describes the architecture and the operation of a linear processor array for iconic image processing. The fundamental data structure are two dimensional matrices. The elements of these matrices are essentially gray scale intensity data (256 levels) but also distance data which are obtained by a laser scanner(triangulation). The parallel computer system is built up by 64 array processors. An array controller supervizes these processors. A micro-programmable address sequencer devides the input images into 64 windows. GVS can operate either in the SIMD or in the MIMD mode. It is mainly conceived for three-dimensional object descriptions which are needed for the part handling of an assembly robot. A first step is done, to combine intensity and distance data in order to get integrated intrinsic images. Internal 3-D models are based on linear octtrees and are generated by the aid of distance data.

* This work was supported by the Bundesministerium für Forschung und Technologie of the Federal Republic of Germany.

1. INTRODUCTION

The operations of iconic image processing influence predominantly the architecture of a vision system. The characteristic features of this kind of image handling are a high data rate ($10^5 - 10^7$ samples/sec), an explicite parallelism (high data independence), a nonconfigurable topology (static) and an operation mode which is SIMD-oriented, /5/. Operations which are more complex and which have to be performed by symbolic image processing require configurable structures (dynamic) and the use of a MIMD mode.

GVS (Grayscale Vision System) is mainly drafted for iconic image processing. But, the architecture of a vision system should also fulfil the different structural requirements of the image processing algorithms, /4/. Appropriate processing structures should be disposable at the different stages of the pattern recognition. For this reason, GVS operate also in the MIMD mode. However, we are just starting to develop some operations for the symbolic image processing (e.g. list processing, tree search).

The complete system is built up by standard units. It is based on MOBIP /9/, and it includes some new features (e.g. laser-interface, pseudo colour, new host computer, X-FORTH). The communication pathes of the processors are defined by a linear array. Each of the 64 array processors is individually programmable in order to get a maximum of flexibility. There is no use of special chips for the destination of workpiece orientation, for the evaluations of convolutions etc.

The calculation of intrinsic images is improved if they are computed simultaneously, since the problem of boundary conditions is solved easier. This system combines two types of sensors in order to calculate intrisic images. Distance data are directly measured (laser scanner) and not by the aid of two cameras. These data are also mainly used for the calculation of the three dimensional symmetry of the parts. Intensity data are get by a conventional CCD-camera.

2. HARDWARE ORGANIZATION

GVS operates in connection with a EUROCOM II host computer, which is based on the 8 bit microprocessor Motorola 6809. This host computer includes an operating system and performs the complete communication with the end users (standard peripherals). The main building blocks of GVS are the microprogrammable address sequencer (AM 2910), the array controller (6809) and the 64 array processors (6809), Fig. 1. The data exchange and the synchronization between these units are supported by three buses.

The picture data bus (5 M Byte/sec) transfers private data from one of four picture memory banks (à 64 k Byte) to an individual array processor. This bus is also used for the data exchange between the video I/O and the picture memory. The video I/O is connected to a camera or to a laser scanner in order to get intensity and distance data. The unidirectional instruction bus connects the array controller with the array processors in order to transfer the instructions to each individual processor (global data). In addition, this bus can also be used for a parallel down-line loading of the array processors by the array controller. The data which are transfered by this operation are for example weighting coefficients of a 3 x 3 convolution. The exchange of neighborhood data between adjacent array processors is performed via the interprocessor bus.

2.1 Address Sequencer

The partition of the input image (256 x 256 pixels) as well as the scheduling of the transfer of these data to the array processors are performed by the address sequencer (cycle time: 200 ns). This unit also controls the activities of the picture data bus. Usually, for local image operations the input matrix is devided into 64 columns. The width of a column is one or more pixels, depending on the image size, Fig. 2. The adjacent columns are transferred via the interprocessor bus.

The address of the picture memory and the address of an array processor are both generated by the address sequencer. The data transfer is done in blocks. The block length
corresponds to the number of array processors. The picture data bus and the array processors are asynchronously connected via I/O registers. Therefore, the input/output of
data and the evaluation of the iconic image processing instructions can be done simultaneously. The addressing of the I/O registers is done by a counter. Therefore, the
array processors can only be sequentially addressed.

2.2   Array Controller

The array controller supervizes the operation of the array processors in the SIMD mode.
It is also connected to the host computer and to the address sequencer. The memory of
the array controller is a global memory (14 k Byte). It stores the program code for
each of the 64 array processors. The operands of the image operations are stored in the
local memories of the array processors. Program branchings which are caused by the local data of the array processors are not allowed.

The main functions of the array controller are:

- selection of the operation mode (SIMD, MIMD)
- preparation of the next instructions (SIMD mode)
- control of the array processors
- transfer of data between the host computer and the array controller
- transfer of data between the address sequencer and the array controller (via picture
  data bus).

2.3   Array Processor

The main parts of an array processor are (Fig. 3):

- CPU MC68B09
- 2 k Byte RAM (local memory)
- Interprocessor (IP) register
- drivers for the instruction bus.

The 64 array processors can either operate in the SIMD or in the MIMD mode. In the
SIMD mode the instructions for the individual processors are stored in the global memory of the array controller. The control signals of the instruction bus supervize the
read operations of the array processors in the global memory.

To prepare the MIMD mode, the private data and the different instructions of the array
processors are transmitted from the picture memory to the local memory of each array
processor via the picture data bus. The array controller starts the program evaluations
and waits for the flag signal of each array processor, indicating the end of a program
execution. Special synchronisation tools like hardware semaphores are not available.

3.   SOFTWARE ORGANIZATION

The realization of the GVS-Software can be described by a four level approach. Fig. 4
schows the software organization. Level 4 represents the user interface. The image processing operations are primarily combined with X-FORTH, /2/. In addition, the languages
PASCAL, C and FORTRAN 77 are available. Level 3 is defined by the FLEX-Operating System
of the host computer. Level 2 is described by the macroassembler for the host computer.
Level 1 is represented by the two macro assemblers for the array controller and the
array processors, and by the horizontal micro code for the address sequencer.

The operations of the iconic image processing are usually devided into three programming packages which correspond to the three hardware units. These programs are running
in parallel. The programs are:

- control program of the image processing routines (host computer)

- micro program for the address sequencer. This program generates the accesses to picture memory. In addition, it establishes the connection between this memory and the array processors
- array controller program to start the array processors.

3.1   User Interface

All GVS-operations are included into the operating system of the host computer. They can be activated by special commands which are embedded for example in X-FORTH. This language is well suited to build up new image processing operations by a given set of standard service routines, which are available as moduls of a function library (macros or subroutines). The operation sequence, storing of an input, image loading of the array processors, performance of the Sobel-operator, transfer of the resulting image to the picture memory and finally the transfer to a monitor can be written in short (X-FORTH) as following:

    OPERATION 1, IMAGE, LDA, SOBEL, STA, DISP;

defining the new image processing function OPERATION 1.

All GVS-operations, which are invoked by the host computer are supervized by a special service program. This program starts and controls all activities which are needed for the execution of the GVS-operations. Therefore, it controls the address sequencer and the array controller by memory mappings. For the quick performance of time critical operations, there exists also an assembly language (host computer).

3.2   Programming of the Array Processors

The array processors as well as the array controller can be programmed by the supporting software of the host computer, since all these three blocks have the same CPU. This means, that for example the array processors can be programmed by X-FORTH. However, in practice there is a specialized assembler for the array processors. It is supported by a library of macros.

In the MIMD mode the array processors are all programmed as independent units. But, the global registers (I/O-, interprocessor registers) and the global memory of the array controller can not be used in this mode. For data exchange, the MIMD mode must be switched to the SIMD mode.

In the SIMD mode all global units can be accessed. However, program branches, which are caused by the local data of the array processors are forbidden, in order to guarantee a constant instruction stream. This problem can be solved by a linearization of the branching control instruction (address calculations).

3.3   Programming of the Address Sequencer

The programming of the address sequencer is supported by a microprogram assembly language, which is resident at the host computer. The word length of the micro code is 40 bits. The programming of new, basic GVS-operations presumes a thorough knowledge of the operation mode of the address sequencer. For this reason, a conventional user can not program this part of GVS.

4.   PERFORMANCE

GVS can be used for a wide range of iconic image processing. Typical tasks are filtering (linear, non linear), histogramming, image enhancement and restauration, grayscale statistics and convolution. Table 1 summarizes the performance data for some operations. The image size is 256 x 256. There exist 256 grayscale levels. The complete system (64 array processors) operates with 30 MIPS.

Operation	Window	performance
High pass	3 x 3	62 msec
Sobel		82 msec
Median	3 x 3	770 msec
Hadamard		580 msec

Table 1: GVS execution time

The range data are obtained with the aid of a triangulation laser scanner, /11/, which is controlled by GVS. This control implies the random adjustment of each laser spot (tracking) and the setting of the scanned image size (68^2, 128^2 or 256^2). In addition, it is possible to define the structure of the projected light (point, grid or line). The camera is used either for the generation conventional intensity data or as an optical sensor for the measurement of the distance data. The calculation time for each distance point is performed by GVS and is about 50 msec.

5.   APPLICATIONS

The intrisic parameters which are calculated in the first step are based on the distance data. These descriptions are the distances according to the laser (central projection) or to the laser/camera plane (parallel projection) defining the location and orientation of a workpiece, the local surface normals (gradients) and the Gaussian curvature. But, the determinations of these characteristic object features from one signal source is very often not unique. Fig. 5 shows a camera picture of a gear and the corresponding distance picture.

It is obvious, that specular surfaces can scatter the impinging laser beam in directions, which can not be observed by the camera. As a result, the distance picture shows black locations (e.g. shaft of the gear). Therefore, the extraction of intrinsic images should be performed by a combination of distance and intensity data. We are now starting to perform these mutual corrections. The common basis of this approach is the topographic primal sketch of Haralick, /6/. The advantage of this approach is the unified treatment of the grayscale images (intensity = g(x,y)) and the range images (distance = g(x,y)) on the base of the Hessian matrix. We calculate at each point, the principal curvatures, the gradients and the projection of the grandients into the directions of the two principal curvatures.

With the help of these two types of data, the intensity data are used to correct the erroneous and uncomplete representations which are generated by distance data. The resulting intrisic image is called an integrated image. In addition, this integrated intrisic image is used to differentiate between different types of edges (e.g. reflectance or distance discontinuities).

The resolution of a camera is not high enough, in order to determinate exactly the reflectance of an object surface. For these reason, we measured in previous experiments the differential reflectance functions of some materials which are often used, like sheet metal, plastic, paper, etc. by the aid of a photomultiplier, /7/. It has been shown, that materials and surface defects can often be determined, be the measurement of the reflected irrandiance. For this reason, a multisensorial vision system should also incorporate a reflectance sensor (e.g. photomultiplier), in order to define exactly the intrinsic parameters of object reflectances.

For robot vision three dimensional object representations are demandatory. For example, this is true for the definition of the gripper points and for the modeldriven interpretation of partial views. The merging of different partial views of an object is performed by distance data, /1/. These data are more suited for the merging operations as the intensity data, because only four corresponding points in two different views must be found (coordinate system and its origin) instead of one hundred or more indentical points, which ared neeeded for stereo vision. Distance data minimize the correspondence problem. The internal representation of the merged, 3-D models is given by linear oct-trees, /3/.

The symmetry definition of three dimensional objects are performed by the evaluation of the moments of inertia and of the Gaussian curvatures, /8/. The volumetric elements which can be described till now are rectangular parallel epipeds, ellipsoids, cylinders, cones, pyramids, spheres and cubes. The three principal axis of inertia for a cuboid and an ellipsoid are different and define symmetry axis. But, for example a cylinder has only two different eigenvalues of the inertia ellipsoid. For a sphere, all three eigenvalues are equal. This symmetry classification can be made unique, if also the Gaussian curvature is evaluated. The calculations of the inertia ellipsoid and the definition of other three dimensional features (e.g. center of gravity) is performed by the partition of the observed area into volumetrical cubes (voxels). For this reason, the distance data are also calculated in the view of a parallel projection.

In a first approach, the applications reported in this chapter have been implemented at an Intel 8086 system (PASCAL). At present, there is an effort to transfer these operations, as fare as the generation of the integrated intrinsic images is concerned, to GVS. The main reason to evaluate these algorithms in a first step at a conventional machine and not directly at GVS is caused by the difficult, parallel programming of this system. For each task, the decomposition into parallel subtasks which have to be allocated to the different array processors and the composition of the sub-results into a task solution by an extended synchronisation mechanism have to be performed. Note, to think in sequential steps is for a programmer much more easier than to do it in parallel.

6.   CONCLUSIONS

GVS represents just a small step toward a vision system which is more general. Such a system should be a multi-sensorical system (increase of reliability) and it should generate integrated intrinsic images at the base of additional knowledge. These knowledge sources should incorporate the illuminating conditions, the generic descriptions of the image generating system and it should be able to produce synthetical images (picture interpretation), as it has been pointed out by Nagel, /10/. At the present, GVS does not exploit these knowledge sources.

These additional physical levels of knowledge can be profitably used, to estimate on the base of intensity data the parameters of intrinsic images more precisely than it can be predicted by GVS. For example, the exploitance of global illuminating conditions and intensity discontinuities (local edge boundaries) can refine the evaluations of surface normals, /12/.

ACKNOWLEDGEMENTS

The author is gratefully indebted to the gentlemen P. Martini and G. Nehr for the reali zation of the basic block of GVS (MOBIP). H. Stiefvater installed the hardware extensions of MOBIP. P. Fenkart has performed the new programming of GVS and its connection to X-FORTH. To both colleagues I am very much obliged for these extensions. I would also like to express my gratitutde to the gentlemen R. Dienst and B. Wetzel (Daimler-Benz AG) for the disposition of several industrial parts (e.g. gears).

7.   <u>REFERENCES</u>

/1/   Biwer, F. "Verschmelzung von Mehrfachansichten zur Volumendarstellung von 3D-Objekten", Diplomarbeit, Lehrstuhl Prof. Dr.-Ing. U. Rembold, (April 1984)

/2/   Eaker, Ch. "X-FORTH User's Manual", Frank Hogg Laboratory, Inc., (1981)

/3/   Gargantini, I. "Linear Octtrees for Fast Processing of Three- Dimensional Objects". Computer Graphics and Image Processing 20, pp. 365-374, (1982)

/4/   Gemmar, P., Ischen, H. and Luetjen, K. "FLIP: A Multiprocessor System for Image Processing". In: Languages and Architectures for Image Processing (eds. Duff and Levialdi), pp. 245-246, (1981)

/5/   Giloi, W.K. "Innovative Bildverarbeitungsarchitekturen", Proceedings of the workshop: Parallele Rechnerarchitekturen in der Mustererkennung und Bildverarbeitung, Karlsruhe, (November 1983)

/6/   Haralick, R.M., Watson, L.T., Laffey, Th.J. "The Topographic Primal Sketch". The International Journal of Robotics Research, Vol. 2, No. 1. pp. 50-72, 1983

/7/   Levi, P. Weirich, E. "Differential Reflectance Functions and their Use for Surface Identification". Proceedings of the 13th International Symposium on Industrial Robots (ISIR), Chicago, pp. 17.61-17.77, (1983)

/8/   Levi, P. "Bestimmung dreidimensionaler Symmetrieeigenschaften auf der Basis von Trägheitsmomenten", 5. DAGM Symposium, Karlsruhe, VDE-Fachberichte 35, pp. 101-106, (1983)

/9/   Martini, P., Nehr, G. "MOBIP: Ein modulares Bildverarbeitungssystem mit Parallelrechnern". Elektronische Rechenanlagen, Nr. 25, pp. 55-65, (1983)

/10/ Nagel. H.-H. "Über die Repräsentation von Wissen zur Auswertung von Bildern". Informatik Fachberichte Nr. 10, pp. 3-21, Springer-Verlag, (1979)

/11/ Stiefvater, H., Vatja, L., Levi, P. "Triangulations-Laser-Scanner: Dreidimensionales Sehen für Industrieroboter". Elektronik, Nr. 26, pp. 119-122, (1983)

/12/ Stuth, B.H., Ballard, D.H., Brown, Ch.M. "Boundary Conditions in Multiple Intrinsic Images". Proceedings of IJCAI 83, Karlsruhe, pp. 1068-1072, (1983)

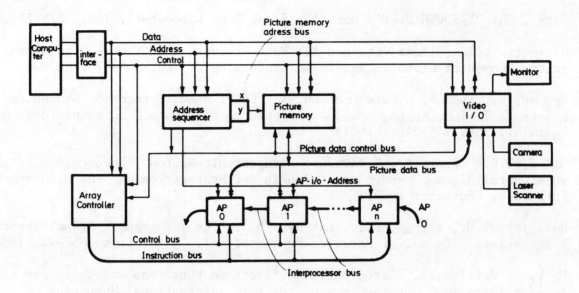

Fig. 1. Hardware organization of GVS

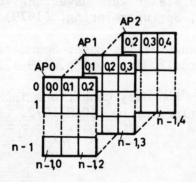

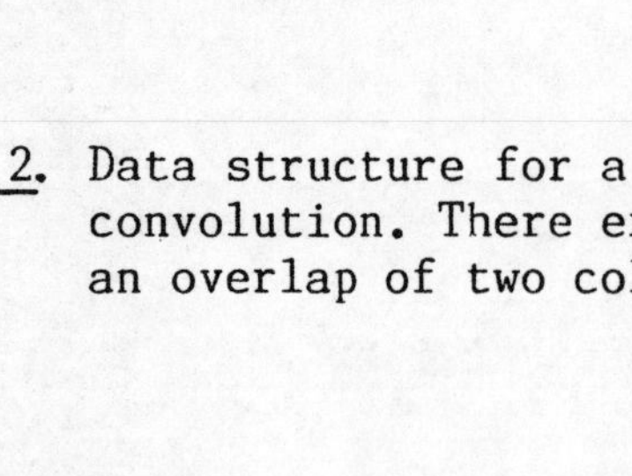

Fig. 2. Data structure for a 3x3 convolution. There exists an overlap of two columns

XFORTH combined with image processing operations		
Pascal combined with MOBIP library		
C		
FORTRAN 77		
FLEX-Operating System		
6809-Macro-Assembler		
6809- Assembler for the Array- Controller	6809- Assembler for the Array- Processors	horizontal Microcode for the Address- Sequencer

Fig. 4. Software hierarchy of GVS

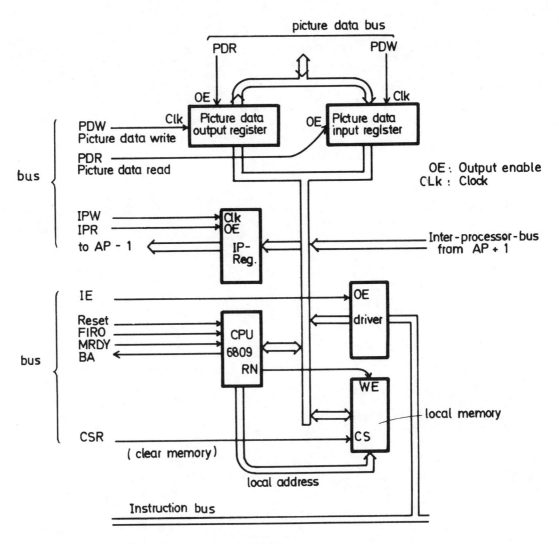

Fig. 3. Organization of an array processor

Fig. 5. Intensity picture of a gear and the corresponding distance picture

Vision – New Sight and Foresight

H. A. Laird
and
D. McKeag
Ulster Polytechnic, UK

Throughout Industry there are many processes which depend on human skills for their successful completion. It is in areas such as these where craftsmen use their skill and experience that the use of artificial eyes is being encouraged - the objectives being to improve consistency, increase efficiency, relieve monotony and reduce human fatigue caused by repetitive tasks and also to obtain information for subsequent analysis.

This paper looks at the components of vision systems and those which are commercially available, with particular interest in their suitability for metal forming operations. One area within this field where vision systems are applicable is in the analysis of the straining caused by deformation processes - a time consuming exercise when the conventional grid circle analysis techniques are used.

Ultimately vision systems will acquire the versatility of the human eye and become the new eyes for Industry.

1. INTRODUCTION

During the last decade the development of new equipment has been accelerated in metal processing and, with the recent advances in machine vision,(1) automation is percolating into the area of metal forming. This automation incorporates the advancing technologies of industrial robots, intelligent vision systems, computer aided design and computer aided manufacture.

Metal forming techniques have been developed by engineers with a practical knowledge of presswork. In the past this approach has sufficed but a greater theoretical understanding is now desirable. There is no doubt that the knowledge accumulated by engineers with industrial experience of deep drawing, has much to offer towards the solution of problems in this area. However, the application of their skills is lacking in a technological approach and there is a lack of scientific definition of the skills.

The forming limit diagrams (Figure 1) of Keeler (2) and Goodwin (3) have found applications where the problem is predominantly one of stretch forming and they have been shown to be useful for the development of deep drawn products in the first stage of the forming process. However, in deep drawing, as the size and complexity of the form of the component increase, the number of draws with interstage annealing increases and hence the forming problems increase.

One approach to this problem is to extend the Forming Limit Diagram technique in conjunction with an Image Analysis System to obtain a quantifiable record and an accurate indication of the approach of material instability in draws between interstage anneals. Given suitable software the Image Analysis System could provide automatic warning of impending metal failure.

2. WHAT IS AN IMAGE ANALYSIS SYSTEM?

An Image Analysis System may be described as a system capable of giving a quantitative analysis of information presented in a pictorial format. Basically it is a vision system which possesses a certain analogy with the sight capability of a human being. In a vision system the retina of the human eye is replaced by a photosensitive array which can receive optical images and convert them into electrical signals. The system also has a "brain" which is a computer for the processing and evaluation of the image data.

A typical vision system (Figure 2) includes an electronic camera with suitable illumination, a computer with necessary software and peripheral equipment such as visual display units, control panels and interfaces for communication. These components are grouped into two main functional blocks:-

Block 1 - data acquisition and conversion which consists of the camera and the illumination

Block 2 - data processing, interpretation and decision making which covers the preparation of data for analysis and the analysis thereof, ie the computer hardware and software.

2.1 Block 1

2.1.1 Illumination The purpose of the illumination provided by a variety of light sources, mirrors and lenses, is to allow information to be obtained from the image of the object and also to provide adequate distinction between the background and the object. The illumination may be either Through Illumination or Reflected Illumination. The former allows objects to be distinguished from their background easily, but only binary images are produced. The latter may be influenced by extraneous light sources but more information may be obtained with the multi-grey level images.

2.1.2 Cameras The cameras currently available for vision systems include thermionic tube cameras and solid state cameras, which may be either line scan or two dimensional array cameras. The solid state cameras are more compact and rugged than the thermionic tube type, however the line scan variety require relative motion of the object across the line of view of the camera. The self-scanning two dimensional array cameras are more expensive but they are particularly useful for gauging applications because of their dimensional accuracy and stability. Regardless of the type of camera, the optical image is projected onto a photosensitive array which is then converted into an electrical signal. The magnitude of this signal varies with the luminance of the image points.

2.2 Block 2

2.2.1 Hardware The hardware of the data processing, interpretation and decision making block comprises a series of microprocessors, a signal processing unit and related electronic equipment. The microprocessors for the data analysis are in the form of a general purpose computer. Peripheral equipment such as visual display units and interfaces are used in conjunction with the computer for communication with the system and any external devices being controlled by it.

2.2.2 Software Vision systems have the same basic hardware components, but the software with which they are provided depends on the potential applications of a particular system. A system designed for use in cell biology requires different software from one which is designed to control an arc welding robot. The computer programmes in the data processing, interpretation and decision making block have two main functions:-

(a) To control the operation of the system.

(b) To prepare the data for analysis and carry out the analysis using suitable algorithms.

Standard programming languages are used for vision system software except in cases where speed is critical. In these cases assembler language may be used.

3. PRESENT APPLICATIONS OF VISION SYSTEMS

The technology of vision systems is now such that systems may be employed in manufacturing environments, present applications including inspection, control of operations and control of components. Other fields in which vision systems are used include metallurgy, biology, zoology, pathology and medicine.

3.1 Inspection

Vision systems are in use in industry for both qualitative and quantitative inspection. Qualitative inspection covers the inspection of components for cosmetic appearance, for surface finish, for flaws or broken parts, for completeness and pattern recognition. Quantitative inspection includes measurement of angles, lengths, areas and distances and also counting applications. In the other fields mentioned above vision systems are used mainly for inspection.

One example, in day to day living, of the use of vision systems for inspection is in the reading of the bar codes which are appearing on food packaging. A system has been set up in one Scottish supermarket to read these bar codes at the check out and produce a list of the items purchased, together with the total cost. This speeds up the whole process of grocery shopping and simultaneously provides an efficient means of stock control, as the computer can record what has been sold.

Reading bar codes is one of the potential applications of the Vidiscan System which has the ability to count items, take measurements, recognise an incorrectly positioned object and reject one which is sub-standard. Examples of more dedicated inspection systems are Keysight, from General Motors Research Laboratories and the Capsule Checker from Fiji Electric. Keysight is designed to verify the presence of valve spring cap keys on enginee cylinder heads and takes one second to inspect each assembly. The capsule checker is an automatic inspection system to detect faults in medicinal capsules.

3.2 Control of Operations

Vision systems are used in the control of operations such as welding or paint spraying, where the manipulation is carried out by a programmable device such as a robotic arm. These vision systems tend to be specialized and therefore require limited and dedicated software.

3.3 Control of Components

Belt picking, bin-picking, orientating, sorting and positioning are areas covered by control of components.

In belt picking operations the system is required to determine the position and orientation of a component relative to some reference point. This information is then used to control a robotic arm which positions the component correctly for subsequent operations. For sorting operations the system is required to recognise a variety of components and select all those which are similar using a number of grippers (4).

Although there are many applications of vision systems in the control of components, there are still a number of technological difficulties to be overcome before systems can be used in bin-picking. Another area where vision systems are unsuitable is in the recognition of non-rigid objects, which cannot be defined in terms of a defined two-dimensional pattern.

4. METAL FORMING TECHNIQUES

Traditional evaluations of formability are based on both fundamental and simulative tests. Tool design and material property requirements for successful stampings are determined either from an accumulation of many past trial and error attempts on similar stampings or from long statistical correlations with press performance data.

4.1 Forming Limit Diagrams

The amount of deformation which sheet metal will undergo before fracturing, depends chiefly on the drawing quality of the material, the design of the tooling, and the frictional conditions between the tool and the work piece. It is desirable to use the cheapest grade of metal having a drawing potential sufficient for the application and it is also desirable to achieve maximum deformation at each stage of forming. Forming Limit Diagrams are useful in determining how much deformation can be incurred without causing necking or fracture.

The Forming Limit Diagram relates the major and minor strains at the onset of necking in sheet metal samples, subjected to biaxial stretching. The curve produced by Keeler (5) was limited to predicting failure for all strains positive. Goodwin (3) developed the curve to include failures in the tension-compression region.

4.2 Grid Circle Analysis

A grid circle analysis system is used to obtain Forming Limit Diagrams. The method currently used is that described by Welch (6). The grid pattern is etched into the blank and a black deposit replated into the grid lines. Such a grid remains visible after severe forming operations and does not induce stress concentrations.

On forming the grid circles become distorted, generally taking the shape of ellipses. In order to determine the strains in the material the major and minor axes of the ellipses are measured and the strains calculated from the differences between these and the diameter of the original circles.

4.3 Measurement Techniques

Measurement of the axes of the ellipses can cause problems. In some cases a pair of dividers and a scale ruled in 100ths of an inch may be used. Although using dividers and a scale is the simplest method, Ayres et Al (7) claim that it lacks resolution at

58

strains less than 5%. They suggest the use of a toolmakers microscope as a more precise, but time consuming alternative.

One method used in the current research programme requires acetate sheet, acetone and a travelling microscope. A piece of acetate sheet is dipped in acetone and placed over the area of the gridded metal where strain measurements are required. This acetate is allowed to dry. The acetate paper takes a replica of the grid pattern during the drying process. When dry the acetate can be peeled off and placed on glass slides, ready for measurement. A travelling microscope is used to measure the major and minor axes of the ellipses.

Another technique used requires a camera to be mounted at a fixed distance from the metal to be measured and take the dimensions from the image projected onto the screen, using a ruler. The replication technique described above is more accurate than this.

4.4 Improving the Techniques

Metal forming methods are widely used in the manufacturing industry, specific examples being the aerospace industry and the automotive industry. For reasons of economy, it is necessary to optimize on the cost of the metal and the cost of the manufacturing process. Accurate analysis of biaxial strain enables maximum formability to be obtained from a metal in any situation. In a single stage forming operation, the cheapest material to fulfill the function can be readily determined. In multi-stage forming operations it is necessary to minimize the number of forming operations and interstage anneals. The obvious solution to this problem is a vision system capable of pattern recognition, measurement and analysis.

5. THE USE OF VISION SYSTEMS IN METAL FORMING APPLICATIONS

In metal forming operations it is difficult to observe the deformation process and monitor the straining of the material. The difficulties encountered in any forming problem where the Forming Limit Diagram technique is used to monitor the process are:-

(a) the time and inconvenience involved in determining a data base for the Forming Limit Diagram of the material being used;

(b) the determination of the onset of material instability in an actual pressing ie the approach of the material to its forming limit.

The use of a computerized vision system to reduce the repetitive nature of the task involved in (a) and concurrently improve the accuracy of the ensuing results is an obvious and straightforward application, however, the determination of the onset of material instability is a more involved problem.

5.1 Determination of the Forming Limit Diagram

In order to obtain a meaningful Forming Limit Diagram, it is necessary to deform a minimum of seven specimens, and to measure both necked and fractured areas on each specimen (Figure 3). General Motors Research Laboratories have developed an Automated Grid Circle Analyser (7). This is a special purpose vision system, designed to measure the distortion of small circles, printed on the surface of sheet metal, as the result of a stamping operation. Forming Limit Diagrams produced by this method are as accurate as those produced using a toolmaker's microscope and may be used as data bases for future forming processes.

5.2 Determination of the onset of Material Instability using a Vision System

In forming processes, particularly deep drawing, more than one forming operation may be required to produce the end product. Generally speaking, the more complex the shape to be formed, the greater the number of stages required to produce it. During the forming process the material strains and if this straining is excessive then necking occurs, the material becomes unstable and ultimately results in fracture if the forming process is allowed to continue. To prevent this happening and to obtain maximum forming from the material, the drawing process must be stopped prior to the onset of material instability. The material should then be subjected to an interstage anneal before the next stage of forming, and so the process continues until the finished form is achieved.

It is possible to obtain an indication of the approach of material instability by comparison of the strains in the material at any given time with those represented by the Forming Limit Diagram. This involves stopping the forming operation and measuring the strains, using the method for determination of the Forming Limit Diagram. The problem which arises is knowing when to stop the process for the comparison. If forming is allowed to proceed too far the material may have become unstable, in which case failure has already occurred. Alternatively, if the forming process is analysed using an incremental approach, whereby the drawing is stopped at pre-determined depths of draw for comparison of the material strains with the Forming Limit Diagram, the time required and the volume of work increase in proportion to the number of increments. Obviously, the smaller each increment, the greater the volume of work and the more accurate the analysis.

Ideally, what is required to overcome this problem is a vision system which:-

(a) is capable of pattern recognition
(b) is capable of measurement
(c) has a computer which is capable of updating its memory bank
(d) is suitable for use in an industrial environment or in a laboratory
(e) lends itself to metal flow observation
(f) is capable of providing automatic warning of impending metal failure
(g) has the ability to control the forming operation
(h) is cost effective

This real time vision/image analysis system should be capable of obtaining material formability data and translating it into a visual display. It should also be useable as a production development tool, to monitor material formability and predict each drawing stage, thus reducing product development lead times and costs.

5.3 Industrial need for a Vision Analysis System

A local company is involved in the deep drawing of complex industrial forms using an aluminium alloy. To date, their approach to press forming has involved costly and time consuming trial and error methods, however they now recognise the need for a theoretical, as well as a practical approach.

A research programme is currently under way at the Ulster Polytechnic (proposed University of Ulster) in collaboration with this company. This work is aimed at developing the techniques of deep drawing of complex industrial forms. There are three clearly defined areas of investigation:-

a) analysis of the experienced engineers approach to problem solving and the introduction of a scientific system for tool and form development;

b) an examination of the deep drawing physical parameters;

c) an analytical investigation.

It is in the third of these areas that the need for a vision system arises. What is required is an image analysis system for evaluation, initially under laboratory conditions and ultimately during a practical forming process in an industrial environment. The system should have the facility to be programmed to recognise certain patterns, obtain certain measurements and perform an analysis of the data. The system will also be used in the formulation of mathematical models of empirical relationships arrived at as the result of concurrent investigations into material properties and deep drawing parameters. This information will be used subsequently as a means of computer analysis and prediction of form development.

5.4 Selection of a suitable System

In the search for a vision analysis system to carry out the tasks, a survey of literature and hardware pertaining to organisations responsible for the research, design, development, production and/or marketing of vision systems was carried out. This survey covered material originating from companies in the United Kingdom, Europe, Japan and the United States of America, together with information from local Businesses, Hospitals and Educational Establishments (1,8).

A number of the systems looked at are commercially available, others are still in various stages of development. The systems currently available cover a variety of applications, inspection and robot control being the most popular. Other systems are designed for more specific tasks.

Following a close examination of the systems on the market, it was decided that this industrial need would be best met by a specially developed, dedicated system. The reasons for this decision are:-

a) None of the systems available exactly met the specification.

b) Any of the few systems which could be modified to fulfil the task were ruled out because of financial constraints.

c) The specific nature of the task involved.

5.5 Development of the System

The system which is under development comprises an Apple lle computer with twin disc drives, a dithertizer, a camera and two monitors.

The Apple lle computer was selected because:-

a) It has 64k main memory locations and a further 64k from the 80-column extender card.
b) It is reasonably portable.
c) It is easy to programme and accepts both high and low-level languages.
d) It is a popular computer, with which people are familiar, and therefore should not cause operating problems.
e) There are a number of digitizers on the market designed for use with the Apple.

The digitizer being used is a Dithertizer lle. This is an interface card designed specifically for the Apple lle. The card allows the analogue image from a video camera to be converted to a digital image which can then be displayed as a graphic screen by the Apple. Images may then be saved to disc or printed on graphic paper. The speed of the dithering process is dependent on the number of grey levels (1 to 64) selected. The dithering process takes 1/60th of a second per frame and using 64 grey levels requires overlaying 64 frames.

The camera recommended for use with this digitizer is a Sanyo VCV 1500 S which is not available on the U.K. market. A suitable equivalent seems to be a National Panasonic WV-1500. Two monitors are required - one to display the analogue image, the other for the digitized image.

Software for the system is being developed at present. The results produced to date, in the experimental work, have been put onto the computer. These results include Forming Limit Diagrams, produced using the replication technique, and also some data from the physical parameters investigations. This information is being used to formulate mathematical models for use in the prediction of form development.

Although the hardware for the system has been selected, there is still a need for further software development before the vision analysis system is fully operational.

6. CONCLUSIONS

There are two main categories of vision systems; Non-Robotic systems are those which can indicate the presence or absence of a good component, whereas Robotic systems are capable of providing sufficient information to control an automated manufacturing system or operation. The main differences lie in the computing capabilities of the systems.

Vision systems have applications in a variety of operations, although the most popular areas at present are inspection and the control of robotic devices. The systems available tend to be of a dedicated nature and expensive. The current trend is to introduce vision systems into assembly operations and develop them for use in random operations such as bin picking.

Despite the number of systems on the market, it was felt that there was a need to develop a vision analysis system for use in metal forming operations, in an effort to reduce product development lead time and to make forming a less time consuming and costly process.

The ultimate aim is to have a vision analysis system which will not only carry out measurements and analysis, but will also take control of a forming process. As yet, the ability to control forming processes using a vision analysis system is Foresight.

7. REFERENCES

(1) PERA Report 366 "Vision Systems" April 1982

(2) Keeler, Stuart P.
"Circular Grid System - A Valuable Aid for Evaluating Sheet Metal Formability"
Society of Automotive Engineers, Inc.   Paper No 680092

(3) Goodwin, Gorton M.
"Application of Strain Analysis to Sheet Metal Forming Patterns in the Pattern Shop"
Society of Automotive Engineers, Inc.   Paper No 680093

(4) Baumann, Richard D. and Wilmshurst, David A.
"Vision System sorts castings at General Motors Canada"
Sensor Review   July 1982

(5) Keeler, Stuart P.
"Determination of Forming Limits in Automotive Stampings"
Society of Automotive Engineers, Inc.   Paper No 650535

(6) Welch, E H
"Gridmarking helps solve forming problems in aluminium"
Sheet Metal Industries   March 1974

(7) Ayres, Robert A., Brewer, Earl G., Holland, Steven W.
"Grid Circle Analyzer - Computer Aided Measurement of Deformation"
Society of Automotive Engineers, Inc. 1979

(8) Laird, Heather A., McKeag, D.
"New Eyes for Industry"
Chartered Mechanical Engineer   May 1983

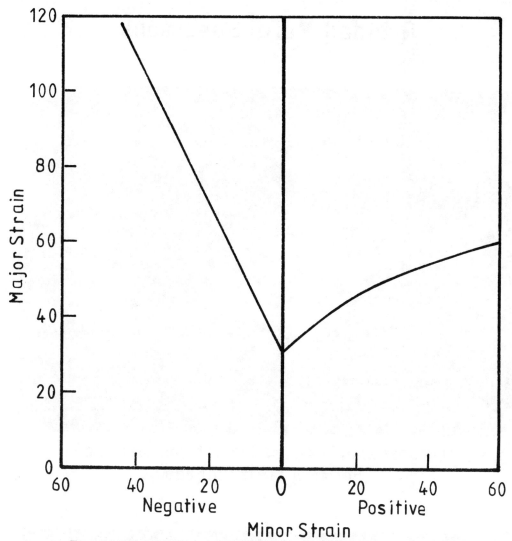

Fig 1. Typical forming limit diagram

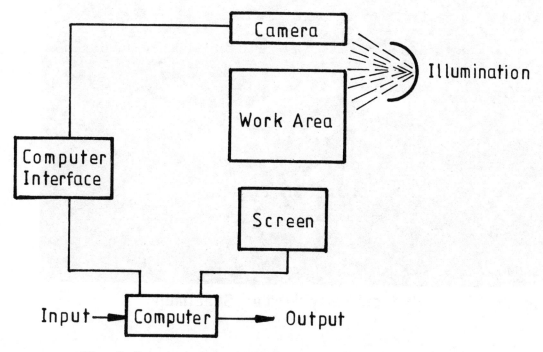

Fig. 2 Basic components of a vision system

Gridded Metal Specimens

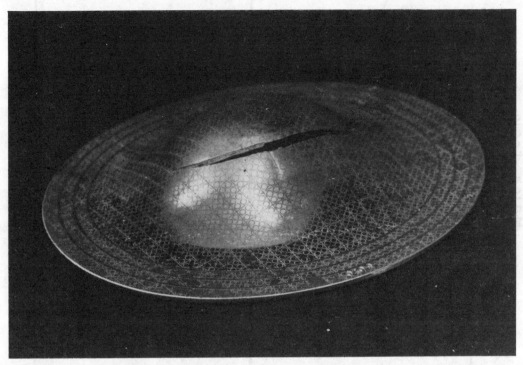

Fig.3a.   Largest specimen used in determination of
Forming Limit Diagram

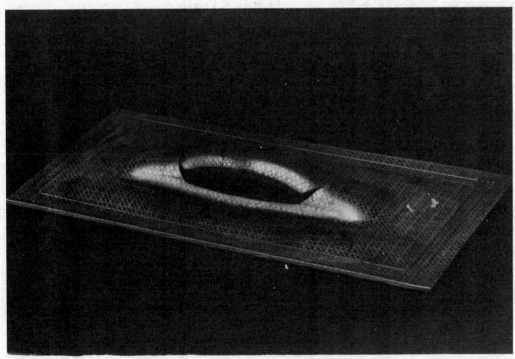

Fig.3b.   Hydraulically Formed Specimen

A set of Shapes for the Benchmark Testing of Silhouette Recognition Systems

L. Norton-Wayne

Leicester Polytechnic

and

P. Saraga

Philips Research Laboratory, UK

Benchmark testing is essential if silhouette image processing systems are to be evaluated and compared objectively, and advance in the state of the art demonstrated.

The authors have devised a set comprising 40 shapes selected to probe for particular weaknesses, which are described and explained. The set tests accuracy in orientation and location, as well as ability to distinguish shapes which are very similar.

1.   INTRODUCTION

Many computing methods for silhouette shape identification have been produced, and a correspondingly large number of systems have been developed for the automated recognition of the silhouettes. Some of these systems have been tailored to perform one specific task. For example some are designed to find the orientation of, and to inspect one specific piece for defects. Others, however, are intended to be applicable with equal facility to a wide range of problems, and are being marketed commercially.

When faced with a problem requiring the processing of silhouette shapes, engineers are often in doubt as to whether equipment which already exists is suitable, or something new must be developed for the task in hand. Even when commercial instrumentation is available, it may be difficult to determine which of a number of alternatives is best, or indeed whether any alternative fully meets the need. Thus, there exists a requirement for objective benchmark tests for instrumentation to process silhouette images.

The British Pattern Recognition Association is setting out to establish some standard benchmark tests. It seems logical to start with tests for systems processing silhouette shapes, since these are simplest. Tests for more complex kinds of pattern, such as those requiring several grey levels for their specification, can then follow. It is important that tests are accepted widely within the pattern analysis community if they are to be useful. The accompanying set of shapes is proposed as a first step; we hope that a process of discussion and compromise will then lead to a set which is accepted widely. BPRA would then act as a clearing house for distributing the set, for receiving and collating comments, and for promulgating the distillation of these comments to users.

In constructing the set, the following considerations have been taken into account:-
    (1) Instrumentation for processing silhouette images is generally required to perform a range of tasks. These tasks include identifying silhouette shapes, determining their orientation and location, detecting and classifying defects in a shape. A good series of benchmark tests must cover all of these.
    (2) Some systems can handle only a few shapes, in extreme cases only one. Others may be required to handle several thousand. For those in the latter category, a facility for efficient self training is a useful attribute. Such a facility would enable the system to learn an additional shape by observing it, and adding it to the shapes already stored.
    (3) There are many attributes which contribute to the overall performance of a system. These include resolution, speed of identification, reliability (expressed by rejection and substitution rates), accuracy of orientation and location, ability to discriminate shapes which are closely similar, and tolerance of minor disturbances such as specks of dust. These attributes are strongly affected by resolution. Thus, specification of performance without including a specification of resolution is meaningless. The variation of performance with resolution is a sound and useful measure.
    (4) The wide variety of methods available for processing silhouette shapes reflects the ingenuity and desire for originality of researchers, as well as the differing needs of the wide range of problems.   Pavlidis[1,2] has pointed out that these methods can be grouped into two classes, which we shall call asymptotic and non-asymptotic. Asymptotic methods, are guaranteed to be able to discriminate any two classes of shape, however close they may be, simply by increasing the resolution of the system. If a 256 by 256 grid is too coarse, then 512 by 512 may well be successful and so on. Characterisation of a pattern by the lengths of radii measured from centroid to boundary, at well chosen angles from some datum such as the longest radius, is one example of an asymptotic method. Non-asymptotic methods, on the other hand, do not guarantee discrimination merely by increasing resolution. An example of a non-asymptotic measure is provided by (perimeter)2/area; one can produce patterns (such as shapes B5 and B5* in the evaluation set) which have the same value for (perimeter)2/-area, and are yet completely distinct. This is an example of the kind of weakness for which the evaluation set must probe.

2. ALTERNATIVE APPROACHES TO THE CONSTRUCTION OF A SET OF SHAPES

Many possibilities exist. One approach is to produce a large set of silhouette shapes, with subsets from many task areas to accommodate the differences in shapes that occur from area to area. For example, engineering components are generally quite different from parts of garments. The latter are much less likely to have boundaries which are straight lines or to have internal holes. An instrument designed specifically for one kind of task might be expected to perform poorly on the other. A drawback to this approach is cost. Large sets of patterns are expensive to reproduce and distribute, and are also time consuming to examine.

An alternative, and more economic approach, is to use much smaller numbers of shapes, each designed to test some particular property or potential weakness of the system. This approach was adopted by Rosen and Gleason[3]. They proposed two series of test shapes. The first contained a number of simple geometrical forms; disk, square, equilateral triangle and rectangle, with and without a single hole centred at the axis of symmetry. The second set contained a set of disks 2" in diameter, containing from 0 to 7 holes, each of 1/4" diameter, disposed symmetrically about the axis. They demonstrated the use of their set to test one particular vision system. One clear merit of their approach is that since the test patterns are simple geometrical shapes, it is not necessary to reproduce them centrally for distribution; a potential user simply draws his own. However, real silhouette objects are normally far from being simple shapes. They are often quite complex, and are rarely symmetric. Therefore, Rosen and Gleason's shapes may not reveal potential weaknesses that would appear when a system is applied to these real shapes.

Our approach lies somewhere between the two extremes described above. The evaluation set that we have produced includes two kinds of pattern. The first is designed to test accuracy of orientation and location, the second tests the system's capacity to discriminate pattern pairs. The samples are designed to test specific properties, such as ability to tolerate internal holes, susceptibility to dust, ability to discriminate left-handed components from right-handed components, and so on. We hope that with a total of some 40 patterns, it will be possible to obtain a reasonable indication of a system's ability to handle satisfactorily substantial numbers of shape classes. We acknowledge however, that complete evaluation of this factor is not possible with our set.

3. PROPOSED BPRA SET OF SAMPLE PATTERNS

3.1 Introduction

In this section, we explain in some detail the properties of the shapes proposed for inclusion in the BPRA evaluation set. The shapes are of four types called A, B, C and D. Types A and B are carefully devised drawings, while types C and D are silhouette photographs of real mechanical parts for which automatic assembly or inspection may be required.

The complete set of shapes is also divided into two categories on the basis of the required task. Shapes of types A and C are intended to be considered individually, and to test for precision in determining location and orientation. Shapes of types B and D are to be considered in pairs, and are intended to test the ability of the system to distinguish one member of a pair from the other.

3.2 Type A Shapes

Shape A1 has been chosen such that the longest radius from centroid to boundary describes an arc subtending a substantial angle (about 90 deg.) at the centroid. This may confuse systems which use the longest radius as a datum for specifying orientation. Shape A2 has two identical longest radii, from the centroid (which lies at the centre of the semicircle) to the remote corners of the rectangle. Naive processing might equally choose one or the other as datum, leading to ambiguity.

3.3 Type B Shapes

Set 1 comprises a pair of patterns (B1 and B1*), which are right-hand and left-hand versions of the same shape.

Set 2 comprises two complex shapes which have a large ratio of boundary length to area. Though clearly distinct, the shapes are of similar construction, and have the same perimeter length and area.

Set 3 comprises two identical ellipse-like shapes. (Actually, they are a pair of semicircular segments joined by two straight lines). Holes have been inserted such that the position of the centroid remains unchanged. The two holes in each shape are different. Thus, although the centroid is at the centre of symmetry of the shape, and there are two identical longest radii diametrically opposed, the two radii are distinguishable. Also, the areas of the shapes (with hole area taken into account) are the same. Thus, the shapes should prove tricky to distinguish. The test could have been even harder by designing the holes to have the same boundary length as well as equal area.

Set 4 comprises two shapes having the same area. The 'bay' has the same area in both shapes, though its shape is clearly different. Methods based on convex hull measurement are likely to find these shapes hard to distinguish.

Set 5 comprises two shapes having the same ratio (perimeter)2/area which are never-theless clearly distinct. An infinity of shapes could in fact be devised having the same ratio of (perimeter)2/area.

Set 6 contains two shapes which are identical except that they are scaled in size by 2% linear dimension. In principle, any system with resolution 64 points by 64 or better should distinguish the pair, particularly if area is used as a feature. In practice, a rather higher resolution will probably be required because of quantisation effects at the boundary.

Set 7 contains two shapes with a slight eccentricity so that a system may find it hard to identify the major axis. The shapes are semicircles joined by small straight lines. In shape B7, the holes lie roughly on the major axis, while in shape B7* they are on the minor axes. The shapes have the same total area and length of perimeter. This pair might equally be included in type A since the ill-conditioning generates doubt in fixing orientation.

Set 8 contains two almost identical complex shapes, one of which has a minor defect inserted. The area of the shapes is unaffected by the error, though the length of the boundary has been increased, but to an extent small compared with likely errors in boundary length introduced by quantisation effects. The test here is to detect the presence of the defect.

Set 9 contains two identical shapes. However, B9* contains two 'specks of dust', of area equal to one notional resolution cell. These are isolated features which the system should ignore. Shapes B9 and B9* should be identified as being the same.

3.4 Types C and D

These shapes are produced by making silhouette photographs of small mechanical parts which are used in the electronics industry, and include such objects as springs, washers, heat sinks, contacts, and component holders. These shapes may cause problems because the ratio of boundary length to area is high, and errors due to quantisation into pixels occur at boundaries. Thus, feature values will be difficult to measure. Also, they tend to lack obvious features which may be used as a datum for establishing orientation or position. It is intended that the variety of shapes should test the versatility and flexibility of a trial system. Some shapes have been included to test the analysis of both internal and external features, and others to investigate the resolution of systems.

As mentioned in section 3.1, the shapes of type C should be treated individually. The task is to determine the location and orientation of each shape. The shapes of type D should be treated in pairs. In this case the task is to distinguish between the members of the pair.

4. SYSTEM ASSESSMENT USING THE SET

Most systems for recognising silhouette patterns start by extracting feature measures from the image immediately after scanning. This reduces the information which must be processed by possibly several orders of magnitude; data containing little information useful for the identification task is rejected as soon as possible. Measures used frequently as features include area, boundary length and so on. It is usual to choose as features measures which are easy to compute, such as total area, which involves merely a count of pixels having an intensity above (or below) a threshold.

It is possible to make some general statements about the intrinsic value of certain individual features. For example, one of the most important sources of error is quantisation noise at the edge of the silhouette. Thus, feature measures such as area and moment values are generally reliable since they depend on cells which are not on the boundary, whereas measures such as perimeter length which depend chiefly on boundary curves are less reliable. Algorithms have been devised[4] for reducing the error in boundary length measurement, but are expensive computationally. Errors in measurements of lengths of radii from centroid to boundary have[5] been shown to obey a binominal distribution, implying mutual independence. This may be exploited to optimise a system for minimum substitution or rejection rate.

However, since the overall performance of a system depends on much more than the type of features used, we suggest that one regards each system as a whole, and that at least the following measurements are made:

(a) Accuracy of orientation as function of resolution.
(b) Accuracy of location as a function of resolution.
(c) Processing time as a function of resolution.
(d) Pair discrimination ability as a function of resolution.
(e) Training time as a function of resolution.
(f) Time to process as a function of the number of patterns stored.

It should then be worth interpreting these tests in the light of the overall processing methodology and features used.

5. CONCLUDING REMARKS

To establish the practicability of the set we have used it to evaluate a system devised to identify silhouette shapes for the apparel industry, and to determine accurately the orientation and location of the shapes. This system has a resolution of 2048 points in the worst direction, and is thus more accurate than most for intended industrial application. It processes about 2 shapes per second, and can store several hundred.

It was however configured specially for performing a particular task, and in conse-quence uses a somewhat unusual processing approach, described in reference 5. When examining appropriate silhouettes (for which internal holes have no consequence and are ignored) their rejection rates are about 0.1%, and their substitution rates too small to be measured. With this evaluation set, however, the substitution rate measured was 2.5% (one pattern gave trouble), and rejection was almost 25%. Shapes B1,B1*,B2,B5*, B6,C3,C5,C8 and C11 were often rejected, and shape B7 was substituted for B7*. Rejection was probably due to shapes having properties with which the recognition system was not designed to cope, and their substitution to ill-conditioning.

On the other hand, the system seemed to find no difficulty in measuring orientation and position to the required 1 degree and 0.008" for any shape, excepting B7* which could not be identified.

This set of patterns is offered tentatively; comment and constructive criticism are welcome. We appreciate that the choice of shapes reflects the authors' experiences, and we would very much welcome suggestions for additional or alternative shapes.

It is imperative that the shapes be evaluated by the widest variety of systems, as soon as possible, and that the quantitative results emerging from such evaluations be fed back to the originators of the set and made available for dissemination to the pattern recognition community at large.

6. REFERENCES

(1) Pavlidis, T. Proc. 4th IJCPR Kyoto, Japan 1978 pp70-85.
(2) Pavlidis, T. "Algorithms for Shapes Analysis of Contours and Waveforms" IEEE Trans. on Pattern Analysis and Machine Intelligence.
(3) Rosen, C.A. and G.L. Gleason "Evaluation of performance of machine vision systems" Robotics International, 1980 paper MS 80-700.
(4) J-D Dessimoz. "Sampling and smoothing curves in digitised pictures" Proc, 1st EUSIPCO Lausanne, Sept. 1980.
(5) Norton-Wayne, L. "A coding approach to pattern recognition" in proc. 1981 Oxford NATO ASI "Pattern recognition-theory and practice" ed. Kittler, Fu, Pau, pub. Reidel, pp93-102.

Type 'A' Shapes

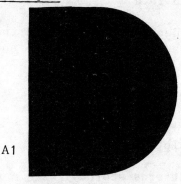

A1

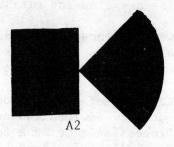

A2

Type 'B' Shapes

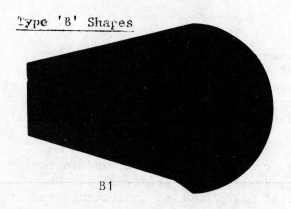

B1

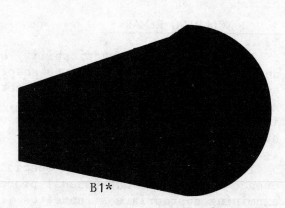

B1*

70

B2

B2*

B3

B3*

B4

B4*

B5

B5*

71

B6

B6*

B7

B7*

B8

B8*

B9

B9*

Type 'C' Shapes

C1

72

C2

C3

C4

C5

C6

C7

C8

C9

C10

73

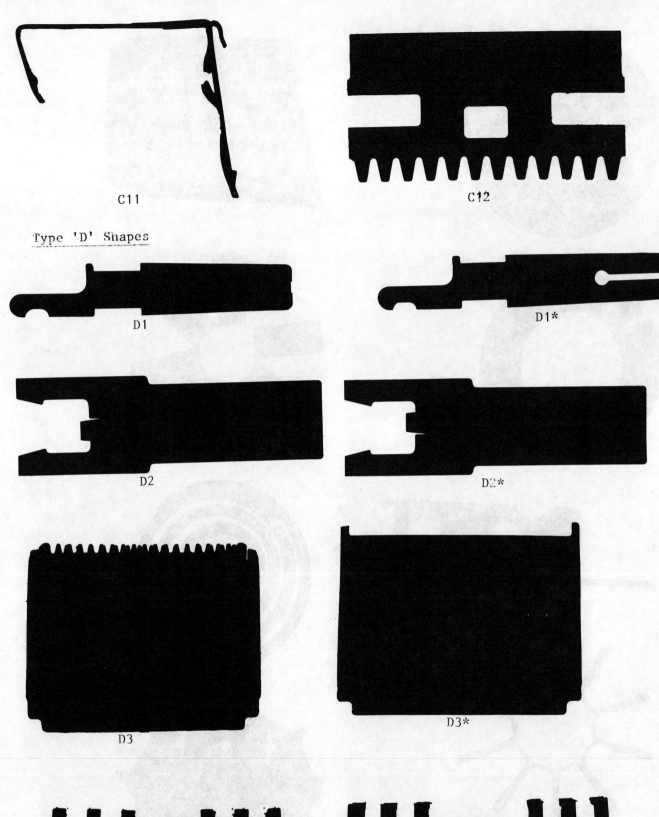

C11

C12

Type 'D' Shapes

D1

D1*

D2

D2*

D3

D3*

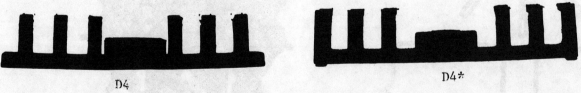

D4

D4*

Solid State Cameras

D. Lake
Fairchild CCD Imaging, USA

ABSTRACT

This paper will discuss the utilization of solid state imagers in complete cameras and camera systems. The discussion will be divided into three sections; the first dealing with Line Scan Cameras, the second, dealing with Matrix Cameras, and the third, dealing with cameras which do not image directly through a lens but rather utilize such things as fiber optics and intensifier tubes. Within each section appropriate critical parameters will be emphasized. Such parameters include resolution, acquisition rate, monochrome versus color, dynamic range, noise, size and ruggedness.

SILICON IMAGING DEVICES

Solid state imaging devices, like most solid state technology, utilizes silicon as the fundamental material. This is because silicon is sensitive to radiation in the visible and near infared portions of the spectrum. The relative response of silicon and that of the human eye is shown in Figure 1. From this it is clear that, with the exception of the deep blue end of the spectrum, silicon response not only can be made equivalent to the response of the human eye, but can also be made to substantially exceed it.

Most solid state imaging devices are constructed from similar individual elements. First there is the photosite itself. This is the area in which the electrons that have been generated by the incident photons are collected. Second there is the transfer gate. This is the mechanism which transfers the electrons collected at the photosite from the photosite to the transfer register. Third there is the transfer register. This is the mechanism which moves the captured electrons from the optical area of the imager toward the periphery of the imager. Fourth there is the output amplifiers. This is the mechanism which converts the electrons into voltages useable by conventional electronic circuitry. There are a variety of techniques used to implement these functions as well as a variety of architectural schemes used to combine them into an imager. The remainder of this paper will concentrate upon the most commonly used techniques.

LINE SCAN CAMERAS

Line Scan Cameras are produced with Linear Imaging Devices (LID) as the inherent solid state imaging device. Therefore both LID and Line Scan Cameras have the identical properties producing one dimensional images via a single line of video information. The block diagram of a LID imager is shown in Figure 2. From this the four basic elements of all solid state imagers can clearly be identified.

There are many properties of solid state cameras, Line Scan or Matrix, which are a function of the imager itself. These will be explained while referring to the LID. The first category of these is that of non-ideal performance.

Crosstalk of Deep Carriers is one of these. This is shown in Figure 3. The longer the wavelength of the incident light the deeper into the silicon the photon travels before generating an electron. Since the photosite capture mechanism is at or very near the surface of the silicon, the deeper into the silicon an electron lies the weaker the capturing field strength is. The lower the field, the more likely the electron will wander randomly and be captured at a photosite other than that which it is associated. It is for this reason that the use of solid state cameras in the near IR must be very carefully, considered in that the image quality degrades significantly. It is also for this reason that solid state cameras normally include filters, typically a BG-38 type, which restricts the response to the visible spectrum. It is also for this reason that Modulation Transfer Function of solid state cameras is very heavily wavelength dependent.

Another major non-ideal property is that of Photoresponse non-uniformity, or PRNU. This is shown in Figure 4. There are two phenomenona shown. First is the low frequency non-uniformity. This is the difference that can be seen from one end of a sensor to the other. The second is the high frequency non-uniformity. This manifests itself as single pixel offset, either positive or negative, that can occur. Fortunately, these two phenomenona are constants in any given sensor and consequently can be easily corrected for if need be.

Yet another kind of non-uniformity is that of Dark Signal. Dark Signal is the electrons that are internally generated from such things as surface states, crystal imperfections, lattice defects and general impurities. The electrons that are generated are, eventually, captured by the photosite capture mechanism. Therefore, the longer the integration time of a solid state camera, the more Dark Signal noise there will be. Unfortunately, this parameter is sensitive to temperature, Dark Signal doubles for every 5 to 10°C increase, and also that the Dark Signal generation rate is not uniform over the entire device. Dark Signal, therefore, generates an annoying spatial noise.

Solid State Line Scan Cameras offer several advantages despite their non-ideal response. A major advantage is that of resolution. Line Scan Cameras are available in two classes of resolutions, the binary progression class and the facsimile standard class. The former provides resolutions of 256x1, 512x1, 1024x1, 2048x1 and 4096x1. The latter provides resolutions of 1728x1, 2592x1, 3456x1 and 5184x1. The camera user is thus less impacted by optics, field of view; and camera location problems with such a range of resolution available to him.

Another major advantage is that of acquisition rate. The major parameter that limits speed performance is the actual number of lines per second that can be outputted from a Line Scan Camera. There are three factors that impact this rate, the speed of the device itself, the length of the line of video, and the degree of parallelism employed. Solid state images can output individual pixels at very high rates; some devices can operate in excess of 20MHz. None-the-less, very high resolution cameras with a single video output stream can produce only 1400 lines per second. Great care must be used in the trade-offs between resolution and acquisition rate.

Fortunately, the inherent advantage of the stability and precision of image spatial geometry help to ease the speed/resolution trade-off. Because the pixel locations are truly real, physical, entities the use of multiple cameras is very feasible. This is due to the fact that once the cameras are installed, all the misalignments can be handled as fixed offsets, offsets that are quite easily corrected for in the system supporting electronics. This property not only is useful in the speed/resolution situation, but also when imaging in separate portions of the spectrum and in using stereoptics.

Yet another advantage of solid state Line Scan Cameras is the wide dynamic range that can be achieved. Dynamic ranges, as defined relative to RMS noise are typically from 2500 to 1 to 5000 to 1. This translates into 11 to 12 bit accuracy when encoding the video output into a digital representation. This is normally much more than is used by vision processing equipment.

Line Scan Cameras are, in general, constructed so as to provide an ideal focal plane behind standard optics and to provide the minimum of electronics necessary for ease of use. The block diagram of the electronics is shown in Figure 5. Basically, the electronics frees the user of having to optimize the operation of the selected solid state imager. When clock and exposure are provided, video and synchronization are outputted. It is then left to the user to provide all of the electronics required to process the video. Since there are no international standards pertaining, to Line Scan Cameras, it is not practical for camera manufacturers to provide any additional electronic integration.

This electronics limitation is also true when using Line Scan Cameras with full color capability. Full color solid state sensors are implemented by depositing appropriate color filters directly on the individual pixel locations of the sensor itself. The technique of implementation on a LID device is shown in Figure 6. Moving from right to left along the photosites the tetrad of green, white, yellow and cyan (G,W, Ye,Cy). This results, on a LID with 1024 pixel locations, in 512 luminance samples and 256 red, green, blue (RGB) triplets. The first order color signal separation is as follows:

Luminance Signal $\quad V_L = V_W + V_G = V_{Cy} + V_{Ye}$

Color Signals $\quad\quad V_R = V_W - V_{Cy}$

$$V_B = V_W - V_{Ye}$$
$$V_G = V_G$$

The ultimate precision of the resultant color is a function of how the individual pixels are processed. Therefore, this block of electronics is not normally included into a Line Scan Camera.

MATRIX CAMERAS

Matrix Cameras are produced with Area Imaging Devices (AID) as the inherent solid state imaging device. Therefore, both AID and Matrix Cameras have the identical properties of producing two dimensional images where one complete image is produced each video cycle. The block diagram on an AID imager is shown in Figure 7. From this the four basic elements of all solid state imagers can clearly be identified.

The discussion of the various inherent characteristics of Line Scan Cameras apply to Matrix Cameras as well. However, because of the increased flexibility of application of Matrix Cameras, there are other factors which take on more importance. This increased flexibility comes from the direct two dimensional imaging capability of the Matrix Camera. One dimensional scenes must be very carefully controlled so as to be assured that the subject of interest does indeed fall into the field of view. The added second dimension significantly reduces the care required for scene control.

One of the major things that is encountered in the easing of the scene control is the variability of lighting. This is normally expressed as wide range of average illumination levels and the occurrence of high intensity "point: light sources, where a point level source is defined as being a small number of pixels, typically less than .1%

The case of widely varying average levels is addressed by the inclusion of Electronic Exposure Control (EEC). The solid state AID sensor includes within it the mechanism to modulate the integration time of the photosites. Therefore, by including within the camera the circuitry to average total video signal with the desired characteristics, the solid state matrix camera can provide optimum response over a wide range of light levels without changing the F-stop of the lens.

The case of the high intensity point light source is addressed by the inclusion of Electronic Anti-Blooming (EAB). There is a finite limit to the number of electrons that can be held with a photosite. When this saturation level is exceeded the excess electrons are free to move through the silicon as noise electrons. Without EAB these noise electrons tend to be captured at adjacent photosites which causes them to appear brighter than they should be. This apparent brightness increase is called blooming. Solid State Cameras with EAB provide a mechanism at each photosite which captures the excess electrons and routes them into the ground buss.

Matrix cameras are, in general, constructed so as to provide an ideal focal plane behind standard optics and to provide the minimum of electronics necessary for ease of use. This is done in two parts, the sense head, as shown in Figure 8, which frees the user of having to optimize the operation of the selected solid state imager, and the camera body, as shown in Figure 9, which provides those functions which are required to produce a complete, stand alone, camera system which is fully compatible with international video standards. This two part approach, where the parts may be separated by several tens of feet, means that a minimum size and weight camera can be located in critical locations, such as the end of a robot arm, while still providing the desired imagery when it is required.

FIBER OPTIC COUPLED CAMERAS

The previous discussion on Line Scan and Matrix Cameras has made the fact assumption that the incident light that falls onto the sensor has come directly from the lens. There are, however, several classes of imaging where this assumptions does not hold. These classes include X-ray image convertors, deep-UV image convertors, streak cameras, and low light level image intensifiers. These cases are handled by providing a general purpose fiber optic interface to the solid state imaging device. A representation of how this is accomplished in the image intensifier case is shown in Figure 10.

The fiber bundle is coupled directly to the sensor, either AID or LID, with an oil interface that is less than 10 microns thick. By utilizing a fiber spacing of approximately 5 microns, each pixel on the sensor receives light from approximately 15 to 20 fibers. The other end of the fiber bundle then is terminated at a faceplate which allows the direct mounting of any of the several different types of spectrum handling apparatus.

SUMMARY

All solid state cameras are produced utilizing a solid state photon to electron conversion device, or solid state imager. Line Scan cameras utilize imagers which are one dimensional in nature. The noise sources in these solid state cameras come from deep carrier crosstalk, photo-response non-uniformity and dark signal. There are major advantages in terms of resolution, acquistion rate, stability and precision of image spatial geometry, and dynamic range. Matrix cameras produce a complete two dimensional image. These cameras have the added advantages of Electronic Exposure Control and Element Anti-Blooming. In addition, direct fiber optic coupling can be used to image other than incident visible light.

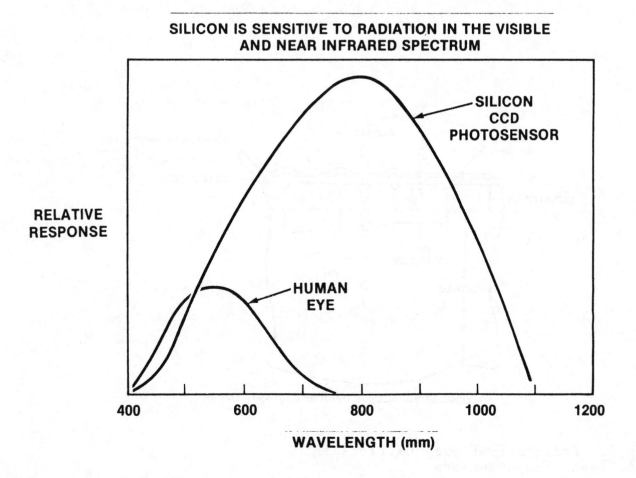

Fig.1. Photon to Electron Conversion Spectral Response Graph

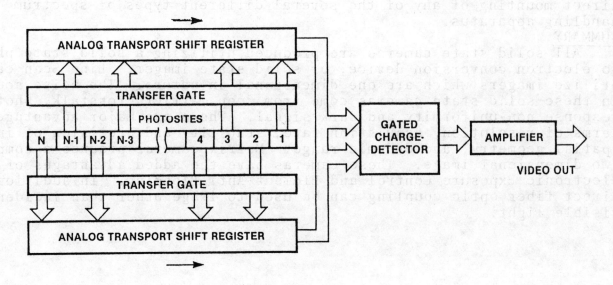

Fig.2.   Line Scan Sensor
         Block Diagram

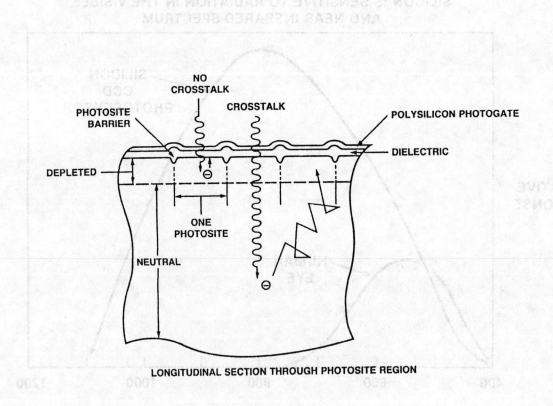

Fig.3.   Crosstalk of Deep Carriers in
CCD Linear Image Sensors

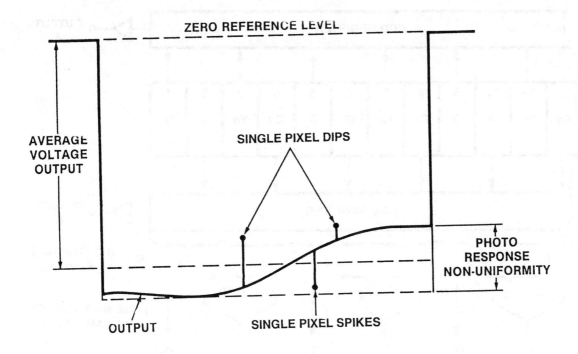

Fig.4. Line Scan Sensor Output
Under Uniform Illumination
Photoresponse Non-Uniformity

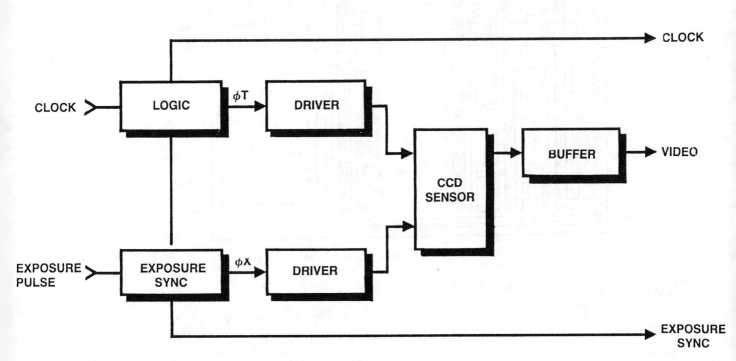

Fig.5. CCD "R" Type Block Diagram

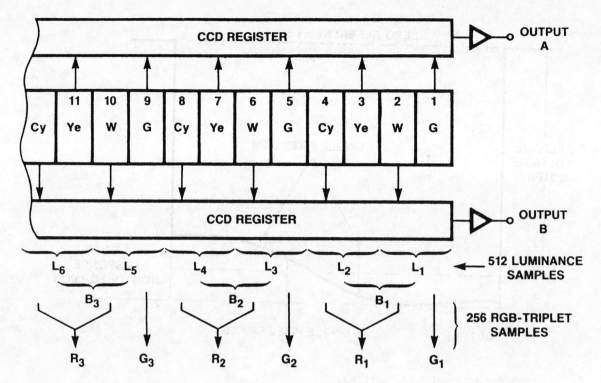

Fig.6. Pixel Order in the CCD133C Color Imager with Diagram Showing Luminance and RGB Signal Derivation

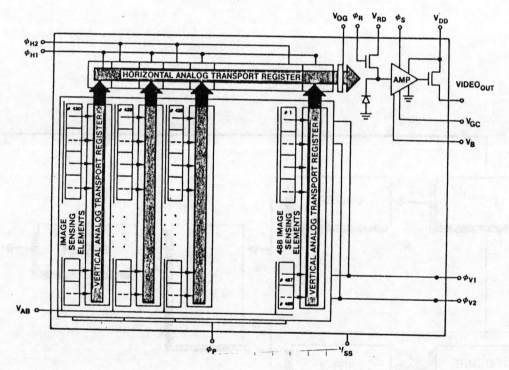

Fig.7. Device Block Diagram

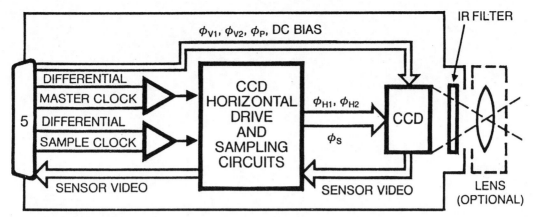

Fig.8. CCD3000 Sense Head Block Diagram

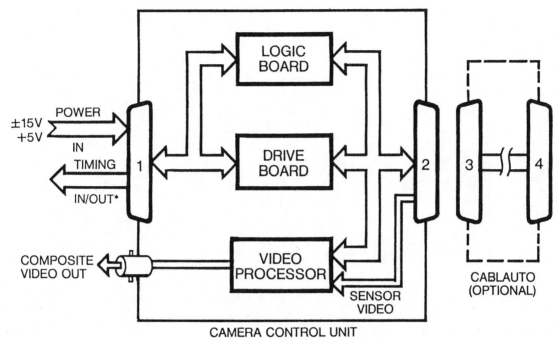

Fig.9. CCD3000 Camera Body Block Diagram

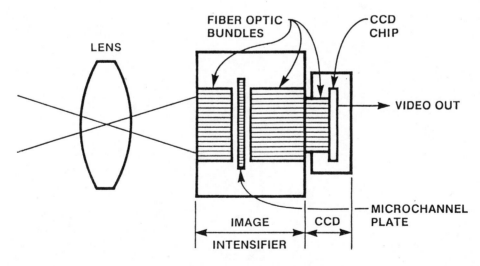

Fig.10. Block Diagram of the Camera Front-End

256×256 Pixel CCD Solid State Image Sensor

G. Boucharlat, J. Chabbal
and
J. Chautemps
Thomson-CSF Electron Tube Division, France

Abstract

A 256 x 256 pixel CCD solid-state image sensor has been designed and developed for shape recognition and image signal processing. The device has square photosites with antiblooming, a pseudo-random access adaptable organization, an electronic exposure time control and a sampled video output. Built with MOS-CCD technology, the sensor possesses all of its associated advantages : small size, reliability, low power consumption and insensitivity to electromagnetic field disturbances. This paper presents the performances and operating modes for the device.

Introduction

The TH X31144 imager presented in this paper possesses a high degree of versatility in combination with the advantages of a photodiode sensor and the compact rugged structure of a charge-coupled device.

Indeed, the design permits many functions and the data output rate adapts to a variety of image acquisition systems. For example, an electronic shutter and a "windowing" function on the TH X31144 make it particularly useful in the areas of shape recognition and industrial checking, where its square format and number of pixels adds a great advantage.

Organization of the device

The TH X31144 image sensor (figure 1) has a 256 x 256 photodiode array and an antiblooming device for each pixel. The pixels have a 29 μm pitch in both dimensions, offering a useful area of 7.4 x 7.4 mm^2, which is compatible with the field covered by the lenses used with TV camera 2/3" tubes. Dark-signal references are supplied by 4 additional rows and 20 columns outside the image zone. The block diagram of circuit organization is given in figure 2. Readout from the photosites is by line transfer[1] and involves :

- a dynamic shift register, successively addressing each row in the diode matrix and authorising the transfer of the charges, integrated by each photosite on a row, towards the line memory. Two identical registers are integrated in the device : they are synchronous, but can be separately initialised, permitting numerous facilities

- an analogue line memory, acting as a buffer between the photosensitive zone and the CCD readout register. This gives improved vertical transfer performance, thanks to two biassing charges[2], Q_0 and Q_1 (figure 3), and supplies the line data to the CCD register in parallel

- a buried-channel CCD readout register which, once loaded via the line memory, sequentially transfers the data packets from each point of the line towards an output device, that performs a charge-to-voltage conversion and feeds the output amplifier.

Operation

The organization of the device permits the following operating modes :

- full 256 x 256 pixel resolution TV. A square format can be obtained on a CCIR standard monitor by supplying two identical frames in succession instead of two interlaced frames, and requires a readout frequency of 7.38 MHz.
Only one address register is required. The performance of the TH X31144 in this mode is described in Table I. A small camera has been built operating at this data rate (figure 4).

- other full resolution TV modes. The maximum readout frequency is 10 MHz and the time taken to transfer an addressed line to the register is of the order of 7 μs. This leads to a minimum of 9 ms for reading out the photodiode array, permitting over one hundred frames per second for a data acquisition system operating at 10 MHz. Naturally, lower readout frequencies are possible all the while the dark signal, which is directly proportional to the frame period, does not reduce the dynamic range by too much. The integration time of the signal for one line, which equals the interval between two consecutive readouts from that line, also increases and raises the signal level.
The sensor can thus adapt to many systems and lighting conditions.

Additional functions

. Antiblooming : a device adjoining each pixel drains away the excess charges caused by a localised overillumination, and thus limits smearing on the nearest neighbours.

. Windowing : The high maximum operating frequency of the address registers (1 MHz) permits very rapid "scanning" of all the lines. The lines to be read can be chosen by selectively applying the addressing voltage on the $\emptyset_{1A}$ and $\emptyset_{2A}$ inputs (figure 2) in synchronism with the address register clocks.

Applying these two capabilities together permits concentrating only a specific zone of the total image, or "windowing". This is achieved by adopting a slow scanning rate and skipping the remainder of the image by scanning at the maximum rate.

This feature has been tried (figure 5) in the TH X31144 camera and allowed just over three zones to be examined in one CCIR frame period (20 ms).

. Variable integration time : Judicious use of the two address registers and the addressing voltages $\emptyset_{1A}$ and $\emptyset_{2A}$ permits reducing the integration time to less than the frame period. This simply requires staggering the start of the two address registers. One same line is then read twice per frame, and the integration time is equal to the interval between the two readings. This technique can be used to obtain an "electronic shutter" accommodating, to a certain extent, variations in illumination, for example, or limiting "trails" caused by the rapid relative motion between the object and the camera.

. <u>Line summation</u> : Conversely, if a loss in resolution can be accepted, information from several consecutive lines may be summed in the line memory before being read out, to raise the video signal output. There is thus a choice of increased apparent sensitivity at the expense of spatial resolution.

Conclusion

A solid-state image sensor aimed at facilitating image pickup and processing has been achieved. Its line transfer organization using two address registers, and its antiblooming protection effective on each pixel, provide for a great versatility in terms of both operating frequencies and illumination conditions. It thus adapts to many configurations for use in industrial checking and shape recognition. The format and number of pixels should satisfy all users.

Acknowledgements

The authors would like to thank Messrs. J.L. BERGER, L. BRISSOT, J.P. BERNARD for their collaboration in designing the device and their fruitful discussions, as well as Mr. A. CUNIBERTI for his skillful technical assistance in operating the sensor. This device would not have existed without engineering competence of Mr. M. BOURRAT and his team.

References

[1] TERAWA, S., et al. "A New Organization Area Image Sensor with CCD Readout Though Charge Priming Transfer", IEEE Elect. Dev. Letters, Vol. EDL-1, pp. 86-88, (May 1980).

[2] BERGER J.L., BRISSOT L., CAZAUX Y., DESCURE P., "A Line Transfer Color Image Sensor with 576 x 462 pels", ISSCC 84, Digest of Technical Papers, pp. 28-30 (February 1984).

TABLE I - MAIN CHARACTERISTICS OF THE TH X31144 IMAGE SENSOR

Number of photosensitive elements... 256 x 256

Dimensions of one pixel.. 29 μm x 29 μm

Area of photosensitive zone...7.4 mm x 7.4 mm

Saturation voltage ....2.0 V

Response (1) ....10.0 V/μJ/cm^2

Nonuniformity of response ....$\pm$ 5 % of Vavg

Nonuniformity of the dark signal ....1 % of V_{sat}

Power consumption..40 mW

Insensitive to magnetic fields

Highly reliable

(1) Colour temperature : 2854 K, plus a BG38 filter

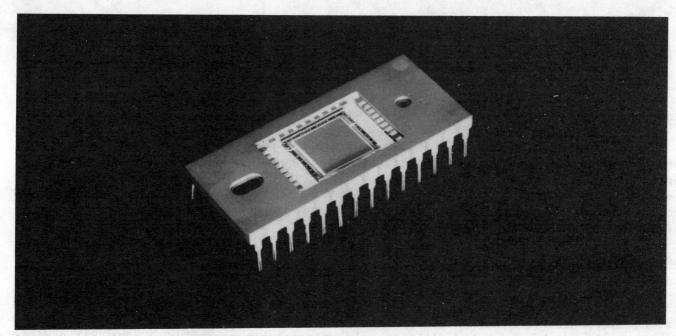

Figure 1 - The TH X31144 image sensor

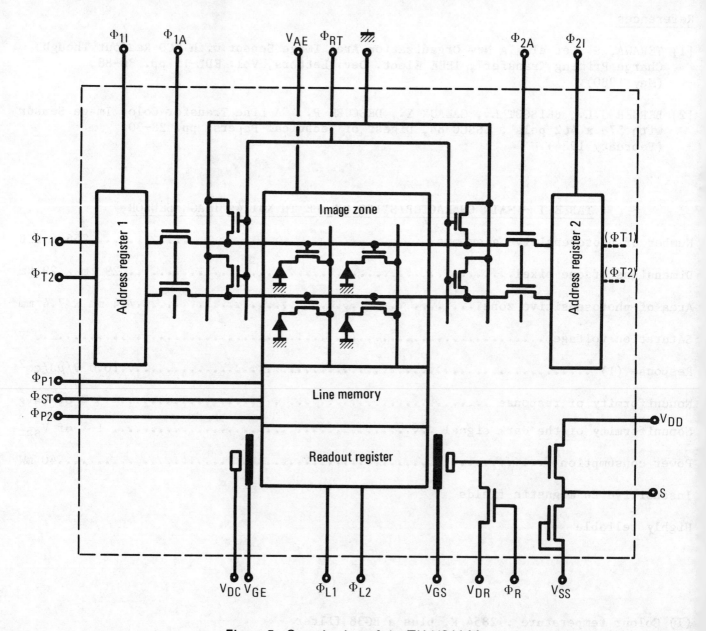

Figure 2 - Organization of the TH X31144

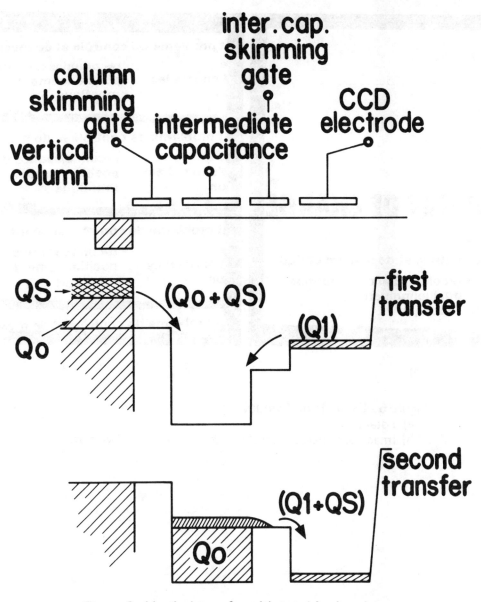

column skimming gate

vertical column

inter. cap. skimming gate

intermediate capacitance

CCD electrode

QS → Qo

$(Q_0 + QS)$

(Q_1)

first transfer

Qo

$(Q_1 + QS)$

second transfer

Figure 3 - Vertical transfer with two biassing charges :
Q_S = signal charge, Q_0 = first biassing charge, Q_1 = second biassing charge

Figure 4 - Miniature camera using the TH X31144 image sensor

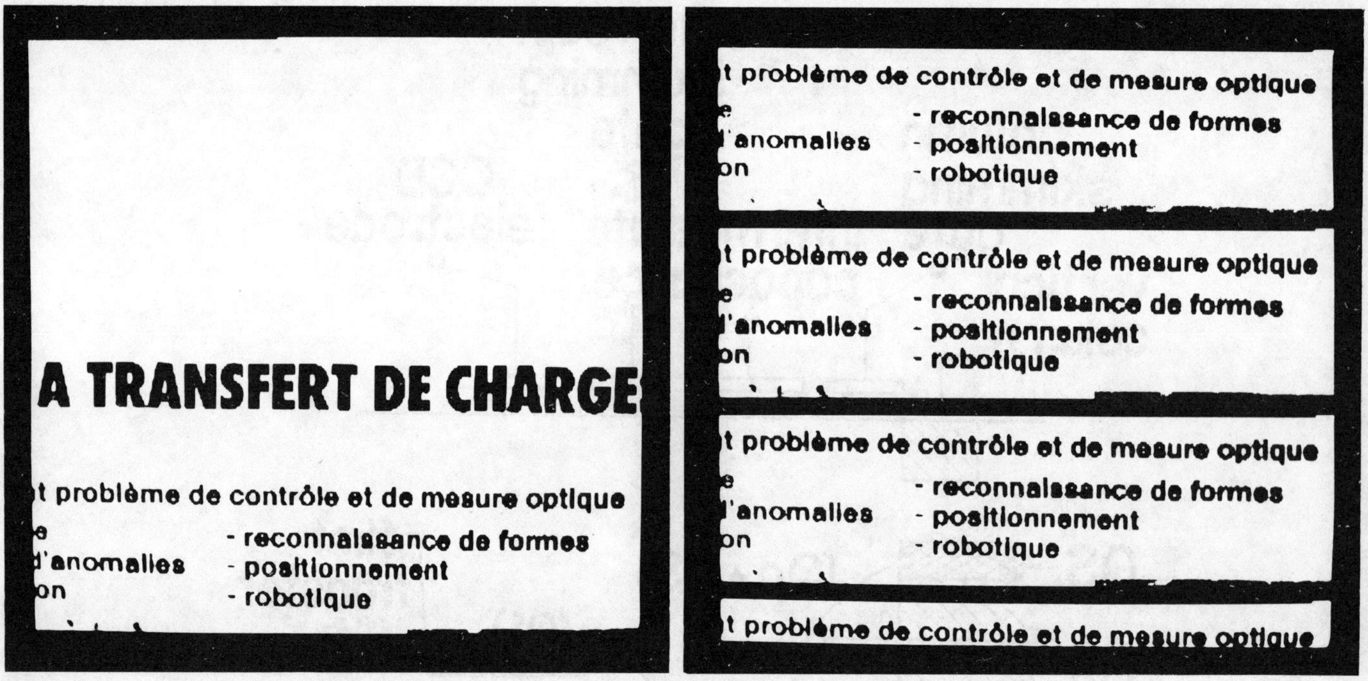

Figure 5 - Example of "windowing" :
   a) Total image
   b) Image obtained by iterative use of a selected "window"

SENSOR GUIDED WELDING

Laser Guidance System for Robots

Z. Smati, D. Yapp
and
C. J. Smith
Central Electricity Generating Board, UK

ABSTRACT

Adaptive arc welding and seam tracking systems rely on the successful combination of various technologies including position sensors, welding power sources, manipulators and industrial robots. Sensors are expected to be a key element in the development of any such adaptive automatic system. In this paper a review of arc based sensors and optical sensors is undertaken. Special consideration is given to laser systems using triangulation.

The development of an adaptive welding system based on a laser range finder is described in detail. This system has been applied to control directly each axis drive of a 4 degrees of freedom manipulative arm to follow complex paths and the algorithms used for image recognition and analysis are described. The application of this system to MIG (Metal Inert Gas) welding of pipes and the use of an industrial 6 degrees of freedom robot for linear tracking are also discussed.

INTRODUCTION

The development of automatic seam tracking and seam characterisation is not new[1]. However, the increasing complexity of welded parts coupled with the desire to increase the productivity of welding processes and improve the control of the quality of the weld have recently stimulated increased activity in the fields of automation and in-process monitoring of welding. A key part of this work is the development and application of suitable sensors and processing systems to provide the basis for adaptive control of the welding process.

Several sensor techniques have been developed[2-8] and some are now commercially available. However, none of the systems presently available can be shown to meet all the requirements of the arc welding process. The requirements which a complete joint characterisation system must fulfil include the following:

91

(i)   the system should withstand the harsh environment of the welding process;

(ii) it should also be usable on the shop floor by a non specialist;

(iii) the system must find its correct starting position; compensation must be
      made for any deviation of the actual starting point from the initial programme
      position:

(iv) the system must operate in real time so that any distortion during
      welding can be detected and compensated for;

(v)   the system must be able to track seams of different geometries and
      provide information on the joint geometry.

    In the present paper, some of the available sensor techniques for arc welding are
reviewed and the practical implementation of one of these systems is discussed.

SENSOR SYSTEMS FOR ARC WELDING

    The techniques which have been considered for welding application include
mechanical, electro-mechanical, ultrasonic capacitive and magnetic sensors. However,
they generally suffer from some overriding disadvantage, such as lack of resolution, or
are insufficiently robust. Tactile probes are attractive because of their simplicity
and reliability; however, it is difficult, if not impossible, to provide information on
the joint geometry. Furthermore, certain weld geometries (e.g. welding into corners)
may preclude the use of contacting sensors. The two methods which appear most promising
are arc- based systems and non-contact optical sensor systems.

Arc-Based Sensor Systems

    In MIG welding, with a constant current power supply, the arc voltage is a function
of the arc gap. Hence, by measuring the mean voltage it is possible to guide the
welding torch at a constant height above the workpiece. If the welding torch is weaved
from side to side, as it approaches the side of the weld preparation, there will be a
change in voltage. Detection of this voltage change at either side of the weld
preparation can then be used to guide the torch along the weld path. Commercial systems
have been produced with simple mechanical oscillators as part of the welding head [9],
by using magnetic deflection at the arc when using robots by weaving the robot arm
itself.

    The chief advantage of these arc-based systems in commercial applications is that
they are relatively simple, the measurements are made directly at the welding torch and
they do not require many additional components to be added to the welding head. They
provide continuous control of the path during welding, and they can cope with reasonably
large deviations from a planned welding path. However, arc-based systems require fairly
well defined edges to the weld preparation and thus cannot be used for welds more than
two beads wide. Furthermore, they cannot cope with a step change of less than 3 mm.
They do not, of course, provide detailed geometric information , but are used to follow
a clearly defined weld preparation. Arc-based systems may have some applications for
single-pass welding but they would still need to be used with a coordinate measuring
system which allowed the approximate welding path to be defined initially.

Optical Sensor Systems

    The simplest optical sensors used in welding provide information about only one
aspect of the weld joint. For instance, equipment is commercially available which
detects a grey level change and can be used to follow a weld path in 2-D where there is
a sharp change in reflectivity. However, for some applications, it is generally
necessary to obtain information both about the welding path to be followed and about the
detailed geometry of the component to be welded. There are currently two basic
techniques which can be applied to obtain information about its detailed geometry. The
first uses the "time-of-flight" of a laser beam to provide distance information

directly, and the second uses the principle of active triangulation either to deduce distance information (using position-sensitive detectors) or to extract dimensional information on key aspects of the component (shadow-cast techniques). Both of these systems gather information some distance in front of the welding torch, although techniques have been developed to allow measurements at the weldpool[10,11].

1.   Laser Range Finders Based on "Time-of-Flight" Measurements

This method is one of the most powerful techniques available, and it overcomes some of the 'hidden part' problems encountered with triangulation, since the transmitted and received beams are usually co-axial. The range is determined from the time needed for the laser light to travel to the target and back to the receiver, (Figure 1). The "time-of-flight" of the beam can be determined directly using a pulsed laser and measuring the elapsed time, or indirectly by using a continuous wave modulated beam and measuring the phase shift[12,13].

The technique of using a pulsed laser and directly measuring transit time, is examplified by the work done by Lewis and Johnson[22]. These authors suggest that a range collection rate exceeding 100 points per second for a 2 cm accuracy over a 1 to 3 metre range was beyond the capability of the instrument they built. In fact under ideal conditions a 128 x 128 range picture would take about 3 minutes to collect. This is still too long for robotic applications. Furthermore a 2 cm accuracy is not good enough for robotic manipulation of small objects.

The second method, using a continuous wave laser and phase detection to measure transit time, is examplified by using the work done at Stanford Research Institute[12]. The authors built an instrument able to capture a 128 x 128 range picture in 2 hours with an accuracy of 7-8 bits and range from 1 to 5 metres. To accommodate the wide dynamic range ($\sim$100 dB) of the reflected energy using a low energy laser (15 mW), considerable integration was required to improve the signal/noise ratio to an acceptable figure. This speed of operation would hardly be suitable for real time robotic work. The authors conclude that much higher energy sources would be required to achieve high speed reliable results; but this would introduce laser hazards to the environments. It thus appears that "time of flight" laser range finders are not yet sufficiently developed as practical devices for robot applications.

2.   Range Finder Based on Triangulation and Shadow Cast Techniques

This technique is the one most widely used at present. It is based on structured light, where a line of light is projected onto the workpiece surface. The intersection of the sheet of light with the surface then represents the profile of the surface as shown in Figure 2. The method has been successfully used in the past to provide adaptive control both of the torch position and welding process[14] and is still under active development for robot applications.

Various camera systems have been used to collect the projected image[7]. In the early work, vidicon tubes were used and the TV image was processed to yield data on joint position, joint gap and weld preparations shape. Most present studies use solid state arrays based on charge coupled devices (CCD) or charge injection devices (CID). In the earlier work structured light was produced by projecting white light past a sharp edge to cast a shadow[14], while later studies[7] have made use of laser light sources both in the visible and infra-red regions, often with modulation or filtering to reduce interference. The 3-D pattern information obtained can be displayed on a TV screen or stored in a frame store. It is possible, in principle, to process the information to reconstruct a true 3-D profile. However, most practical systems extract only key information required for a particular application, such as joint position or joint gap. A most important feature of this selective approach to data extraction is that is it easier to operate the system in real time, i.e. to make measurements and control welding while the weld is being made. This is of obvious benefit in avoiding extra time for a measurement stage, and may also be advantageous in situations where the component can move or distort during welding (e.g. car body manufacture).

Because of these advantages, it is expected that shadow cast or structured light systems using light sheets will be widely developed for future commercial applications.

3.   Range Finders Based on Triangulation and Position Transducers

The basic principle of this technique is similar to that described in the previous section. However, a projected point source of light is normally used together with a position sensitive photo-detector. The equipment is calibrated to produce a measurement of the distance from actual surface to a reference surface.

As shown in Figure 3, the light source shapes the light beam and projects a spot onto the surface to be measured, The scattered light is imaged by a lens onto the position sensitive photo-detector. This device produces a current which is a function of the position of the imaged light spot, and the output is calibrated to remove non-linearities. In order to reduce interference from the arc light, narrow band pass optical filters may be used, and the laser beam may be modulated. In addition, the power of the laser can be regulated in a control loop in order to compensate for reflectance changes. Commercial systems of this type are now available, and a number of practical 3-D vision systems have been based on this technique[15,19].

In order to provide information on surface profile it is necessary to scan the laser beam over the surface of the component, either mechanically or by using mirrors to deflect the beam. The distance information can then be processed, and key features can be extracted (e.g. joint position), and the surface profile can be reconstructed and displayed on a TV screen for operator interaction.

Although the position sensitive detector technique requires a scanning system, it has the advantage that commercial systems can be used and the data can be processed relatively easily. For this reason this technique has been adopted for the present work.

DESCRIPTION OF THE PROTOTYPE SYSTEM FOR PIPE WELDING

The Vision System

The range finder used in the present work is based on triangulations and single axis position sensitive detectors. The scanning of the laser beam is achieved mechanically over a fixed distance of 70 mm once every second.   The data are then sent to a central computer for joint characterisation. This computer carries out all the operations from data handling to path recognition, robot control and welding parameter selection and control. The camera unit is shielded from direct arc light interference. Furthermore, both narrow band pass filtering and modulation of the laser beam are used to suppress most of the noise caused by spatter or by the arc itself. When data from the sensor system are received by the computer, on-line data averaging is performed as a further method of noise filtering.These data are then stored in memory, where data analysis is required to identify the various welding preparations.

Algorithms are used for joint characterisations. The joints considered are those most commonly encountered in pipes as well as in linear welds on plates.

In the simple case of a butt joint, the gap size is deduced by computing the total average of the data collected during one complete scan. The first and last points that exceed this average by a given amount represent the start and end of the width of the butt joint. For further accuracy, a second iteration is necessary. This procedure is shown on Figure 4. For a lap joint, the procedure starts in the same way as for a butt joint. However, only one point is required. This point should correspond to the intersection of the two plates. Figure 5 shows how to determine this intersection. A fillet joint can be considered in the same way as a lap joint, and the same algorithms are therefore used.

For a V groove, the procedure is more complicated. The total average of the data collected during one scan must be found. The intersection of this line with the V

94

groove is then found and the lines are split there. Two "average calculations" are then required. The first set of data comprises all the points up to the first intersection and the second set of data starts from the second intersection up to the last collected data point. This procedure is then repeated until the edges of the weld preparation are detected with the required accuracy. To find the depth of the groove, to a first approximation, the point in the middle of the edges is chosen. All the data around this point are checked until the minimum is found. Figure 6 shows the implementation of this technique.

Additional algorithms can be developed to characterise more complicated weld preparations.

The Four Degree of Freedom Manipulative Arm

When the vision system described above has detected the welding preparation, the welding torch is moved to the required position using the manipulator system in Figures 7 and 8. This consists essentially of a cartesian configuration with a "rotate" movement on the workpiece. The system is driven by four stepping motors, giving a resolution of $\pm 8.10^{-3}$ mm. The mechanical range of the system is approximately:

Torch movement;	vertical range	$= 75$ mm
Torch movement;	horizontal range	$= 70$ mm
Range finder;	scanning range	$= 70$ mm
Workpiece;	rotation	$= 360^{o}$

The motion of the torch and the laser range finder takes the form of straight line, constant velocity movement. The processor is required to calculate the speed of the individual axes as well as the path necessary to maintain the torch over the required welding preparation. Having calculated the motor speed, the value may be loaded into the linear axis interface together with the displacement values.

When the process is first initiated, the controller resets the axes and drives them to a "START" position. The scanning phase is then started where the groove is found, as described above. The torch is then driven to the correct position and welding can proceed. During welding any deviation detected by the sensor system and the welding torch is then moved to the correct position.

The welding conditions are set externally by the main processor and monitored during welding to maintain optimum metal transfer.

Multipass Welding

For a fully automatic welding system, three basic elements are required, i.e. a sensor system, a manipulative arm and a welding model. In the previous section, the processing of information from the sensor system, together with the control of the manipulative arm have been discussed. To provide the third required element, the computer model developed at MEL[20] for the prediction of multipass welding conditions for thick section joints is used. This model simulates the formation of thick section welded joints and can be used to examine the effects of welding variables on weld bead size. Thus it can be used to calculate optimum joint filling conditions as well as the position and size of each weld bead. For example, the procedure for filling a typical vee-preparation, consists of depositing a root run corresponding to an average condition and then proceeding to the second layer by placing a corner bead, calculating the corresponding height of a flat layer, and the number of weld beads required to complete the layer. The number of these beads must be rounded to a whole number. The weld preparation is then filled to within 1 mm of the top surface by adding successive layers. The last 2 layers may be modified by increasing or decreasing the gross heat input to alter the individual weld bead corner geometry and hence the appropriate layer height. This strategy must be worked out prior to welding because the calculations involved are complex and therefore it is not possible to implement this procedure in real time. When these calculations (using the range finder data) are finished, the first bead can be deposited. Any deviations of actual bead shape from that calculated are detected by the vision system and the welding conditions for the next layer can be modified accordingly; this process is fast enough and may be implemented in real time.

The pulsed MIG welding process is used for both the root run and for subsequent passes. A welding model[21] is used to maintain optimum spatter-free metal transfer. This mode allows bead shape to be related to the main welding parameters, as follows:

$$BW = \alpha_1 + \alpha_2 WFS + \alpha_3 TS$$

$$Pen = \alpha_4 + \alpha_5 WFS + \alpha_6 TS$$

$$BH = \alpha_7 + \alpha_8 WFS + \alpha_9 TS$$

with BW, BH, Pen, WFS, TS defined respectively as bead width, bead height, penetration, wire feed speed and travel speed. The α_n are all constants. All the main welding parameters are monitored and changed accordingly to keep the required bead characteristics as well as optimum spatter-free metal transfer.

A Laboratory System for Linear Welding with an Industrial Robot

A similar system has been developed for linear welding using a Cinncinnati Milacron robot[15]. In this case, the "Adjust" function on the robot is used to link the robot to an external computer. The sensor is protected from possible spatter and fume and it is also air cooled to guard against excessive temperatures. Scanning of this sensor is also achieved mechanically. The processing algorithms used here include data averaging and correction in lateral positions. Process parameters such as wire feed speed, travel speed and pulse MIG conditions are also optimised and monitored. During all the tests performed using pulsed MIG on mild steel plates with different V preparation, the grooves were detected within an accuracy of $\pm$ 0.5 mm, which is within the repeat accuracy of the robot.

DISCUSSION

In this paper, we have shown that laser range finder sensor systems can be used to generate the basic data required for automatic welding, and we have discussed some of the simple algorithms which can be used to process the range data in order to generate 3-D profile information. Additionally, we have summarised work reported in detail elsewhere[20,21] on modelling the welding process and on the strategies needed for correct weld bead placement. It is now possible to combine these basic components in order to produce automatic systems. The system will find, measure and track the weld preparation, place all the weld beads in the correct position and also report on how the joint was completed if required.

We have used the prototype systems in this work to track and measure both linear and circumferential weld seams. Multipass welds have been made and the resulting weld structures agree closely with the predictions of the model.

Laser range finder sensor systems are now readily commercially available and in some cases integrated with robot control systems. These developments, in combination with further work on weld models, will allow the production of practical automatic welding systems in the near future.

CONCLUSIONS

(1) Non-contact sensor systems have been reviewed and three techniques identified which have potential for further development; arc-based systems, light sheet systems and scanning laser systems.

(2) A scanning laser system has been chosen as the basis for a prototype robot guidance system and integrated with the control hardware. Algorithms have been developed to fulfil the requirement of joint characterisation.

(3)   A welding model has been developed to predict bead shapes and maintain spatter-free optimum metal transfer conditions for pulsed MIG welding.

(4)   A 4-axis manipulative arm has been designed, built and integrated with the control hardware to enable different sizes of pipes and tubes to be welded.

(5)   A Cinncinnati Milacron robot has been used together with our sensor system to achieve a seam tracking unit for linear welds.

(6)   The main subsystems for a true multipass welding system, sensor unit, manipulative unit and welding model are now operational.

ACKNOWLEDGEMENTS

This paper is published by permission of the Central Electricity Generating Board.

REFERENCES

(1)   Brown, K.W., "A Technical Survey of Seam Tracking Methods in Welding". Welding Institute Report, 1975.

(2)   Brown, M.A. et. al., "Passive Visual Tracking for Robotic Arc Welding". International Conf. on Optical Techniques in Process Control. Paper F2, pp 207-216, (June, 1983).

(3)   Kodaira, N. et. al., "Microcomputer Control of an Arc Welding Robot with Visual Sensor". IEEE, 1982.

(4)   Cook, E.C. et.al., "Robotic Arc Welding: Research in Sensory Feedback Control". IEEE Trans. on Ind. Electronics, Vol. IE30, No.3, pp 252-268, (August, 1983).

(5)   Presern, S. et. al., "Tactile Sensing System with Sensory Feedback Control for Industrial Arc Welding Robots". ROVISEC 1, Stratford-Upon-Avon, U.K., (April, 1981).

(6)   Doerr, A.J.R., "The Anatomy of an Intelligent Robot". Proc. of 2nd Euro Conf. on Automated Manufacturing, Birmingham, U.K., (May, 1983),

(7)   Morgan, C.G. et.al., "Visual Guidance Techniques for Robot Arc Welding". ROVISEC 3, pp 615-624, Cambridge, Massachusetts, U.S.A., (November, 1983).

(8)   Dufour, M. et. al., "Adaptive Robotic Welding Using a Rapid Image Pre-processor". ROVISEC 3, pp 641-648, Cambridge, Massachusetts,U.S.A., (November, 1913).

(9)   King, F.J. et. al., "Seam tracking Systems with the Arc as a Sensor". 2nd Int. Conf. on Advances in Welding Processes, Harrogate, U.K., (1978).

(10) Richardson, R., Private Communication, (1983).

(11) Rider, G., "Control of Weld Pool Size and Position for Automatic and Robotic Welding". ROVISEC 3, Massachusetts, U.S.A., pp 625-634, (November, 1983).

(12) Nitzan, F.J. et. al., "The Measurement and use of Registered Reflectance and Range Data in Scene Analysis". Proc. IEEE, Vol. 65, No.2, pp 206-210, (1977).

(13) Page, C.J. et. al., "Non-contact Inspection of Complex Components using a Range Finder Vision System". ROVISEC 1, Stratford-Upon-Avon, U.K., (April, 1981).

(14) Westby, O., "An Adaptive Controlled Welding Machine". Report No. 1977-66-01, (STF16 A77028), Sintef, Trondheim, (1977).

(15) Smati, Z. et. al., "An Industrial Robot Using Sensory Feedback for an Automatic Multipass Welding System". 6th British Robot Association Annual Conference, Birmingham, U.K., (May, 1983).

(16) Weise, D., "Software Inspection Teams Robots and Laser Sensors". Robotics World, Vol. 1, No.1, (1983).

(17) Pickard, L., "Sensory Feedback for Robot Inspection and Control". Robotic World, Vol. 5, (1983).

(18) Grandstedt, O., "The ASEA IRB in Automotive Industry". "Automan '83", Birmingham, U.K., (May, 1983).

(19) Oomen, G.L., "A Real-time Optical Profile Sensor for Robot Arc Welding". ROVISEC 3, Cambridge, Massachusetts, pp 679-668, (November, 1983).

(20) Alberry, P.J. et.al., "Computer Model for Predication of Heat Affected Zone Microstructures in Multipass Weldments". Metals Industry, Vol. 9, No. 10, pp 419-426, (October, 1982).

(21) Smati, Z. et. al., To be published.

(22) Lewis, R.A. and Johnson, A.R., "A Scanning Laser Range Finder for a Robotic Vehicle", 5th Int. Joint Conf. on Artificial Intelligence, pp 762-768 (1977).

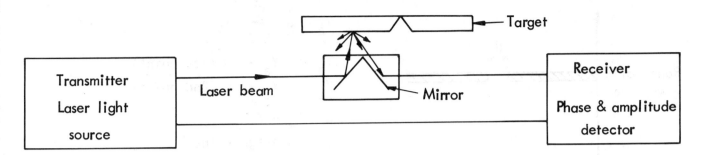

Figure 1. Block diagram of a phase shift range sensor

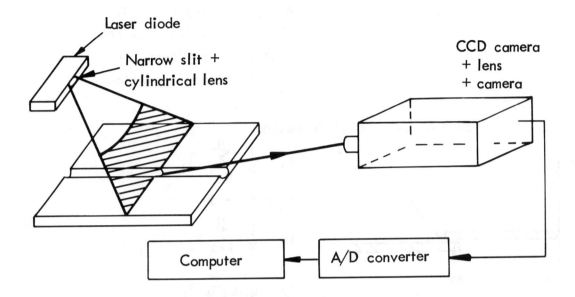

Figure 2. Principle of the shadow cast technique

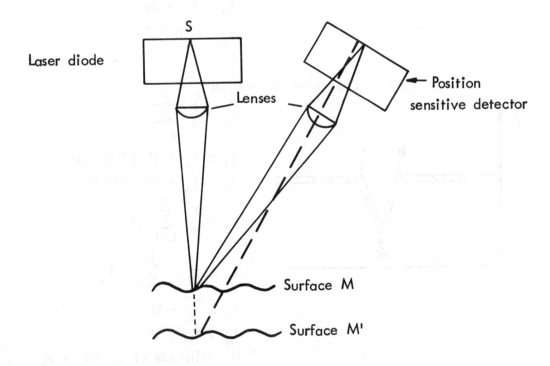

Figure 3. Principle of the scanning laser system

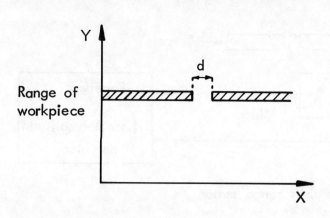

$$Y_a = \frac{1}{N} \sum_{k=1}^{N} Y_k$$

$Y_b \geq Y_a - SD$ (1st point)

$Y_e \geq Y_a - SD$ (2nd point)

$d = X_a - X$

Y_a: average value

Y_k: range data

N : total number of points

Y_b, Y_e: beginning and end of gap

X_b, X_e: positions of Y_b, Y_e

SD: accuracy required

X_1, X_2: first iterations

X_3, X_4: second iterations

Figure 4. Butt joint detection

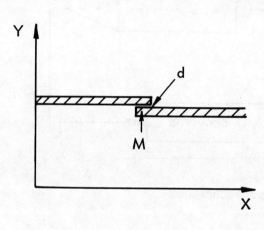

$$Y_a = \frac{1}{N} \sum_{k=1}^{N} Y_k$$

$Y_m \leq Y_a - SD$

$$Y_1 = \frac{1}{M} \sum_{k=0}^{M} Y_k$$

$$Y_2 = \frac{1}{N-M} \sum_{k=M}^{N} Y_k$$

$Y_b \leq Y_1 - SD$

$Y_e \geq Y_2 + SD$

$$X_{av} = \frac{X_b + X_e}{2}$$

Figure 5. Lap joint detection

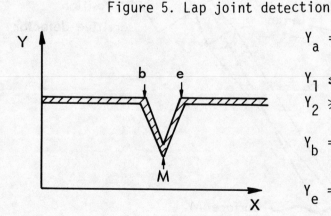

$$Y_a = \frac{1}{N} \sum_{k=1}^{N} Y_k$$

$Y_1 \leq Y_a - SD$ (1st point)

$Y_2 \geq Y_a + SD$ (2nd point)

$$Y_b = \frac{1}{b} \sum_{k=1}^{b} Y_k$$

$$Y_e = \frac{1}{N-e} \sum_{k=e}^{N} Y_k$$

$Y_3 \leq Y_b - SD$

$Y_4 \geq Y_e + SD$

M = minimum of x_k $(x_3 < x_k < {}_4)$

Figure 6. V-groove detection

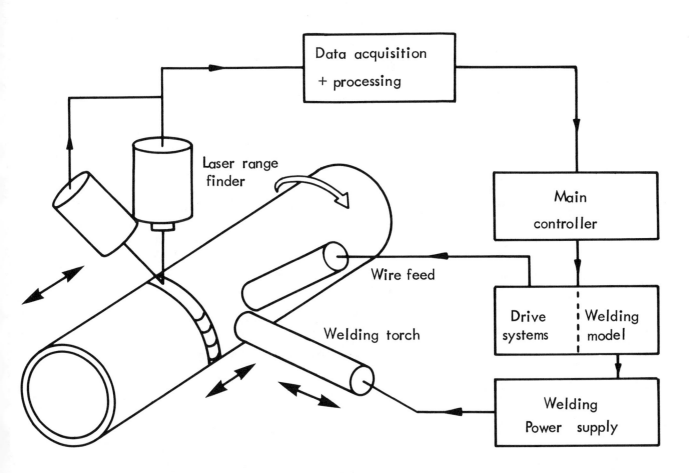

Figure 7. Diagram of the MEL system

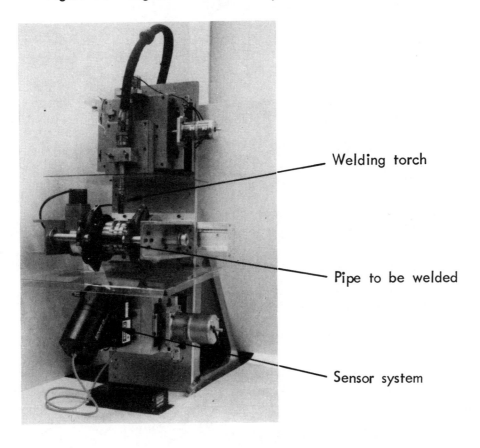

Figure 8. Fully automatic pipe welding system

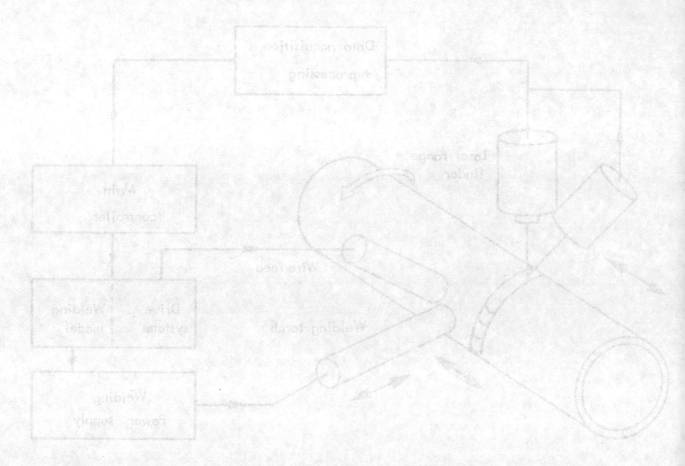

Figure 7. Diagram of the MPL system

Figure 8. Fully automatic pipe welding system

General Visual Sensing Techniques for Automated Welding Fabrication

J. E. Agapakis
and
K. Masubuchi
Massachusetts Institute of Technology
and
N. Wittels
Automatix Inc., USA

ABSTRACT: Welding fabrication invariably involves three distinct sequential steps: preparation, actual process execution and post-weld inspection. One of the major problems in automating these steps and developing autonomous welding systems, is the lack of proper sensing strategies. Conventionally, machine vision is used in robotic arc welding only for the correction of pre-taught welding paths. In this paper, novel uses of machine vision for the determination of the weld bead and joint geometry are presented, developed visual processing techniques are detailed, and their application in welding fabrication is covered. Coding of the required welding knowledge in the form of an "expert" interpretation system is finally proposed.

INTRODUCTION

Work reported in this paper has been performed as part of an ongoing research effort to develop novel vision aided welding systems[1]. Motivation for undertaking this study came basically from two different areas which share the need for sophisticated visual sensing and where the developed technology can be directly applied.

Specifically, the primary motivation has been the immediate need for improvement of factory floor productivity by the large scale introduction of welding robots. In such applications, the major limitations of conventional preprogrammed robots come from the requirement for repeatedly high fixturing accuracies and tight tolerances in part positioning and edge preparation, as well as from the inability to cope with weld distortion and part-to-part variations. These limitations often render the current teach-playback robots inadequate and make the development of novel sensory controlled

103

systems a necessity[2,3].

Equally important to our research interests has been the need to develop capabilities for unmanned construction and repair in remote, hazardous, or unreachable environments. Such capability is a prerequisite for large scale human utilization of outer space or the deep ocean, and for such unconventional applications as the repair of highly contaminated nuclear reactor components or work in other hazardous environments. Under conditions encountered in remote fabrication, even simple manual tasks become extremely complex and costly. In addition, the skills that are required for the successful performance of high quality welding are not likely to be readily available. Therefore, extensive automation, improved sensing strategies, and local intelligence are required for the development of any viable remote welding techniques[4,5]. In fact, series of remote welding simulation experiments performed by our group[1] repeatedly demonstrated the need for improved sensing strategies (particularly utilizing machine vision) both for welding process automation and for human operator assistance.

OBJECTIVES AND APPLICATIONS OF VISUAL SENSING IN WELDING FABRICATION

Welding fabrication invariably involves three distinct sequential steps: preparation, welding process execution, and post-weld inspection. Since the actual welding process requires only a small percentage of the total fabrication time, any efforts to improve productivity, to automate welding, or to develop remote welding fabrication systems should equally weigh each of the above basic steps. Therefore, uses of machine vision in all these fundamental steps must be explored. Such possible applications are summarized below:

(I) Pre-Weld Preparation:

- Sensing of part location and orientation (visual offsets determination)

- Sensing of joint type and shape (necessary for the selection or adjustment of welding process conditions)

- Dimensional accuracy of fit up, edge preparation, and material surface inspection

(II) Actual Welding Process Execution

- Visual Guidance of the Welding Torch (Torch position/orientation control)

    * Single- or multi-pass seam tracking

- Welding Process Control

    * Weld pool and/or weld bead geometry sensing and control

    * Weld heat input sensing and control (utilizing thermal imaging)

(III) Post-Weld Inspection and Quality Control

- Automated visual surface inspection

    * Weld bead geometry (conformity to codes and design specifications)

* Surface defects detection (undercut, overlap, etc.)

- Automated NDT techniques (dye penetrant, X-ray, etc)

Most of these applications are within current technological limits. However, only a few are now actively pursued. Specifically, the vast majority of the work in this area deals with single-pass seam tracking and visual offsets determination. Limited work is also done on weld pool sensing and automated X-ray inspection. In our work, reported in this paper, we chose to concentrate on the development of general techniques for the visual sensing of the <u>weld joint</u> geometry ahead of the welding arc, and of the <u>weld bead</u> behind the molten metal pool. Capability for sensing of the weld bead geometry, in particular, is essential for automated post-weld inspection, planning of multipass welding paths, welding conditions selection, and weld bead geometry control. No other similar research efforts in this particular area have been reported in the literature.

The weld joint and bead features of welding significance that must be recognized include: the <u>root location</u> and <u>wetting points</u> which are needed for the determination of all the weld dimensions, such as <u>leg lengths</u> or <u>throat size</u>; the <u>preparation angles</u> and <u>fill volume</u> of the joint gap and the <u>area and centroid</u> of the deposited bead cross section, which are necessary for the planning of the welding conditions and the torch path during any subsequent welding passes; the <u>wetting angles</u> and exact <u>profile</u> of the bead, which are required in order to detect any surface defects such as overlap, undercut, excessive convexity, etc.

At this point, it is important to note that although we are not sensing the molten metal pool itself or directly viewing the arc, rugged sensor construction and specialized visual processing is required since the system must operate in the very harsh welding environment, in the presence of intense arc light, smoke, flying spatter, electrical noise, and very high temperatures. Furthermore, it is important to mention that widely different image representations and shape descriptions, measurement accuracies, resolutions, and processing times are dictated by the different possible applications which range from slow detailed inspection to real-time welding process control.

GENERAL SYSTEM ARCHITECTURE

Hardware:

As in most of the applications of machine vision to robotic welding, the fundamental objective here is to obtain three-dimensional world information from a two-dimensional image. This can be done in a variety of ways, most of which essentially introduce a necessary extra linear constraint in addition to the ones introduced by the camera model. However, due to the nature of the welding environment and the type of features of interest, methods such as stereoscopy or other "shape-from" techniques[6] are not easily applicable. "Structured lighting"[6, 7], in the form of a projected plane of light, is successfully used. When a plane of light is projected and moved across the scene, stripes appear on the illuminated objects. Depth can then be readily inferred, since each point on a stripe will lie both on the light plane and the line-of-sight determined from its image coordinates.

In our system, the plane of light is created by refracting a 10 mW He-Ne laser beam using a cylindrical lens. The image is detected using a 320x244 CCD array (a Hitachi KP-120 camera). A narrow-band optical interference filter is used to block most of the ambient and arc light. Both the projection optics and the camera are mounted on the wrist of a Hitachi HPR-10A robot, controlled by an Automatix AI-32V integrated robot/vision controller, where all the developed visual processing algorithms are also implemented. Alternative sensor design philosophies have also been examined and considered to be advantageous in some cases. Specifically, a configuration where a laser beam is scanned along the scene and detected using a linear image sensor will in general increase the signal-to-noise ratio - since the beam is concentrated at one spot

- and at the same time will eliminate the problems with multiple reflections that might occur when a plane of light is projected on an angled highly specular surface. A solid state laser could also permit higher powers and lead to a more compact design. The current prototype sensor configuration, however, has been sufficient for a first implementation of the developed visual processing algorithms.

Software:

Visual processing in any artificial vision system can be broken down into imaging, image processing, image analysis, and image understanding and scene interpretation. In the first step, a two-dimensional discrete array of intensities is computed and stored in pixel memory by focusing the light emitted from the scene through the optical system onto an image sensor and by sampling and quantizing the resulting analog video signal. During image processing, operations are performed on the original image in order to reduce noise and highlight features of interest. Further operations are performed on the image during image analysis in order to detect features and to arrive at a symbolic image description. Finally, during image interpretation, the derived image description is used in order to reason about the world task at hand.

For most general description and recognition, one should rely on information from several laser stripes in order to create a geometric model of the workpiece[8]. Although we are also currently pursuing variants of this approach, we realize that the computational burden for a complete real-time geometric modeling would be excessively high even for the most powerful commercial vision system processors. However, due to the rather constrained world we are operating upon, sufficient information for recognition can be extracted from each stripe.

Most of the developed visual processing primitives were coded in Pascal and MC68000 assembly language. Initial algorithm development and most of the user interfacing were done in RAIL, a high-level robot and vision system programming language.

DETAILS OF THE DEVELOPED VISUAL PROCESSING TECHNIQUES

Imaging

As mentioned above, triangulation using a projected sheet of light is employed in order to get world information from the image. The <u>transformation</u> from image to world space is formulated as a matrix equation of the form:

$$s[x,y,z,1]^T = M[u,v,1]^T \tag{1}$$

between world and image homogeneous coordinates. Matrix M is calculated during camera and projector calibration[6,7]. A special calibration fixture is used for this purpose. This procedure has to be slightly modified if the camera is moved by a robot. Then, calibration has to be done in tool (torch) space with respect to which the camera is fixed. Transformation to world coordinates is computationally expensive. It is only, therefore, done in the final steps of our processing scheme. Feature detection and description is primarily done in pixel coordinates.

Image Processing and Feature Detection

The laser stripe, which is the only object of interest in structured light images, is readily detected by <u>intensity thresholding</u>. However, since the bright pixels corresponding to the laser stripe are usually much fewer than the ones corresponding to the background, the intensity histogram of the picture is not bimodal and a midgray threshold cannot be easily found. Cumulative thresholding techniques[6,9] can be successfully used in this case, by taking advantage of the known thickness of the laser stripe. Furthermore, adaptive threshold selection is necessary. Primarily, this is required in order to cope with the wide variations of stripe/background contrast that are caused by varying arc light intensities in the different areas of the image. Secondly, adaptive thresholding can help alleviate the problem of artifacts generated when the laser stripe is projected on a highly reflective object with irregular or highly angled surfaces, such as the weld bead. In this case, minor reflections will occur on the small facets adjacent to the main area illuminated by the laser line. The gray levels of these areas will vary randomly around the values within the stripe. For different thresholds the shape of the stripe might be different and artifacts can be introduced as in Figure 1. Locally adaptive thresholding can be very conveniently performed on each scan line (vertical or horizontal) and can be combined with a thinning operation which must follow.

If intensity thresholding is used, the detected laser stripe will have a thickness of several pixels. The exact dimensions will depend on the actual size of the laser beam, the magnification of the optics and the threshold value. However, for the determination of the 3-D object geometry, a thinned version of the stripe (one pixel wide) is required. Although general thinning algorithms are available[10],[9], particularly simple thinning techniques can be applied here. Specifically, if a single laser plane is projected in a roughly vertical (or horizontal) direction on most typical objects, the resulting laser stripes will cross each row (or column) of the image only once. So, one can proceed line-by-line (or column-by-column), detect runs of pixels above a specified threshold, and obtain the center pixel for each run. This method will break down in some cases. Then both vertical and horizontal "slices" through the stripes have to be made. However, this was not found to be necessary in our applications.

The ability to detect the laser stripes can be enhanced if the above thresholding/thinning scheme is modified so as to reject any run that is not within some thickness limits and if thresholding is combined with raster tracking[9]. The latter technique takes into account the spatial continuity of stripes and searches for stripe pixels only in a neighborhood of the pixels detected in the previous line (or column). In the case of a discontinuity, such as when illuminating a lap weld, a stripe will not be found in this area and search has to be performed on the whole line (or column). If the laser is projected on a highly specular object with angled surfaces, such as a V-groove joint, then for some camera and projector orientations, multiple reflections of the stripe might be observed (Figure 2). Then thresholding will pick up both the stripe and its very bright reflections. Separation is impossible with simple thresholding, since the reflections can be equally (and sometimes, more) bright than the stripe itself.

An alternative technique to intensity thresholding is spatial filtering with a filter which would give an identified (extremal or zero) response at the center of the stripe while suppressing other brightness sources (noise, reflections etc.). The main feature that can be used for this discrimination is the stripe thickness. In the majority of the scenes illuminated with structured light, the stripes are roughly vertical or horizontal and an one-dimensional such filter can be employed. A discrete mask is convolved with each row (or column) of the image and the maximum response on the given row (or column) corresponds to a point on the stripe centerline. It can be demonstrated that the required filter must be a bandpass filter in the frequency domain. The only such filter that combines space and frequency domain localization is the Laplacian of a Gaussian which can be readily approximated by an easily implemented difference of Gaussians filter[11]. A zero order approximation of such a filter is essentially a second difference operator which has been widely used in the literature[9] for line and streak detection. This filter has been successfully implemented in hardware by other investigators in this field[12]. For efficiency, in our implementation we combined spatial filtering with raster tracking. A correctly tuned filter can indeed easily discriminate the laser stripe (as in Figure 3) but the computational burden of the software implementation is rather high. Roughly one second was required for convolution in a 100 x 100 pixel window in our initial software implementation (for a 4 pixels wide stripe). However, since even spatial filtering will sometimes fail to discriminate between the laser stripe and its reflections, it was decided not to implement any such scheme in hardware, at least not before alternative projector/sensor

107

configurations (that would possibly eliminate reflections) are tested.

Image Description

Irrespective of the actual technique used, the output of stripe detection is a set of points located on the stripe centerline. Several such stripes along the last part of a single fillet weld are shown in Figure 4(a). However, representations richer than a set of (x,y) coordinate pairs are needed in order to recognize the essential weld features from the stripe data. Descriptions of the stripe shape must be computed. Several hierarchical levels of description are actually possible, but depending on the application and the time available, all, or only part of the power of the developed recognition primitives should be used. Multifeature and multilevel representations, based both on significant orientation changes along the stripe and on curve segmentation and fitting with elementary segments, are used for recognition in our system. In order to obtain these representations, further processing of the stripe centerline data is first required.

Specifically, gap filling and smoothing are necessary. The stripe centerline description might be sparse if the feature extraction software fails to recognize pixels along the stripe or might contain isolated noise pixels. In order to fill the gaps, Bresenham's line generation and drawing algorithm[13] was adapted from computer graphics . Use of this algorithm resulted in a significant (at least 50%) time saving as compared with simple straight line interpolation. Filtering can be necessary in order to smooth the stripe and remove noise. One dimensional low-pass nonrecursive filtering (essentially a running average filter) and median filtering have been implemented. Mean filtering results in loss of high frequency detail and can introduce significant artifacts around noise spikes. Median filtering on the other hand is a nonlinear processing technique which suppresses noise by substituting a value with the median of its neighbors in a window. The most important characteristic of a median filter is its ability not to affect step or ramp functions while discarding noise pulses. However, sorting needed for the median calculation, is very time consuming. To overcome this limitation, a fast median algorithm[14] is successfully adapted in our one-dimensional filtering problem. In this implementation, the input values are stored in a histogram and sorting is only performed for the first window values when the histogram is formed. Changes to the histogram can be performed very rapidly. So, as the window slides, the histogram is only modified rather than being recalculated.

Changes of orientation along the stripe correspond to image features that must be recognized. For example, in a fillet weld, the wetting points where the weld separates from the joint side are such locations. A second difference operator applied on the stripe coordinates would result in a maximal response at such locations of slope change. However, this response would be lost in noise due to the stripe quantization. An improvement of this idea is the use of central differences. An operator of the form:

$$S(x) = (y(x+k) + y(x-k) - 2y(x)) / 2k \qquad (2)$$

can successfully eliminate the noise problem by essentially low pass filtering. It can be readily shown that the response of the operator is zero on a straight line and proportional to the slope difference when evaluated at a point of an abrupt slope change between two straight lines. (Figure 5) This operator has been used for edge detection in stripe images by Sugihara[15]. In our work, we extend this idea by evaluating this operator over different scales[1]. Specifically, by varying the interval k in the above equation, different amounts of smoothing can be introduced before the differences are calculated. Minor or local changes of orientation along the stripe will only manifest themselves as local extrema in the response of operators with small values of k. More global changes will be evident in the response of operators with both low and high values of k. This behavior is illustrated in Figure 6. We are currently investigating the possibility of generalizing this approach and applying in our problem the concept of curvature primal sketch of Brady[16]. The basic idea there is that step or other significant changes of orientation along a contour can be detected using operators that are used for detection of similar changes in intensity images.

<u>Approximation of the stripe with a set of elementary segments</u> is necessary in order to generate a compact representation and as input to the recognition heuristics. Representation by straight line segments was found to be sufficient for our purposes. A polygonal line ("polyline") representation permits significant data compression by requiring storage only of the breakpoints between the segments. In addition, area calculations that are required in other stages become straightforward. Several line fitting algorithms were implemented and evaluated[1]. The "hop-along" split-and-merge algorithm by Pavlidis[10] was selected and adapted to our application. The algorithm sequentially examines separate subsets of pre-specified (hop) size. These subsets are iteratively split or merged, based on the result of a collinearity test performed on all the points in the subset. Fine-tuning is possible by selecting an optimal hop size. Collinearity checking is performed by evaluating a distance measure for all points from the endpoint line. The performance of the algorithm can be enhanced if other measures such as the number of sign changes of the pointwise error or the area between the point set and the end line, are included in the collinearity test. The accuracy of the representation can be increased by tightening the collinearity test parameters. On the other hand, if the test is relaxed, then a representation expressing the global character of the stripe can be obtained.

Recognition and Interpretation

Based on the coarse description of the significant orientation changes along the stripe, the general character of the joint or weld bead can be inferred. Furthermore, combining the information from the finer detail levels with the polyline representation, the joint sides and the weld bead can be identified. Heuristics for the determination of the desired weld joint or weld bead features can then be developed. For example, in a fillet weld, after the sides of the weld joint are found, the root can be located by extrapolation (Figure 4(b)). Having the root and the wetting points, leg lengths and throat dimensions are easily computed. Examination in some detail of the actual profile in the neighborhood of the wetting points can give information about the wetting angles, undercut, overlap etc.

All the dimensions obtained from a single stripe refer to the section of the weld joint or bead defined by the light plane. If this cross section is perpendicular to the weld line, then these dimensions and features are the same as the ones specified in welding design, and measured during manual inspection. If this condition is not met, however, then corrections or further geometric modeling have to be performed. We solve this problem, to a first approximation, by utilizing information from two closely spaced stripes as in Figure 7. The two stripes can be obtained by moving the same line projector along the weld and combining the pictures taken sequentially. Since this might introduce additional errors due to the robot positioning accuracy, a dual laser line projector has been designed where a single laser beam is separated into two parallel beams using a beamsplitter/mirror combination. Repeating the previously detailed processing steps the joint root and the wetting points can be found for both stripes. Weld cross-section parameters can be then computed. For example, the leg lengths can be readily obtained by projecting the wetting points onto the segment defined by the two root points.

After the weld features are recognized, further interpretation is needed before this information is useful in welding fabrication. This additional processing requires knowledge and reasoning specific to welding and the task at hand. For example, in the case of weld inspection, knowledge of the structural design specifications and applicable codes is required in order to reason on the adequacy of the measured weld dimensions and the significance of the detected weld defects. Or, in multipass welding, the sensed weld bead profile between passes can be effectively combined with knowledge about the location and formation mechanisms of interpass defects, in order to avoid them. Other information and knowledge about the intended service of the structure or the anticipated loading conditions is also essential in this step. So, for example, undercut detected in the toe of a weld bead might not be significant for static strength but very detrimental if the weld is subjected to oscillating loads and fatigue is a consideration. In addition to the above, further quantitative knowledge about the process is necessary if corrections to any observed problems are to be proposed. So, process models or other information relating the weld conditions, the joint geometry and the attained weld bead shape and size are necessary in order to suggest appropriate corrections of the welding conditions based on the sensed bead geometry.

Although some simple interpretation heuristics can be readily included in a particular implementation, development of a general system is very difficult if conventional implementation techniques are used. We are currently examining the feasibility of an alternative approach and attempting to build a first prototype of a more general visual interpretation module which will encompass, or have mechanisms of capturing, most of the required welding expertise[1]. Such technology is particularly essential for the development of autonomous remote welding fabrication systems[4]. It is expected that knowledge in this first prototype will be coded in the form of production rules[17]. Using the inference mechanisms and the knowledge representation available in a standard production rules language (YAPS[18]), the development of this system can be significantly simplified.

APPLICATIONS, CONCLUSIONS, AND FURTHER WORK

In the work reported in this paper, general visual processing techniques have been developed in order to permit new applications of machine vision in all three phases of welding fabrication (preparation, welding process execution, and inspection). Using these techniques, weld joint and weld bead features that are of significance in welding can be accurately recognized. Availability of such sensory information is absolutely essential in the development of autonomous welding systems and in novel applications (such as welding process conditions selection and adjustment, adaptive and multipass welding, or post-weld inspection), which we are currently pursuing.

Selection of welding process conditions is currently more an art than a science requiring the welding engineer to iteratively determine appropriate adjustments to the welding conditions based on the interpretation of series of procedure qualification tests. Most of the efforts to automate conditions selection have been based on descriptive or normative models that attempt to predict the weld bead geometry in an open loop fashion. The empirical descriptive models require a significant amount of experimentation for their derivation and are very difficult - if at all possible - to generalize. The analytical normative models, on the other hand, are usually very simplistic and impossible to apply in any practical situation. We believe that availability of the developed visual sensing techniques permits (by closing a feedback loop), an alternative approach, where the system can essentially be self-trained just as the welding engineer does, using a sparse database for an initial guess and simple rules and models for iterative adjustments. Similar arguments hold for real-time weld bead geometry control.

The available weld bead and weld joint geometric information can also be used in multipass welding applications during the planning of welding passes other than the first (where conventional seamtrackers invariably fail). Finally, this sensory input can permit automating welding inspection and quality control, which are now performed manually. Development of such an inspection system can lead to productivity improvements during manual inspection and can also provide a record of the weld shape, much in the same fashion as an X-ray record is now kept during non-destructive radiographic examination.

In conclusion, therefore, we believe that continuing efforts for the development of improved visual sensing strategies, along the directions presented above, can certainly lead to significant advancements in automated robotic welding fabrication, and can make the development of remote welding systems possible.

REFERENCES

1.   Agapakis, J.E., Vision Aided Remote Robotic Welding, PhD dissertation, M.I.T., Fall 1984.

2.   Wittels, N. and Libby, C.J., ``Vision Aided Arc Welding,'' AutoWeld Conference, RI/SME, Southfield, MI, November 16-17 1983.

3.   Agapakis, J.E., Masubuchi, K., Koreisha, N., and Somers T., ``Basic Research on the Introduction of Welding Robots to Commercial Shipbuilding,'' Tech. report to the Maritime Administration, Contract No. DTMA 91-83-C-30030, M.I.T., August 1984.

4.   Masubuchi, K., Agapakis, J.E., DeBiccari, A., and Von Alt, C., ``Feasibility of Remotely Manipulated Welding in Space - A Step in the Development of Novel Welding Technologies,'' Tech. report to the Office of Space Science and Applications, Innovative Utilization of the Space Station Program, NASA, Contract No. NASW-3740, M.I.T., September 1983.

5.   Masubuchi, K., Agapakis, J.E. and Von Alt, C., ``Remotely Manipulated Welding; Space, Underwater, and Other Applications,'' 65th American Welding Society Annual Convention, Dallas, TX, April 11-13 1984.

6.   Ballard, D.H., and Brown, C.M., Computer Vision, Prentice Hall, Englewood Cliffs, N.J., 1982.

7.   Nitzan, D. et al, ``Machine Intelligence Research Applied to Industrial Automation,'' Tech. report No. 12, SRI Project 2996, S.R.I. International, January 1983.

8.   Kremers, J., et. al., ``Development of a Machine-Vision-Based Robotic Arc-Welding System,'' 13th International Symposium on Industrial Robots, Chicago, Ill, April 17-21 1983, pp. 14-19 to 14-33.

9.   Rosenfeld, A. and Kak, A.C., Digital Picture Processing, Academic Press, New York, NY., 1982.

10. Pavlidis, T., Algorithms for Graphics and Image Processing, Computer Science Press, Rockville, MD., 1982.

11. Marr, D. and Hildreth, E., ``Theory of Edge Detection,'' Proceeding of the Royal Society (B), Vol. 207, 1980, pp. 187-217.

12. Morgan, C.G., Bromley, J.S.E., Davey, P.G., and Vidler, A.R., ``Visual Guidance Techniques for Robot Arc-Welding,'' 3nd International Conference on Robot Vision and Sensory Controls, Cambridge, MA, November 7-11 1983, pp. 390-399.

13. Bresenham, J.E., ``Algorithm for Computer Control of Digital Plotter,'' IBM Systems Journal, Vol. 4, No. 1, 1965, pp. 25-30.

14. Huang, T.S. et al, ``A Fast Two Dimensional Median Filtering Algorithm,'' IEEE Transactions on Acoustics, Speech, and Signal Processing, Vol. ASSP-27, No. 1, February 1979, pp. 13-18.

15. Sugihara, K., ``Range-Data Analysis Guided by a Junction Dictionary,'' Artificial Intelligence, Vol. 12, 1979, pp. 41-69.

16. Brady, M. and Asada, H., ``Smoothed Local Symmetries and their Implementation,'' The 1st International Symposium on Robotics Research, MIT Press, Cambridge, MA, 1984.

17. Hayes-Roth, F., Waterman, D.A., and Lenat, D., Building Expert Systems, Addison-Wesley, Reading, MA., 1983.

18. Allen, E., ``YAPS: Yet Another Production System,'' Tech. report CS TR-1146, University of Maryland, February 1982.

FIGURES

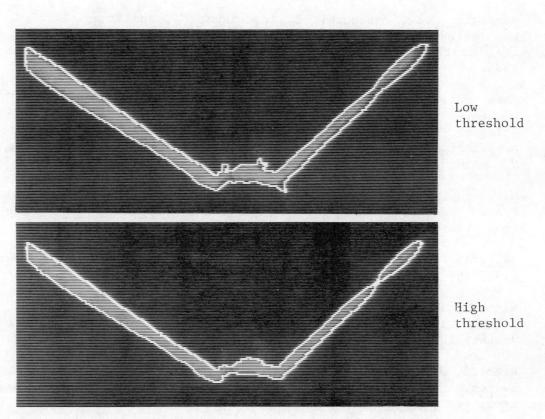

Low
threshold

High
threshold

Fig.1. Effect of threshold selection on the shape of the stripe.

Laser stripe   Reflections   Reflections   Stripe

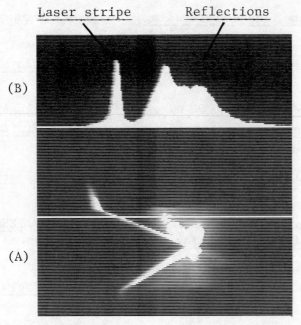

(B)

(A)

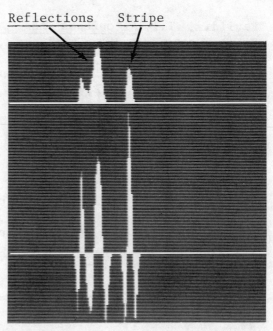

Fig.2. (A)Laser stripe and reflections on an aluminum V-groove joint. (B)Intensities along the highlighted scan line.

Fig.3. Response of the spatial filter for the reflections and the laser stripe.

112

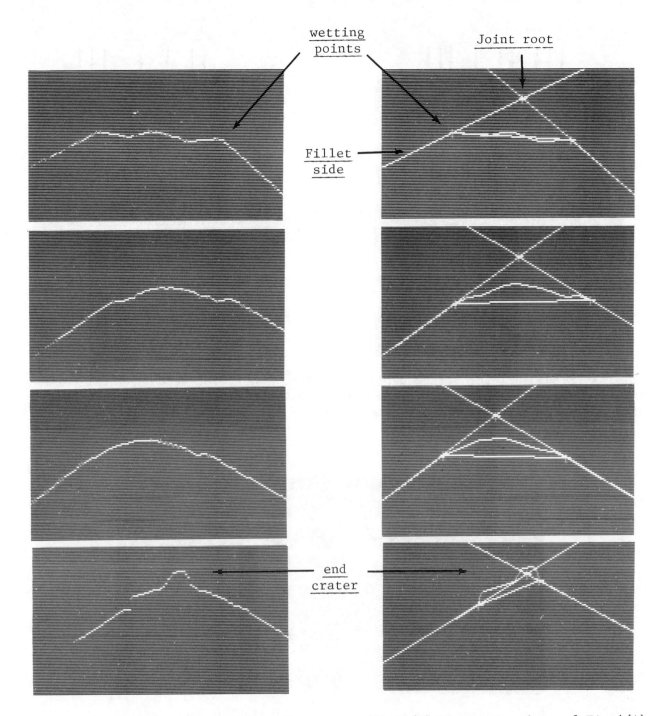

Fig.4(A). Stripe centerlines across the last part of a fillet weld (close to an end crater)

Fig.4(B) The stripe data of Fig.4(A) after processing and recognition of the joint sides.

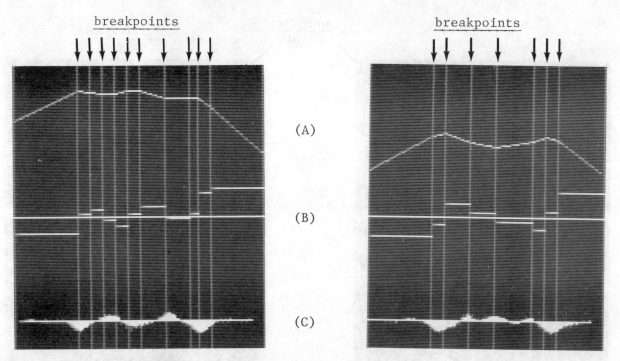

Fig.5. (A) Processed laser stripe centerlines from two fillet weld profiles (B) Slopes of fitted straight line segments between the breakpoints (shown by vertical lines) (C) Second central differences operator evaluated along the stripes.

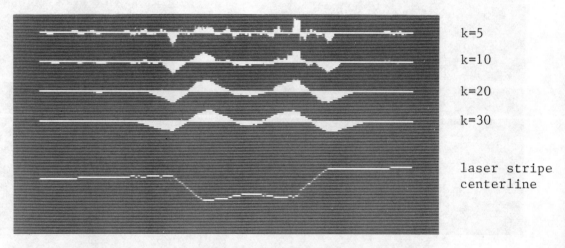

Fig.6. Second central differences evaluated over various scales along the profile of a multipass V-groove butt weld.

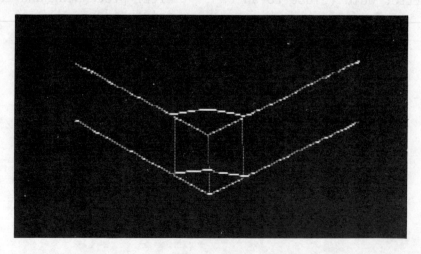

Fig.7. Processed image of two laser stripes projected on a fillet weld.

Flexible Eddy Current Sensors for Industrial Applications

H. Clergeot, D. Placko
and
F. Monteil

Laboratoire Electrotechnique Signaux et Robotique, Ecole Normale Superieure de l'Enseignement Technique, France

High performance eddy current sensors have been designed, the primary destination being imaging of a metallic surface.

Individual sensors achieve good precision and stability, with fast digital acquisition and multiplexing. The combinaison of fast, accurate, rugged sensors with the flexibility of the digital treatement makes this arrangement very interesting for a variety of industrial uses.

Applications are given for distance measurement in very adverse conditions (high temperature), for fast measurement of the thickness of metallic layer.

A more sophisticated application is given using a linear array of sensors for imaging of the profile of a seam. A numerical deconvolution has been perfomed to improve the image. This sensor is currently being tested for seam tracking and automatic arc welding.

I - Introduction

In a previous communication[1], we presented an imaging system using eddy currents. The performance of the individual sensors and the versatility of the electronic scheme designed for this purpose makes this measuring system suitable for a number of industrial applications including individual multiplexed distance, thickness or conductivity measurements.

We first discuss the principle of measurements through eddy current sensors, then we describe the electronic circuits used and show how they can be accomodated for various applications. We then give the lasted progresses concerning the application to imaging by the use of equispaced sensors.

II - Measurement through eddy current sensors

1°) Principle

The basic sensor uses a core with high permeability and an excitation winding driven by a sine wave signal.

If a high enough frequency excitation is used the penetration of the field inside a conductive material is negligible. In such a case, a metallic plate near the sensor just modifies the geometry of the path of the magnetic field (Fig. 1) and gives a modification of the reluctance independent of the physical properties of the material, which is very interesting for distance measurements.

The exact expression of the penetration of the field is given by the skin depth $\delta = \sqrt{\dfrac{2}{w\, \mu_0\, \mu_r}}$. But for magnetic materials, a δ-wide path in the metal is roughly equivalent to a $\mu_r \cdot \delta$-wide path of air : a metallic plate is "seen" by the sensor at a distance

$$\delta' = \mu_r \delta = \sqrt{\frac{2 \mu_r}{w \mu_o}}$$

(1)

behind the actual surface of the plate (Fig. 2). Beside the reluctance modification, the second effect of a metallic body is to introduce losses. The reluctance and losses may be measured separately, from the current in the coil, using synchronous detection with quadrature or in phase reference signal (with respect to the driving voltage).

2°) Sensitivity and stability

At long distances the magnetic field may well be represented by a dipolar field wich varies as D^{-3} (or D^{-2} for a linear array of sensors[2]) as a function of the distance D on the axis. The consequence is a very sharp decay for the sensitivity of the reluctance to small variations of D (see Fig. 3).

In order to increase this sensitivity, we have used a metallic shield to reduce the leakages, and a high permeability core with large poles.

Due to the decrease in sensitivity, the sensor is more and more sensible to noise and drifts when D increases, and this will set the pratical limits in which the sensor can be used with the proper accuracy. The metallic shield helps keeping the interference noise to a very low level, even in severe industrial environnement. In order to cut off drifts, and mainly thermal drifts, a balanced structure has been used (Fig. 4). The magnetic flux is divided into two parts : an active path outside the sensor and a reference path inside. The signal output is collected on a second winding making the difference between these two flux. This disposition permitted a reduction by a factor of ten of the drift (Fig. 5). The residual unbalance is connected to the small volume allowed for the reference path, and its drift due to dilatation or conductivity variation.

We may note by the way that the balanced structure gives a better linearity and reduces the requierements on the dynamic of the analogue preprocessing circuits. The massive metallic body used for the magnetic shield makes easier and more efficient an occasional air or water cooling system.

III - Electronic circuits

The circuits already described in a preceeding communication /1/, are sketched in Figure 6. The synchronous detector may be switched for in phase or quadrature measurements ; it has good perfomances up to several MHz.

The digital linearisation is very flexible ; the calibration curve may be registred in a RAM corresponding to the proper application or to different kind of sensors. For a given type of sensor individual disparities with the standard curve are accomodated using digital offset and gain corrections.

IV Sample of possible applications

1°) Multiplexed independent distance measurements

With the balanced structure (Fig. 4), in-phase detection is used. Usually the tests were done at the frequency f = 300 KHz. In the case of steel on iron sheets, we have operated up to 2 MHz in order to reduce the equivalent air depth (1).

We have represented (Fig. 7) a sensor designed for distance measurement in the range 0 to 7 cm with a resolution of 0,25 mm. It was tested in an industrial plant at an ambient temperature of about 200°C; using air cooling, the bodies of the sensors was around 70°C. At this temperature, in spite of the long wires used(about 12 m)the sensors exhibited good long term stability and no sensitivity to noise. Without correction at a distance of 5cm, the thermal drift was equivalent to 6mm in reading from 20°C to 70°C due to the residual unbalance of the temperature coefficient (see fig.5) ; this unbalance could be reduced by a more carefull design of the sensor or corrected automatically using a temperature sensor. The minimum acquisition time is about 20μs for one sensor, coresponding to a maximum measurement frequency of 50kHz.

2°) Touchless thickness measurement

If the thickness of a metallic sheet is of the order of the skin depth, a variable fraction of the magnetic flux can cross the sheet, so that, at a given distance, the reluctance and the losses are a function of the thickness.

In view of a particular industrial application we had to measure the thickness of aluminium samples in the range 0,5 - 5mm. To achieve a large enough skin depth the frequency was 1kHz. We have represented figure 8 the variations of the reluctance, that has to be registred on the linearization memory to get a direct reading of the thickness

3°) Imaging through a line of sensors

This application uses a linear array of equipaced sensors, each one giving the distance of the point of the metallic surface in front of it. This gives a two dimensionnal picture of the profile of this metallic surface in the plane of the sensor. A three dimensional picture can be obtained by using a translation of the sensor, or of the metallic body (see figure 9).

The advantage of eddy current sensors[3] over optical systems[4,5] is their ruggedness and their good immunity to noise and temperature.

The main limitation occurs from the spread of the field of the sensors : the image of the profile is "smoothed" by a spatial window which reflects the limited resolution of the sensors. In our previous communication we have explained how the profile can be restored with a good accuracy using a digital deconvolution /1/. In the next paragraph we detail some improvements and new results on this method.

V Imaging using eddy current sensors

The purpose of the digital treatment used to restore the image is to "invert" the smoothing operator associated to the sensor, so that the first point is to get a good description of this operator (subparagraph V 1). Due to the non linearity of the relation, as already discussed else-where[1] , the deconvolution is better achieved using an iterative algorithm ; we introduce in paragraph V 2 a more general recursive algorithm more appropriate for a real time implementation. We then give some exemples of images in different situation.

1°) Modelisation of the relation profile/image

From a theoretical study of the field, an analytical model for the smoothing filter can be justified[2] in the form :

$$h(y,x) = (a^2+y^2)/(a^2+y^2+x^2)^{3/2} \tag{2}$$

Let $y(x)$ be the exact distance of the surface and $Y(x)$ the image given by a sensor. For simplicity we way use the distance between two sensors as unit lenght for x, so that for each sensors, x is an integer corresponding to the index of this sensor in the linear array. The image is then approximated by the expression :

$$Y(x) = C.\sum_{x'} y(x') \, h(y(x'), x' - x) \tag{3}$$

where C is a normalization constant computed so that :

$$C.\sum_{x'} h(y(x'), x' - x) = 1 \tag{4}$$

A comparaison between the measurement image Y and image $\tilde{Y}$ computed according to (3) is given Fig. 10.

2°) Recursive deconvolution

Let us denote $\underline{Y}$ the set of observations corresponding to the N sensors,

$\underline{Y} = (Y(1),\ldots, Y(N))^T$. In the timevarying case this vector depends on the time and is denoted $\underline{Y}_t$ where t will be considered as an integer corresponding to the index of the time sample. We denote in the same way $\underline{y}_t$ the exact profile and $\underline{\widetilde{Y}}_t$ the image computed by (3). Let P be the exact operator giving $\underline{Y}_t$ as a function of $\underline{y}_t$ and $\widetilde{P}$ the operator corresponding to (3).

The recursive algorithm proposed may be represented by the block-diagram Fig. 11. As can be seen, from the previous estimate $\hat{y}_{t-1}$ of the profile, the estimate $\hat{\widetilde{Y}}_t$ of the image is computed by the relation (3) (operator $\widetilde{P}$) and compared to the real image $\underline{Y}_t$ given by the sensor. The error is used to correct the estimate of the profile, via the gain operator K.

Obviously, if convergence occurs, then $\hat{\underline{Y}}_t = \underline{Y}_t$ and $\hat{y}_t = \widetilde{P}^{-1} \underline{Y}_t$, so that if $\widetilde{P}$ is a good estimate of the true operator P, the restored image $\hat{y}_t$ will be close to $\underline{y}_t$.

The gain operator K controls the speed of convergence, or the tracking performance, and controls the filtering of the measurement noise, as will be discussed now.

3°) Discussion on the choice of K.

A good insight on the effect of K may be obtained under the hypothesis that P, $\widetilde{P}$, K are linear operators, which is true for small deviations of y(x) from an average $\overline{y}$, and the hypothese of spatial stationnarity by neglecting the end effects on the sensors. Then the operations of linear filtering turn into simple products by the use of spatial Fourier transformation. For a given spatial frequency, let p, $\widetilde{p}$, k be the gain of the operators P, $\widetilde{P}$, K and let $(\hat{y}_t)$ be the amplitude of the corresponding spatial component of $\hat{y}t$. The algorithm of figure 11 reduces to the scalar recursion :

$$(\hat{y}_t) = (\hat{y}_{t-1}) + k \left\{ (Y_t) - \widetilde{p}\, (\hat{y}_{t-1}) \right\} \tag{5}$$

whence :

$$(\hat{y}_t) = k \left\{ (Y_t) + r\, (Y_{t-1}) + \ldots + r^n (Y_{t-n}) + \ldots \right\} \tag{6}$$

using :

$$r = 1 - k\widetilde{p} \tag{7}$$

Note that, the filters associated to P, $\widetilde{P}$, K being symetric p, k, $\widetilde{p}$ are real. We will assume that $0 < k\widetilde{p} < 1$. Then (6) is a low pass time filtering relation with the time constant τ given by :

$$\tau = \frac{1}{1 - r} = \frac{1}{k\widetilde{p}} \tag{8}$$

and the static gain :

$$k \left\{ 1 + r + \ldots + r^n \ldots \right\} = \frac{k}{1 - r} = \widetilde{p}^{-1} \tag{9}$$

Thus $(\hat{y}_t)$ will track the value $\widetilde{p}^{-1}(Y_t)$ with the time constant τ. In presence of white noise on the observation Y_t, the effect is characterized by the equivalent gain gb :

$$gb^2 = k^2 (1 + r^2 + \ldots + r^{2n} + \ldots) = \frac{1}{1 - r^2} = \widetilde{p}^{-2}\, \frac{1 - r}{1 + r} \tag{10}$$

$$gb = \widetilde{p}^{-1}\, \frac{1 - r}{1 + r} = \frac{\widetilde{p}}{\sqrt{2\tau}} \tag{11}$$

The effect of a large time constant (k small) is a degradation of the tracking performance, but also a decrease of the noise by averaging. We will discuss now some particular choices of K.

The simpler choice is K = the identity operator. This was the choice made in[1] For our application $\widetilde{p}$ is a decreasing fonction of the spatial frequency, equal to one at the zero frequency. In such conditions the time constant is very small for low frequencies and increases for higher spatial frequencies. This may be interesting because higher frequencies are more corrupted by noise. Instead of the unity, a constant less than one may be choosen for K, to provide some noise reduction even at the zero-frequency.

Another possible choice would be $K = \tilde{P}^{-1}$; this would provide quick convergence for all frequencies, since then $r = o$. To achieve the same non zero time constant Θ, the choice would be $K = \frac{1}{\Theta} P^{-1}$. But it is necessary to introduce at least a cut off filter for the higher frequencies to reduce the noise.

Note that the filter K could be used too to provide interpolation between the measurement points (at an intervall smaller than the sensors spacing in the linear array).

4°) Exemples of restored profiles

Typical results of the imaging system are presented in figure 12. The sensor was constituted by 24 elements, with 6 mm spacing. As can be seen the restoration is quite accurate, but for the sharp edges that exhibit somme smoothing or overshoot.

VI - Conclusion

Our imaging system previously presented[1] has been improved to give a better restoration of the profile and to meet the requirements of real time operation.

The circuit used proves to be usefull too for a variety of idustrial applications, using multiplexed measurements on independent sensors.

Acknowledgements

This work was supported by the CEA and it is a pleasure to acknowledge J.M. Detriche for its active collaboration. Application concerning distance measurements and imaging are protected by a patent CEA-ENSET[6].

References

1 - Clergeot H, Placko D, Monteil F "Imaging using eddy current sensors" Pocedings ROVISEC (November 1983).

2 - Placko D, "Dispositif d'analyse de profil utilisant des capteurs à courants de Foucault" Thèse de 3ème Cycle, Université Paris Sud, ORSAY (April 1984).

3 - Detriché J.M., Marchal P, Cornu J., "Self adaptive arc welding by mean of an automatic joint following system" Proceedings 4 th RO.MAN.SV, Varsovie 1981.

4 - Morgan CG, Bromley JSE, Davey P.G., Vidler A.R., "Visual guidance for robot arc welding", ROVISEC 3, Boston, November 1983.

5 - Sami Z, Smith CJ, Yapp D, "An industrial robot using sensory feedback for an automatic multipass welding system" BRA Conference, May 1983.

6 - Clergeot H, Detriché JM, Monteil F, Placko D, "Dispositif pour mesurer la proximité d'une surface métallique conductrice" Brevet CEC N° 83/5323 Septembre 1983

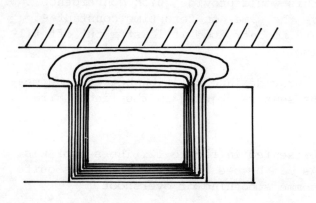 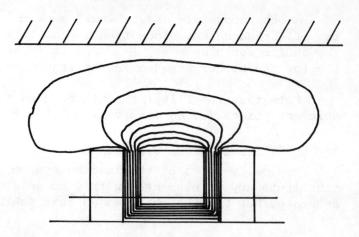

Figure 1 : Modification of the field path by a metallic plane (digital simulation)

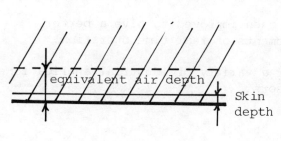

Figure 2 : Equivalent penetration in a magnetic materiel

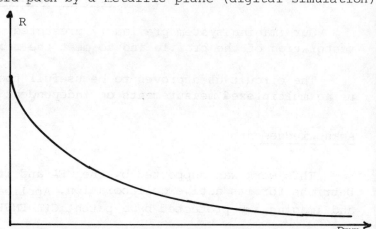

Figure 3 : Variation of reluctance with D

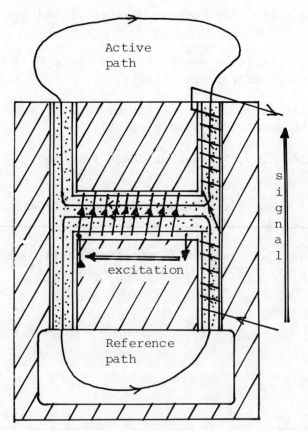

Figure 4 : balanced structure for the sensor

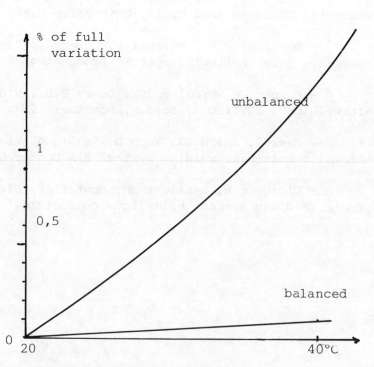

Figure 5 : thermal drift for balanced and unbalanced sensors

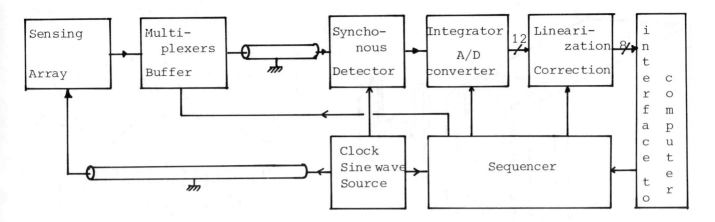

Figure 6 : Electronic circuits

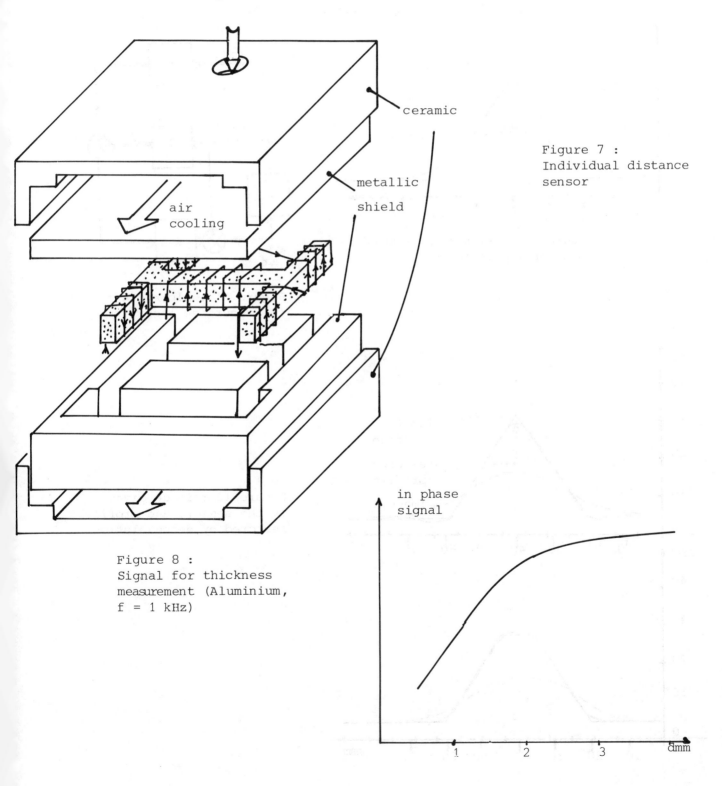

ceramic

metallic

shield

air
cooling

Figure 7 :
Individual distance
sensor

Figure 8 :
Signal for thickness
measurement (Aluminium,
f = 1 kHz)

in phase
signal

121

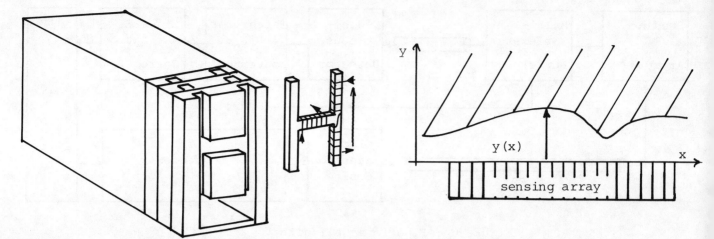

Fiugre 9 : Profile imaging through a sensor array

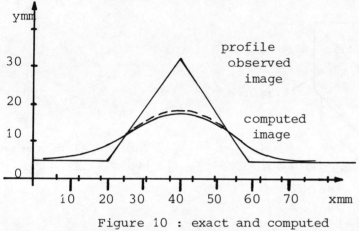

Figure 10 : exact and computed
image of a profile (scale 1)

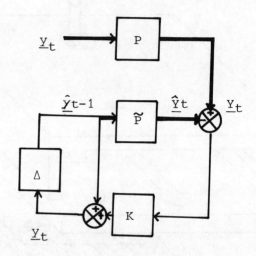

Figure 11 : block diagram of
the recursive algorithm

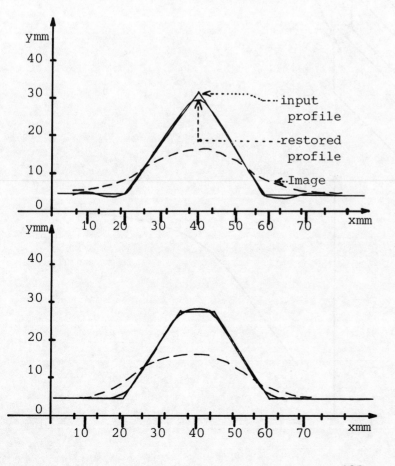

Figure 12 : image given
by the sensor, and profile
restored by deconvolution

Robot Plasma Welding with Integral Arc Guidance

R. V. Hughes, C. J. Page
and
R. P. Walduck
Coventry Polytechnic, UK

Abstract - A prototype robot welding system has been developed which is based on the plasma arc process. Results obtained so far have shown that high-speed welding of exotic materials such as stainless steel, nimonics and titanium down to thicknesses as small as 0.25 mm is practical. An inertialess weaving system is under development and a prototype unit has demonstrated the feasibility of the technique for compensating for weldment fit-up inaccuracies during high-speed welding.

INTRODUCTION

The process of plasma arc welding covers a wide spectrum of currents, ranging in magnitude from as low as 0.01 A up to 500 A or more. The development of this welding technique allows the joining of metals in the gauge range 0.01 mm to 20 mm or more and extends to a wide variety of material specifications.

PRINCIPLES AND CHARACTERISTICS OF PLASMA WELDING

Plasmas occur in all electric arcs, existing as high temperature regions of ionized gas. Plasma welding exploits this phenomenon by constricting the arc to form a high-intensity, collimated arc stream. The resultant "pencil beam" arc makes the process very much less sensitive to variations in arc length than the gas-tungsten-arc (G.T.A.) process, in which small changes in arc length bring about large variations in the impingement area on the workpiece, and consequently in heat input per unit area. With plasma welding, over a wide range of arc length, the heat input density is virtually constant. Comparison of arc power intensity shows it to be of the order 10^7 watts/cm^2 for plasma arcs and 10^4 watts/cm^2 for GTA welding. The plasma arc welding process can operate in either transferred or non-transferred arc modes. With the transferred mode the arc is formed between the non-consumable electrode and the workpiece, whilst with

123

non-transferred mode the arc is established between the electrode and the torch nozzle. The majority of plasma welding of metals is performed with the transferred mode. Non-transferred arc is particularly suited to welding of the thinnest gauges of metal, where accurate control of very low heat inputs is important. Furthermore, since with a non-transferred arc, the workpiece need not be electrically conductive, non-metals can be joined.

WELDING TECHNIQUES

The plasma process can operate in two distinct fusion forms. The melt-in technique is similar to that encountered with GTA welding - arc impingement causes melting to some depth to form a weld puddle, but will only produce full penetration on thin gauges of material. Plasma gas flow is reduced to a level where deformation of the weldpool does not occur. The alternative method of operation is "keyhole" welding; under favourable conditions of plasma gas flow, arc current and travel speed, a relatively small weldpool is produced with a hole penetrating through the complete work section on the joint line. Known as the "keyhole", this phenomenon is unique to plasma amongst the arc welding processes. In stable keyhole operation, molten metal is displaced to the top bead surface by the plasma stream to form the characteristic keyhole. As the plasma torch is mechanically traversed along the weld joint, metal melted by the arc is forced to flow around the plasma stream to reform and solidify at the rear of the weldpool.

Figure 2 illustrates a typical keyhole condition, however the size of the hole is slightly larger than during actual welding, due to the difficulties involved in producing very rapid arc extinction and consequent weldpool freezing.

ASCERTAINING OPERATIONAL PARAMETERS

In broad concept, robotised welding portrays many characteristics of the more conventional means of torch manipulation. This similarity provides a useful function for simulating the process conditions under which robotic welding can be accomplished, but free of problems directly resulting from interfacing the welding equipment with the robot. Accordingly, the initial work to establish the operating conditions for welding was performed on a traversing unit equipped with a variable speed motor. The results of initial tests indicated butt-welds of satisfactory appearance and metallurgical structure could be produced at welding speeds up to 17 mm/second. It should be noted that this compares to speeds of about 4 mm/second under normal manual operations. The test programme also highlighted the importance of joint match-up and showed that special design of fixturing was necessary to minimise the extent of heat-affected zone and thermal distortion.

Overall, the results appeared to provide a sound basis on which to commence development of a robotised plasma welding system.

ROBOTISED WELDING

The robot used in development of the system was a Unimation Puma 500 series machine. Initially this was connected to a B.O.C. Sabre-arc microplasma unit with a maximum current output of 15 A. Using fixturing and operational data obtained during mechanised welding tests, many welds were performed with the robot in materials ranging from 316-S12 grade stainless steel to IMI 318 grade titanium alloy. The autogenous butt-welds produced from the outset were of equal quality to any weld produced by mechanised welding - with excellent top and underbead appearance. Horizontal "down-hand", semi-vertical and even vertical "uphand" welds all portrayed the same quality characteristics.

A major point highlighted when using the robot was that attainable welding speed is directly related to the welding current available. In response to this, a Messer

Griesheim Euromatig plasma unit, with a current capacity of 250 A was obtained, such that welding speeds even greater than those achieved with the microplasma unit (17 mm/second) could be reached. With the more advanced weld program control available with this system, welding speeds of 60 mm/second were easily reached before the draught effect caused by rapid torch movement caused a deterioration in the inert gas shielding of the weld area. Substantially higher welding speeds than this can, however, be achieved by supplementing workpiece protection with blanket shielding derived from the fixturing itself.

WELD FIT-UP AND ITS INFLUENCE ON JOINT INTEGRITY

The successful production of autogenous butt-welds using the plasma process, particularly the keyhole mode, is very dependant on the quality of joint fit-up. A large number of welding tests have confirmed that in general, a gap between abutting joint faces of greater than 10 to 15 percent of sheet thickness, independant of material specification, will produce a lack of fusion on the joint line more akin to plasma cutting than welding. It is important to appreciate that this rule applies to other autogenous welding processes, since it is essentially a feature of metal flow conditions, but is acutely noticeable when attempting keyhole welding. The problem becomes of most significance where very thin sheet material, of say, below 1 mm gauge is concerned, as accurate preparation of sheet edges is almost impossible.

Joint fit-up is further complicated when it is realised that to minimize the effects of thermal distortion it is desirable to produce as narrow a weld bead as practically possible. For a fixed set of welding parameters, increasing the welding speed produces a lower heat input per unit length of weld, with consequential narrowing of the weld bead. This can be clearly seen in Figure 3, where for fixed welding conditions, the beads resulting from three different welding speeds are compared.

The normal technique employed to alleviate the fit-up demands of the joint is the continuous addition of filler metal (homogenous welding), to the weldpool, as it progresses along the joint. This assists the "bridging" of the joint gap and provides material re-inforcement. Where significant gaps exist between adjoining edges eg. Where the gap is greater than the sheet thickness, it is mostly necessary to invoke a weaving motion with the welding torch to assure satisfactory bridging and bead re-inforcement. Weaving becomes most significant when high-speed welding is considered.

WEAVING LIMITS THE WELDING SPEED

Welding robot manufacturers were quick to realise the necessity of providing a weaving ability in the robots repertoire of movements. Most robots have the facility for programs to be written in which both the frequency and amplitude of the weaving motion, superimposed on the actual weld line, can be specified. There is an inherent weakness in all these systems however, in that all rely upon mechanical oscillation of the robot arm. High weaving frequencies, that is, weaving above 10 Hz is not feasible because the inertia of the arm and torch assembly causes "overshoot" and dynamic instability.

Depending on the type of joint, and the rate of filler wire addition, this maximum weave frequency of 10 Hz limits the attainable welding speed to between 20 and 30 mm/second. If the welding speed is increased above this level, lack of fusion and voids within the joint will become apparent.

What is needed is a method of weaving which in practical terms possesses no inertia.

MAGNETIC ARC DEFLECTION

The ability to influence the path of an electric arc by the controlled application of a magnetic field has been used for some considerable time as an aid to weldpool stirring.

Research to date has shown that other arc control features are possible using this technique.   A system of magnetic arc deflection has been devised which allows very high frequencies of weaving to be achieved at the weld joint, entirely independant of the robot.   Figure 4 shows the prototype system, whilst Figure 5 gives a detailed view of the arc deflector mounted in place on the welding torch.   The deflector operates on the principle that, when placed perpendicular to a magnetic field, a current carrying conductor (in this case the arc column), will be subject to a force causing motion at right angles to that field.   The direction of arc movement is dependant on the polarity of the magnetic field and the arc polarity.

The deflector in Figure 5 is essentially an electromagnet, with specially shaped soft-iron pole pieces.   When the assembly is arranged such that the poles are perpendicular to the direction of welding, transverse oscillation of the arc column i.e. weaving, can be induced.   An alternating waveform, usually sinusoidal or triangular, derived from a signal generator is directed to a power amplifier.   The amplified signal is fed through a coil assembly which results in the generation of a similarly alternating magnetic field at the pole pieces.   This field acts upon the arc column, forcing the arc impingement spot on the workpiece to move back and forth across the weld joint.   As there is virtually no lateral inertia in the arc column, it is possible to attain very high weaving frequencies indeed, although practical welding requirements are unlikely to demand arc oscillation above a few hundred Hertz.

Figure 6 shows the effect on the weld bead when a 2 Hz triangular waveform is combined with a welding speed of 10 mm/second.   The difference between this weld and those shown in Figure 3 is clearly apparent, with the bead width being increased by some 60 percent compared to a normal weld conducted at that speed.   Increasing the deflection frequency to 20 Hz whilst welding at 10 mm/second has the effect shown in Figure 7.   The flow pattern of material in the weldpool becomes much smoother, producing a more nearly straight-edged weldbead.

AMPLITUDES OF DEFLECTION OBTAINABLE WITH THE ARC COLUMN

Experimental work has shown that a peak to peak deflection amplitude on the workpiece of approximately 1.5 times the arc column length is achievable before undesirable metal flow characteristics such as "undercutting" manifest themselves.

As already described, with the plasma arc process, a columnar arc form exists, which permits large variations in torch standoff without significant effect on the workpiece. The length to which the arc column can be drawn during welding is essentially a function of three parameters:-

1.   Arc current
2.   Plasma gas velocity
3.   Plasma gas type

Using argon plasma gas at a typical velocity of 400 m/s, a 100 A arc column can be drawn to a length of about 25 mm and maintain stability.   With a 250 A arc and flow velocity of 700 m/s, the arc column can reach as much as 100 mm in length.

If it is assumed that under normal welding conditions, an arc length of 20 mm is used, the implication is that a weaving amplitude of up to 30 mm is practicable.   This has been borne out by experimental data.   Figure 8(a) illustrates the type of bead profile attainable when bridging a 3 mm gap between 304-2B stainless steel of 0,44 mm thickness.   It will be noted from Figure 8(b) that the underbead of the weld is of similar quality to the topbead shown in Figure 8(a).   The overall bead width is approximately 5,5 mm.

FURTHER DEVELOPMENTS

In addition to the many possibilities offered by the system described for the purposes of controlling the arc path, on two axes if necessary, a potentially far greater advantage is suggested.   This is in the arena of weld seam tracking.

At present, real-time seam tracking limits welding speeds to around 20 mm/second (see reference 1).   This is largely due to the time taken to carry out complex position-correction computations within the robot computer.   If the output from a conventional seam sensing device - optical devices generally prove most resilient in the arc welding environment - is coupled to a signal conditioning unit which generates an output proportional to the sensed seam deviation, a DC current of the appropriate magnitude and polarity supplied to the coil of the magnetic deflection assembly will cause the arc column to "bend", following the desired, rather than the pre-programmed path.

It is this area towards which research is now being directed.   Because in most practical environments, seam placement deviations of only a few millimetres are encountered, it is envisaged from the work undertaken to date that magnetic arc positioning will be able to meet tracking requirements.   The advantages offered by a hardwired seam compensation system are that firstly, welding speeds would be higher than those attainable when using software based systems operating on the robot itself, and secondly, the equipment is not limited to any particular manufacturers robot.

REFERENCES

(1)   Morgan, C.J. Bromley, J.S.E., Davey, P.G., and Vidler, A.R.
       "Visual guidance techniques for robot arc welding".
       Proc. 3rd   RoViSec,   Nov. 1983, Cambridge, USA,   pp. 615-624.

(2)   N.T. Dick.
       "The application of magnetic fields to T.I.G. welding arcs - part 1 - transverse A.C. magnetic field" Welding Research International 1972   Volume 2 No. 1.

(3)   G. Rider.
       "Control of weldpool size and position for automatic and robotic welding"
       Proc. 3rd   RoViSec,   1983   pp. 625-634.

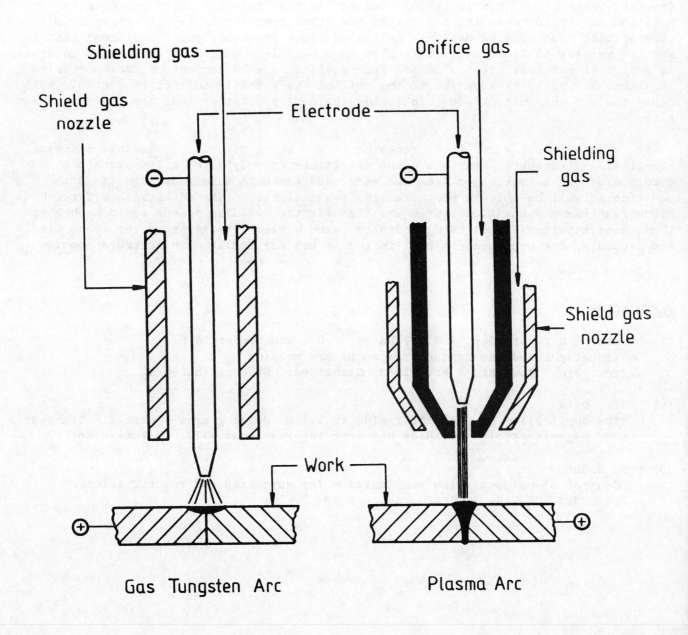

Fig. 1 Arrangement of G.T.A. and Plasma Welding Systems.

128

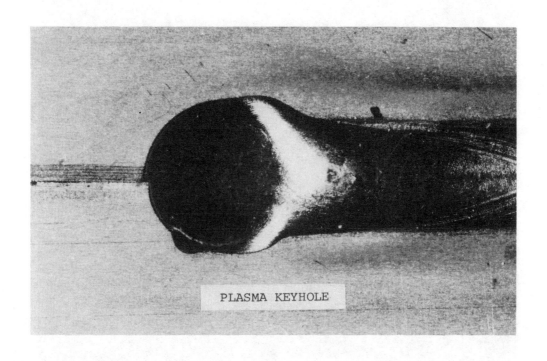

PLASMA KEYHOLE

Fig. 2 Typical plasma keyhole in BS.1449 Pt 2, 304-2B
stainless steel, 1,2 mm gauge.

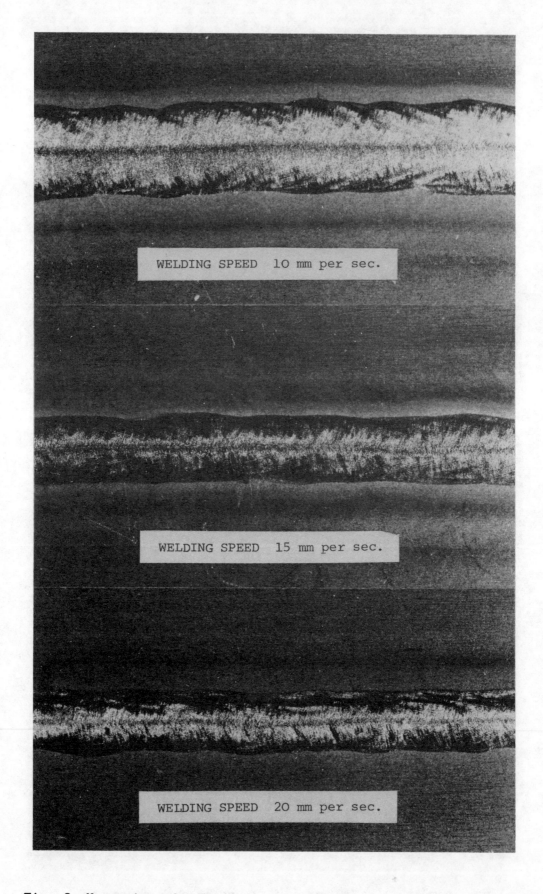

WELDING SPEED 10 mm per sec.

WELDING SPEED 15 mm per sec.

WELDING SPEED 20 mm per sec.

Fig. 3 Narrowing of weld bead due to increase in welding speed.

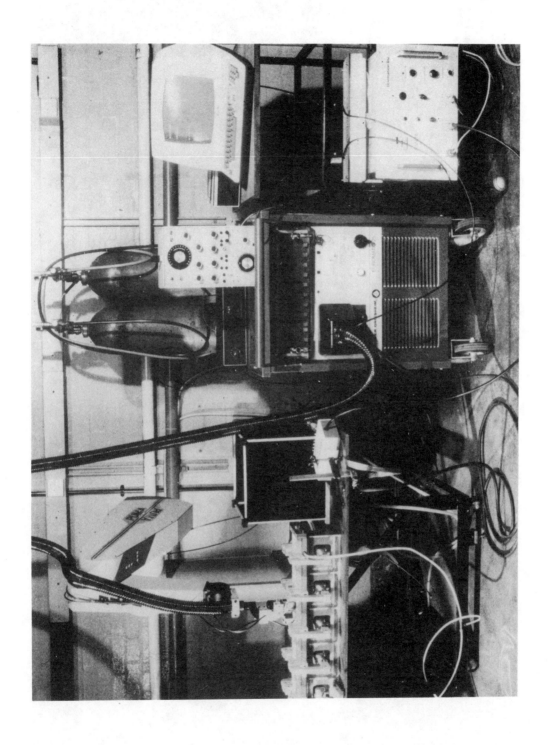

Fig. 4 Robotic Plasma welding system.

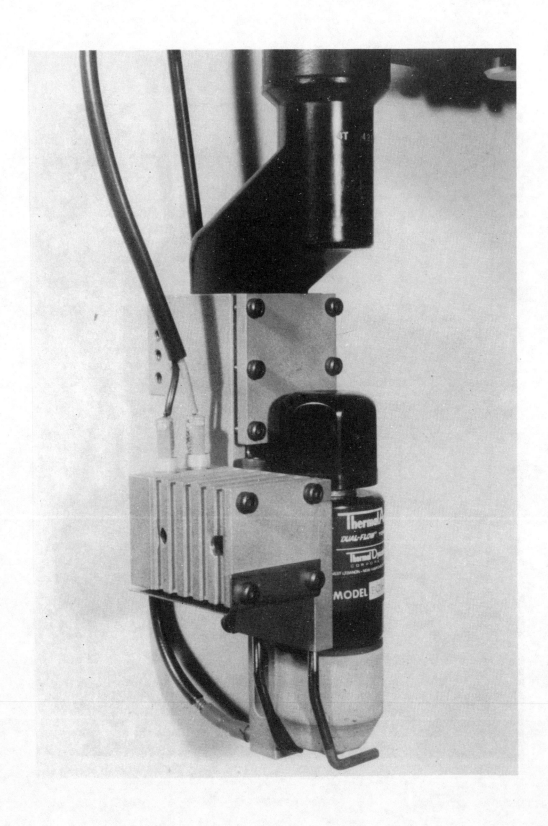

Fig. 5 Magnetic arc deflector fitted to commercial machine torch.

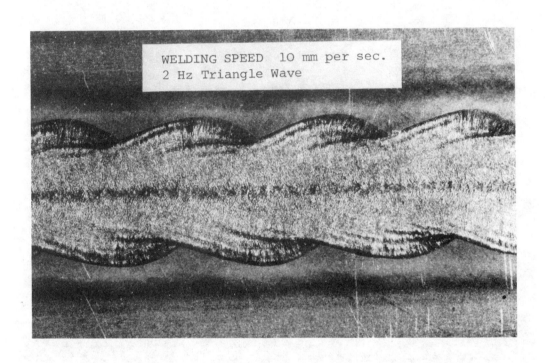

Fig. 6   Weld produced using low frequency magnetic arc deflection
to illustrate principle of the technique.

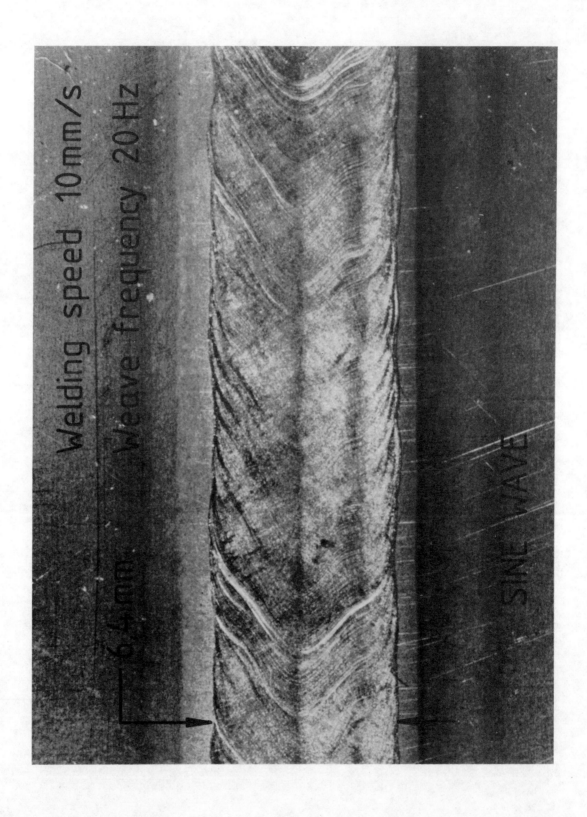

Fig. 7 Large increase in weldbead width possible using magnetic
arc deflection.

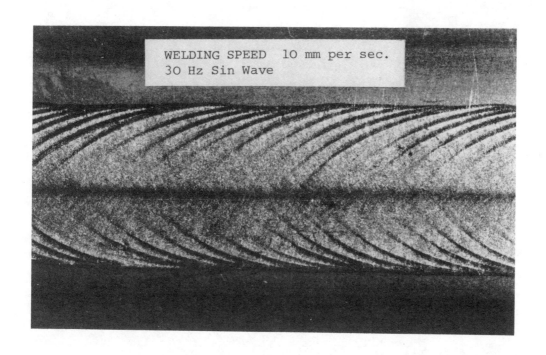

Fig. 8(a)

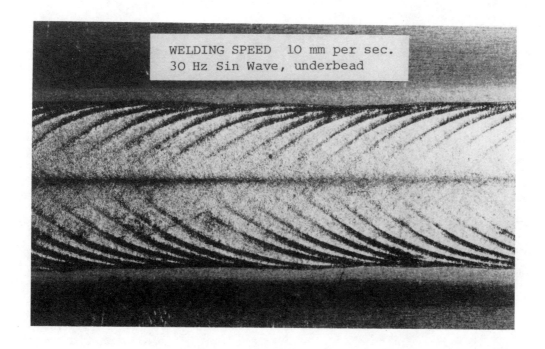

Fig. 8(b)

Butt Weld produced in 0.4 mm gauge 304-2B stainless steel
with 3 mm gap between joint edges.
Filler material: Stubs 316L
Bead width   5,5 mm

Machine Vision Algorithms for Vision Guided Robotic Welding

N. R. Corby Jr

General Electric Company, USA

ABSTRACT

Robotic welding systems in the past have primarily been taught a specific trajectory to be welded, along with appropriate process parameters for welding that path. This has been satisfactory for less demanding applications but in most situations, part fit-up and in-process variations have caused less than satisfactory results.

Real-time machine vision feedback control has been suggested as a way to compensate for static and dynamic geometry variations as well as to control the welding process parameters.

Robust forms of visual processing must be used to deal with the difficult visual environment. In this paper, techniques for enhancing the scene, developing navigation data, and temporally processing the scene to derive process control data are described, along with descriptions of the processing hardware employed.

INTRODUCTION

The development of robotic arc welding systems has progressed rapidly over the last three years. There is widespread research interest and much commercial interest in the topic as evidenced by the number of research and development projects underway. The number and type of welding sensors continue to grow daily.

There are many review papers dealing with defining the various methods of arc welding and the characteristics of each type. Also, papers exist [1] which review adaptive control for automated arc welding. This paper will deal primarily with tungsten inert gas welding (TIG or GTAW). This type of welding is a precise, high value added type of welding employing a non-consumable electrode. An arc is struck between the electrode tip and the workpiece and the resultant arc supplies heat sufficient to melt the workpiece metal. The volume of metal produced by melting the parent parts is usually insufficient to completely fill the gap between the parts. Filler wire is usually fed continuously into the molten weld pool to remedy this, as well as to introduce some specific metallurical properties to the resultant joint, if desired.

The "optimum" weld is generally described as one that permeates the metal of all constituent workpieces in a prescribed manner and which fills the original gap in a specified way e.g. with a specified amount of "drop-through" and a specified amount of "positive reinforcement".

Simple robotic welding systems are taught a desired trajectory for the torch tip based on the specific part or object to be welded. Robots which use point-to-point control need many such points. In addition, extra taught points are commonly required to allow for smooth motion at difficult points such as sharp corners. More sophisticated robots are able to get by with fewer taught points due to software which allows for proportional or interpolated movements. The number of taught points required in these systems is still large however.

After a taught path has been specified and verified, the weld schedule is created. This is a schedule, stored in the robot controller, which contains the weld process parameters and is usually specified as a function of path segment. These process parameters are such variables as current, frequency and duty cycle for example. This schedule is keyed to specific segments of the taught path.

Most welding systems now are capable of compensating for gross positional errors through the use of global static positional offsets to the taught path. In many systems, the offsets are determined by simple sensory touch probes which contact prescribed points on the workpiece in prescribed directions. Such sensors are incapable of compensating for varying fit-up and distributed plate mispositioning.

While there exist some real-time sensors capable of determining the centerline of the joint to be welded, there are very few sensors which can determine process related information. Approaches which attempt to decouple the navigation task from the process control task do not generally produce resultant seams of satisfactory quality. In general, most systems use apriori weld process schedules to vary the weld process variables while adaptively folowing the weld path segments. Some recent experimental systems, have demonstrated non-coupled active joint following and sensor-driven process control but only during relatively benign segments of the path.

In the General Electric system, the navigation and process loops are coupled since for the execution of a "proper" weld, the electrode may need to be positioned at a point which is not directly over the joint centerline. The navigation system determines the topography of the plates immediately (4-8mm) ahead of the electrode and the process measurement system determines the condition, shape, and position of the weldpool relative to the electrode. Other information such as metal type, thickness, desired penetration and joint width are known to the control system and are factored into the control algorithms. An integrated closed loop control system then determines appropriate coordinated steering, velocity and process commands to cause a properly centered weld with the specified penetration to occur.

SYSTEM DESCRIPTION

The major components of the GE vision guided welding robot are shown in Figure 1. The balance of this paper will concentrate on the imaging and image analysis portions of the system. For a more global view, the reader is referred to [2] in which systems and control related aspects are discussed.

The primary components are the Thru-Torch sensor (an integrated imaging welding torch), a programmable laser pattern projector (designated "laser"), a GE TN2500 CID solid-state imager, and a distributed multiprocessor for real-time image analysis (shown as a collection of circuit cards above the box labeled Adaptive Welding Control System). The imaging torch is carried by a General Electric process robot. The Adaptive Welding Control System (AWCS) recieves the navigation and process monitoring data at a rate of approximately 10-15 Hz from the WeldingVision multi-processor and integrates this information with an extensive analytical model of the weld process. Based on a minimal taught path (as little as three or four points) the AWCS then computes integrated, real-time trajectory commands along with computing real-time weld process commands. The system also contains the intelligence for fully automated control of startup, weld and shutdown.

Integrated imaging welding torch

A development prototype of the integrated imaging welding torch is shown in Figure 2. Initial research [3,4] has indicated the value of obtaining a visual image of the weld pool itself. The apparatus that has been used to obtain these views in the past has been rather awkward to use in practice. A coaxial viewing geometry is preferred, which results in an "aerial" view of the weld pool and the surrounding plates to be joined. A custom TIG welding torch was designed that allowed a clear view down the gas cup of the torch providing a view coaxial with the electrode. A high resolution coherent fiber bundle with an appropriate optical system is mounted above and coaxial with the electrode. The system thus relays a full spectrum image of a circular area centered on the electrode to a GE TN2500 CID imager. Suitable neutral density filters were used to adjust the intensity of the pool image initially. As described later, the neutral density filters were later replaced by specific narrow bandpass filters. To increase the range of geometries that could be accessed by the torch, the circular gas cup was later changed to an elongated design approximately 25mm long by 13mm wide. Numerous changes having to do with shrinking the overall envelope of the torch have also been carried out. The current result is a very compact torch specifically designed for robotic manipulation. The depth of field for the current design exceeds 8mm and is adequate for the range of tasks currently under test.

Programmable Light Pattern Generator

In order to accomplish the navigation function for automated welding, the system must be able to determine the position and attitude in space of the parts to be joined. It is generally sufficient to determine this information relative to the robot (or electrode tip).

The prototype system uses the principle of structured illumination to obtain mutltiple profiles of the joint area immediately ahead of the electrode. In principle a planar sheet of light is allowed to impinge on the surface to be measured. When the intersection of the light sheet and a planar surface is viewed from a position not in the plane of illumination, a straight line image will result. If the surface to be profiled is not a plane, then the image of the intersection will not be a straight line. The deviation from the straight line case can be processed to yield height information for the surface.

The primary task of the robot mechanism is to guide the electrode tip along the joint to be welded. For maximum stability in the control process and for minimum constraints on the path to be welded, it is desirable to determine the plate topography as near to the electrode tip as is possible. Past systems have attempted to measure plate topography at a point well forward of the arc (50 - 100mm)in order to avoid the problems of dealing with the intense arc in the field of view. Past attempts have placed the illumination source (flash lamp, laser diode etc), pattern aperture and projection optics on the torch which in general results in a bulky package and a static illumination pattern.

After evaluation of arc spectra and the imaging system properties, it was determined that a low power Helium Neon CW laser would provide suitable illumination energy. Other wavelengths were found to possess characteristics that will be useful in some of our future welding robots.

It is desirable to be able to dynamically alter the structured illumination pattern spatially and to be able to temporally modulate it. Different patterns can be adopted at different points in the task, each one specifically designed to assist in determining the required geometric information. An illumination source (in this case a 7 mw HeNe CW laser) is formed into a ray and the ray is then moved thru space by a two-dimensional positioning system (for example, a galvanometer driven x-y mirror system or an acousto-optic modulator) and is imaged onto one end of a coherent fiber bundle. A dynamic, adaptive sequence of beam positions can be computed in real-time by the Vision Multiprocessor or a precomputed set can be supplied from a RAM or ROM. A suitable set of projection optics at the exit end of the coherent fiber bundle projects the pattern onto the surface to be profiled.

The actual pattern generator used in the prototype system conveys multiple patterns, some dynamic and some static. The illumination for each pattern is a small CW HeNe laser. This programmable light pattern generator forms an integral part of the imaging welding torch described above. The resultant compact torch contains no active circuitry and is able to:

1) project multiple dynamically varying light patterns into the zone immediately forward of the electrode
2) image those patterns in such a way as to enable a structured light determination of plate topography
3) image the weld pool area using the same single camera

Vision Multiprocessor

The Vision Multiprocessor analyses the images delivered from the imaging torch in real-time at a rate in our laboratory prototype of approximately 10-15 Hz. Commercial prototypes in our product laboratories are able to run significantly faster: approximately 30 Hz for navigation and 15 Hz for process sensing. The research oriented Vision Multiprocessor is currently composed of five 68000 single board computers arranged on a Multibus. A GE Vision Memory System (PN2150) is interfaced on the Multibus. This set of cards acquires a roughly 256 by 256 pixel image with 128 gray levels per pixel directly from the digital port of the GE TN2500 CID camera. The weld scene image for the CID camera is delivered by the coherent fiber bundle connected to the Thru-Torch imaging welding torch. A 256 x 256 color graphics board on the Multibus provides colored graphic overlays which are superimposed onto the weld pool scene and can be viewed on the system monitor. In this way the operator is provided with indications of system function and performance. Both serial asynchronous interfaces and 16 bit parallel interfaces are also present, in addition to EPROM boot cards.

One 68000 is designated overall master and is responsible for arbitration of resources, control of vision related peripherals (pattern modulation for example), and interlocked communication with the Adaptive Welding Control System.

Two types of information are supplied to the Vision Multiprocessor-startup information and run time information. The startup information allows the Vision Multiprocessor to configure itself optimally for the upcoming conditions. Appropriate algorithms and parameters are chosen from libraries stored in the Vision Multiprocessor. The real-time data is used to select alternate algorithms appropriate to the conditions that are about to be encountered, such as when changing from butt to lap welds along a joint. In addition, parameters can be adjusted based on the world information. For example, torch attitude and orientation is continually requested from robot controller and the AWCS.

The other 68000 processors are designated as slaves and are assigned dynamically varying duties within dynamically changing regions as designated by the master at the beginning of each processing cycle. The number of slave processors can be easily altered. The result of adding more processors is higher processing speed.

VISION PROCESSING FUNCTIONS AND ALGORITHMS

This section will outline the types of processing that occur within the Vision Multiprocessor. Program were developed on either a VAX 11/750 or on a 68000 based UNIX system in "C". Once the algorithms were defined and debugged, the code was partitioned and then cross assembled for the 68000 processors. For testing in the lab, the resultant machine code is electronically "downloaded" to the Vision Multiprocessor. For floor prototypes, the code is burned into EPROMS and the system software is loaded automatically when the system is powered up.

This section will be divided into two sub-sections: navigation and weld pool processing. Figure 3 is a photograph of one of the typical video images that are processed. Figure 4 offers a schematic view of the same scene. The dark spot near the center of the image is the electrode. Surrounding the electrode spot is the bright, crescent shaped arc flare. Filler wire is shown being fed in from the upper edge of the scene. At the left side, the two laser stripes are shown with the plate joint indicated by a darkened segment of each stripe. The weld pool is the dark, roughly elliptical area extending to the right of the electrode. The actual weld pool extent is about 5mm by 10mm.

Navigation

The section of the weld scene of processed for the navigation function is the left third of Figure 3. For the purposes of illustration, two parallel sheets of light are employed here. Other light patterns have been evaluated and certain patterns are currently being added to the navigation library. The image shown is for the case of a butt weld. Different topographic parameters are determined for other types of joints such as lap or fillet joints. Appropriate strategies for these situations are stored in the navigation library and can be retrieved when needed while traversing the weld path.

The shape and brightness of the "stripes" can vary greatly over the course of the weld, as can joint width. The information about the plates are such things as: height and tilt for each plate, width of the joint at the "signposts" and the position of the edges of the joint.

In addition to the "normal" optical degradations and distortion caused by the lenses, filters and windows etc, the coherent fiber optic bundles introduce some unique distortions. This is true of both the image forming path and the pattern forming path. One of the distortions that the bundles exhibit is a periodic lateral dislocation of the image that, for example, transforms vertical straight lines into a "square wave" patterns. Such fixed distortion is measured during calibration so that its effects may be corrected for at run-time.

The torch tip is positioned perpendicular to a flat reference surface and at a known height above it. An image is taken and processed to determine stripe centerline and stripe brightness. The laser intensity is adjusted by the Vision Multiprocessor to provide a nominal desired brightness. If filler wire is to be used during the welding run, the wire feeder is commanded to advance the wire to its nominal extended position. The attitude and position of the possible interference with the stripe patterns is detected, measured and stored. The filler wire is then retracted. The robot then moves the torch to the start of weld.

Two modes of processing are provided - cold start and warm start. Cold start processing is used when starting the weld or after an image that could not be processed. Warm start processing is performed on any cycle which was preceded by a cycle that yielded reliable data. Each stripe is processed by its own processor with a dynamic processing window. All processing is gray scale in nature.

The cold start window extends vertically over the whole height of the stripe image and horizontally a small distance on either side of the stripe. Using the calibration data, a search is made downward along the stripe. Each row of the windowed image is normalized and scanned by a least mean square slope detection filter which filters out the effect of detector and image noise and provides a filtered image whose value at each point is equal to the smoothed slope of the original. The positive-to-negative zero crossing of the filtered waveform determines the peak of the luminance wave form (which is the center line of the stripe).

At the joint, the stripe centerline may shift due to the height change encountered at the plate junction or the stripe may be absent due to shadowing or to no bottom being present at the joint. The criteria for determining when an edge is encountered is a jog (relative to the calibration data) that exceeds a specified amount or failure to detect the joint centerline over a specified number of rows. For the case where the plate surfaces are not nearly parallel, the joint center is determined by intersecting lines fitted to the plate surface measurements.

For the more usual case of a warm start, the correlation over time of the process is exploited. In this case, the windows are made much smaller than the cold start case and a search for joint edges made in the vicinity of the region where the edge occurred on the last cycle. Additional critera in this case are consistency of joint width and heading. Estimates of stripe brightness made on eack cycle allows the system to adapt to changing surface reflectivies.

The raw navigation data is transformed to a common coordinate system, passed to the master processor and is then transmitted to the control system along with estimates of the validity of each piece of data. The closed loop adaptive welding controller then determines navigation commands and process commands.

There are a number of other estimation procedures that continuously monitor the condition of the navigation image and can alter the nature of the algorithms as well as the parameters used by the algorithms. Global storage areas in the master CPU store current feature values for transmission to the AWCS and thus indicate the current state of the Vision Multiprocessor system.

Process Measurement

The weld pool boundary is the liquid-to-solid boundary of the molten pool of metal directly below the electrode. The appearance of the weld pool covers an extraordinarily wide range. The surface of the pool under normal conditions is globally convex, but may have local flat spots or depressions caused by materials on the molten surface or by cover gas flow. The nature of the surface shape as well as the shape of the perimeter varies greatly with velocity of the torch and with tilt of the workpiece surface. Pool shapes encountered during startup and shutdown or during sharp turns have characteristic shapes which are quite different from those encountered during the bulk of the run. Electrical and thermal imbalances can create quite unsymmetrical and distorted pool boundaries. If filler wire is not used, the pool is slightly better behaved and often has a concave or flat surface with a fairly regular shape.

The average reflectivity of the surface is very high, but the surface can have local areas of much lower reflectivity. This can occur in the formation of transient "oxide" rafts. Slag "rafts" can also form but in usual practice these lead to unacceptable welds. These rafts can move and circulate due to electromagnetic and thermal circulation effects. The thermal distortion due to filler wire introduction can also distort the pool boundary. These are only some of the variations that one can expect when trying to determine weld pool boundaries in a real welding situation.

Some pool boundary detection approaches have tried to make use of the ambient arc light to detect pool boundary. The nature of the arc and its instabilities make this approach a difficult one. The presence of filler wire can also influence the arc as can variations in the cover gas flow. Electrode tip geometry has a pronounced effect on arc position. An external source of illumination was used in order to acquire a satisfactory image in which the weld pool could more easily be detected. A low-power Helium-Neon CW gas laser provided a suitable illumination source. The neutral density filter used earlier was replaced by a 3 nm bandpass interference filter centered at 632.8 nm. Most arc light is therefore blocked from the camera. Appropriately shaped static laser patterns delivered by the Programmable Pattern Generator, provide a considerably enhanced pool image.

The processing of the weld pool images proceeds as follows (Refer to Figures 3 and 5). The image in sampled radially at a predetermined angular and radial resolution relative to the electrode tip. The electrode forms a fixed point relative to the camera and serves as the origin. The angular extent of sampling is approximately plus and minus 100 degrees relative to the angular orientation of the weld pool. The "zero" angular ray varies from frame to frame due to rotations of the image caused by the welding torch following the joint with a preferred filler wire orientation. Image sample intensities are non-linearly transformed and smoothed. If desired, an exponentially weighted sum of current and past samples can be formed for processing by later steps. This time-weighted series exploits the frame-to-frame time correlation of the pool scene.

Each one-dimensional radial image intensity waveform is processed syntactically to determine possible occurances of patterns that indicate a true pool boundary. Figure 5 shows the radial sampling pattern (light gray dots), a forbidden processing zone centered on the electrode (large brighter gray dot) and potential edge locations (small bright dots).

This set S' of possible boundary points is then scanned in order to delete isolated points and obvious false boundary points. The resulting set of waveforms is called S. Those rays of S that lie within plus and minus 30 degrees of the "zero" axis of the pool are then selected and all potential edge points fitted in a least squares sense to a three-degree of freedom parabola. The extremum or "peak" of the fitted parabola occurs roughly where the tail of the pool occurs. The same subset of S that was within plus and minus 30 degrees is rescanned using the first parabola as a rough reselection template. This has the effect of selecting only those points that lie with a fraction of a standard deviation of a two-dimensional curve. The resulting parabola usually is a very good approximation to the true rear edge of the pool. The extremum of the parabola indicates the position of the rear extremum of the pool.

The determination of the top and bottom edges of the pool proceeds as follows. A vertical search window is established at a point approximately midway along a line segment L joining the electrode center and the above determined rear pool point. The pool image is sampled perpendicular to L in two windows centered on the expected location of the pool top and bottom edges. The columns of these two windows are processed by the same algorithms used to process the rear pool edge data. The set of possible locations of the upper edge SU and the set of possible locations of the lower edge SL are each subjected to a weighted majority vote. The result is an estimate of the lower and upper pool extent relative to L. Figure 6 shows the result of the pool edge extraction process. The single rear dot is the estimate of the position of the rear edge extremum of the pool. The double dot at the top and the double dot at the bottom edge are the result of the detection/majority vote process. In addition, the position of the joint edges for each laser stripe is indicated at the left edge of the figure.

A two parameter curvilinear fit is made based on the upper edge point and the rear edge point. The fitted form of the curve is based on empirical studies and analytical models. The same is done for the lower edge point and the rear edge point. The upper and lower half curves form a composite boundary on which estimates of the shape and extent of the pool can be based. These features are then determined and sent to the master for formatting and transmission to the AWCS.

CONCLUSION

This paper has described a prototype system for visually guiding in a closed-loop manner the trajectory and process parameters of a robotic TIG welding system. The prototype system is undergoing evaluation and extension by General Electric. Commercial versions of the system are under design at the Robotics and Vision Systems Department of General Electric, Orlando, Florida. Research aimed at increasing the robustness of processing and the breadth of welding techniques and situations is also continuing. With the advent of applied expert systems, the possibility exists for very robust systems capable of handling the myriad of details encountered in a complete system solution of automated welding.

REFERENCES

(1)   Richardson,R.W. "Review of the State-of-the-Art of Adaptive Control for the Gas Tungsten and Plasma Arc Welding Processes". Center for Welding Research Report 529501-82-5, Ohio State University.

(2)   Sweet, L.M., Case, A.W., Corby, N.R. and Kuchar, N.R. "Closed Loop Joint Tracking, Puddle Centering and Weld Process Control Using An Integrated Weld Torch Vision System", Control of Manufacturing Processes and Robotic Systems, ASME Winter Meeting, November 1983.

(3)   Richardson, R.W., Gutow, D.A. and Rao, S.H. "A Vision Based System for Arc Weld Pool Size Control", Measurement and Process Control for Batch Manufacturing, ed. D.E. Hardt, ASME, November 1982.

(4)   Richardson, R.D. and Richardson, R.W. "The Measurement of Two-Dimensional Arc Weld Pool Geometry by Image Analysis", Control of Manufacturing Processes and Robotic Systems, ASME Winter Meeting, November 1983.

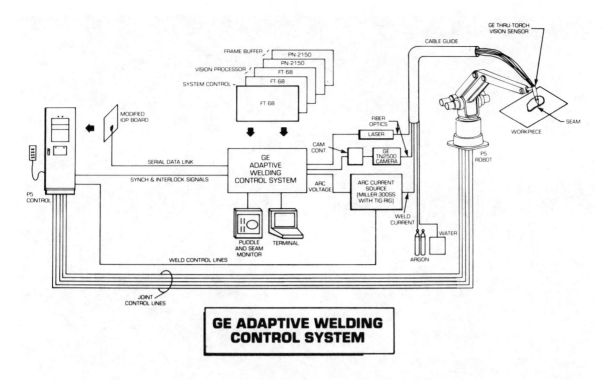

Figure 1   Schematic of closed loop welding system

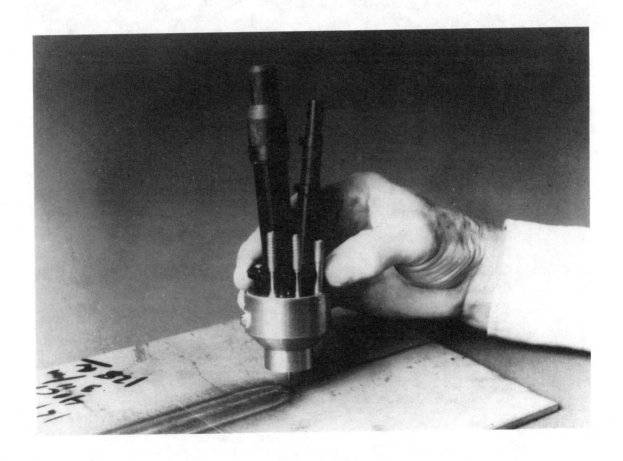

Figure 2   Thru-torch integrated welding torch sensor.

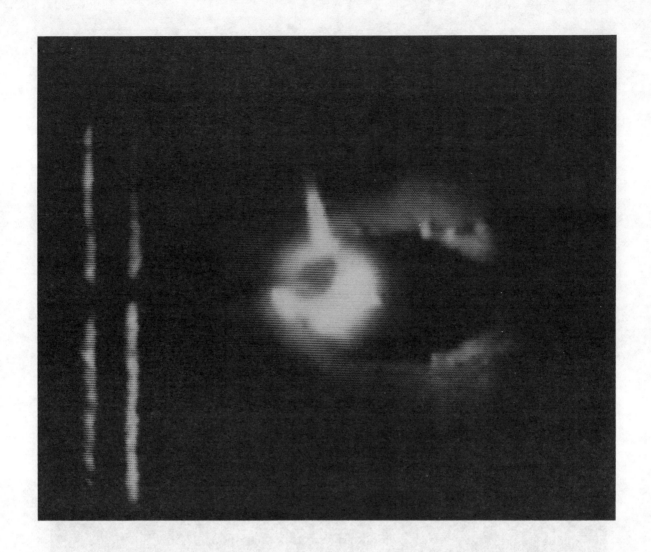

Figure 3   Unprocessed sensor image of weld-pool scene.

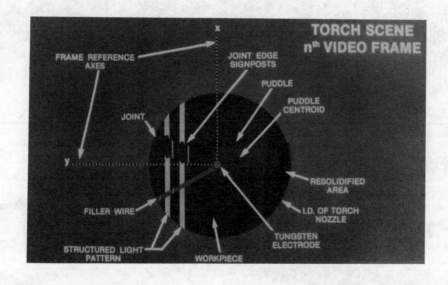

Figure 4   Schematic of features in vision sensor image.

146

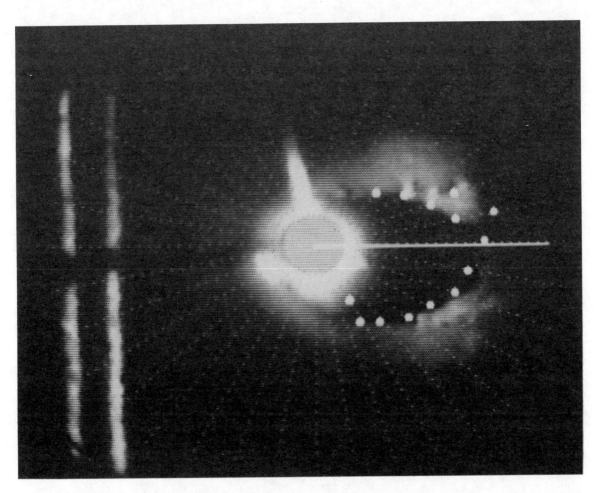

Figure 5   Sensor image with sampling points and
raw extracted features.

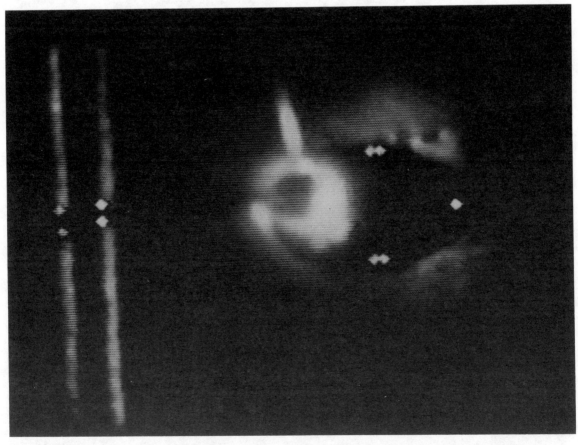

Figure 6   Sensor image with final pool boundary
points and navigation points.

A General Purpose 3D Vision System

G. Betz

Robotic Vision Systems, USA

ABSTRACT

A general purpose 3-D vision system will be described and several applications will be discussed. The system uses structured light and optical triangulation to digitize the surface of the workpiece. The vision system is an industrial product which has been installed and operated in both assembly line and batch manufacturing operations. This paper shall describe the sensor and how it is modularly constructed such that it can be easily configured to implement the turn-key systems described. Further, a more advanced 3-D vision system which is not yet in commercial production will be described.

INTRODUCTION

The widespread use of robotics will to a large extent depend on the integration of robots with various types of sensors (i.e., 2-D vision, 3-D vision force, torque, etc.) and intelligent sensory processors. This field is as of now very immature, however, some systems have been developed and are in use. In the area of 3-D vision, RVSI has produced a series of sensor systems, the first of which was installed in 1981 and is still in operation. Since then, the 3-D vision sensor has evolved and it is now a modular product which can be configured to accomplish a variety of tasks.

The core of the Company's 3-D Robo Sensor® vision systems is the electro-optical Vision Module sensor (Model 210 Series). This sensor, introduced in the spring of 1984, is designed to be easily reconfigurable so that it may be optimized for each task without an extensive design effort. Another major element of the Vision Module is the vision Processor Unit, which contains dedicated vision processing hardware as well as an embedded microcomputer. The microcomputer is a Motorola 68000 which can be easily configured to perform a variety of operations by simply adding standard VERSAbus™ or VME printed wiring board.

The last and most likely the most significant element of the modular system is the applications software. It is the software which determines the function of the vision system, that is whether the system will be used for inspection, welding, assembly, material handling, etc. The software itself is modular with modules dedicated to vision data processing, robot interface and system control. Each of the elements of the system will be described in more detail in the following sections.

VISION MODULE

The Vision Module consists of a solid state structured light projector and solid state array camera. The unit projects a plane of light onto the surface to be digitized. The projection of the plane on the surface forms a cross sectional image of the surface which the camera converts into a signal that is processed into digital coordinates of each resolvable point.

An outline drawing of the sensor is shown in Figure 1. The sensor is factory adjustable to provide field of view in the Z axis from 0.5 in. (1.27 cm) to 4.0 in. (9.16 cm) and can be configured to be either bisector normal or projector normal. The sensor mounting plate is on the top of the unit and can be used to mount the sensor directly to the robot tool plate or the sensor can be mounted to an adaptor plate which also mounts the necessary tooling as shown in Figure 2.

VISION PROCESSOR UNIT

The Processor Unit is housed in a half height Nema 12 cabinet as shown in Figure 3. It provides DC power for the sensor, controls sensor operation, processes the camera signal into 3-D measurements of the viewed surface and communicates with the robot controller. Additional communications ports enable communication to other robots, computers and terminals. A Motorola 68000 microprocessor provides sufficient computational power for many applications. High speed signal processing is performed by analog and digital hardware circuitry which presents the results to the 68000 for further processing and external communication.

Depending on the application and the preference of the end user, the vision processor can be supplied in one of the following 2 basic configurations:

- Image Processor - This is the most elemental form of signal processing for the Vision Module. In this configuration, the Vision Module will collect and process data as requested by a host computer. The microcomputer will locate and identify the features requested by the host and return their location and orientation in sensor coordinates.

- Stand Alone Processor - In this configuration, the microcomputer in the processor takes on the function of the image processor and the host computer. The host computer function typically involves tasks such as vision correlation with nominal data bases, robot path generation, user interface and station reporting. The degree of hardware modification to include the host function in the Vision Processor Unit is dependent on the complexity of the task. For simple operations when time is not critical no hardware additions are required because of the existing capability of the 68000 base microcomputer.

SOFTWARE

Although the Robo Sensor® hardware has been designed to be inherently flexible, the greatest degree of flexibility is afforded by the software configuration. The modular software concept has been followed in the design of RVSI's software. An ever increasing library of software modules is maintained by RVSI. Software modules exist for the following functions: image data correction, feature detection, feature measurement, data base correlation, robot path generation, robot communications and user interface.

The various software modules can be combined to be run either in a stand alone vision system or the software may be divided to operate in a combination of a vision processor and a host. To date, the Robo Sensor® and its modular software have been integrated to create inspection, material handling, adaptive welding and other manufacturing processes. A brief summary of some of the key programs follows:

Automated Component Optical Measurement System

The system was designed to inspect engine casting for the Cummins Engine Co., a major diesel engine manufacturer. The equipment is general purpose, however, in that by reprogramming the minicomputer, a variety of large, complex objects may be inspected and reports automatically generated. The system can be scaled to larger or to smaller more accurate versions for inspection of objects differing considerably from the 2' x 2' x 4' (610 x 610 x 1220mm) volume of the engine casting.

During a period of 40 minutes the system automatically scans all six sides of the casting, measuring over 800,000 data points to an accuracy of 0.010 inches (0.25mm). Figure 4 is a photograph of the engine block laying on its side on the transport mechanism that carries the block back and forth in front of the vision sensors that can be seen to the left of the block. The system operates as follows: prior to an engine inspection, the scan mechanism scans a known surface for automatic calibration and verification of system operability. The engine block is transferred to the scan table mechanism which positions the block with a repeatability of 1/4 inch (6mm). The scan table mechanism moves the block in front of the vision sensors which are position automatically in: (1) height to view the appropriate swath; and (2) in and out to place a portion of the engine block optimally within the sensor's field of view. This procedure is repeated until the entire surface of one side is scanned. The data is correlated to datum references which are selected on that side of the block and are visible when viewing adjacent sides. The scan table mechanism then rotates the engine block and the process is repeated until all four sides are scanned. The inspected block is returned to a roll-over barrel where it is rolled over to allow the final two sides to be scanned.

Although over 800,000 data points are measured, the data is processed with regard to a stored design data base to allow approximately 1250 comparisons to critical engine block dimensions to be reported. In order to keep the amount of data the computer must process to a manageable level, a hardware preprocessor under adaptive computer control is used. The hardware preprocessor accepts a total of 600 megabytes of data at a rate as high as 500 kilobytes per second.

The operator can select six different hard copy reports: an Exception Report on out-of-tolerance measurements; a Marginal Report on "close to" out-of-tolerance measurements; and an Engineering Analysis Report on all measured parameters for all features; a Shifted Report on each of these reports with the part reference planes automatically shifted by the system computer to minimize out-of-tolerance measurements, similar to that which a skilled machinist would do in an attempt to make the block useable. In addition, a graphics printout capability of selected features is provided, an example of which is shown in figure 5.

A menu is used to prompt the user for data entry which allows relatively untrained operators to use the system. Once the system begins its inspection process, no human intervention is required and the equipment operates unattended.

Seam Automated Welding System

RVSI developed and installed four vision based adaptively controlled welding robots in the General Motors Janesville, Wisconsin assembly plant. The vision sensors accurately measure the position of the front pillar-roof seams and rear-sail roof seams on the J-cars. The weld paths of Cincinnati Milacron T3 welding robots are then modified to adapt to the measured seam location and cross section. The systems accommodate a position ambiguity of +/- 1/4 inch (6mm) as well as any arbitrary rotation. An identical equipment is also installed in the General Motors Technical Center in Warren, Michigan. A photo of RVSI's 3-D vision Robo Sensor mounted on the T3 robot is shown in Figure 6.

A similar system was installed at the Caterpillar Tractor Co. East Peoria, Illinois. This system uses a PDP 11/44 host computer which can control and coordinate up to 10 robot/vision welding cells utilizing Cincinnati Milacron 746 electronic robot. In this case the welds must be structural, as opposed to the cosmetic welding done for GM. The part material varies from 1/4" (6.3mm) to 1.5" (38mm) and the system is configured to track and process 5 groove cross sections (i.e., fillet, single bevel, V-groove, J-groove and U-groove).

Automated Propeller Optical Measurement System (APOMS)

Historically, U.S. Navy ship propellers of up to 24 feet (7.3M) in diameter have been inspected by a labor-intensive process. RVSI is developing a Robo Sensor based inspection system with sophisticated computer data processing. A three-dimensional digital description of the propeller is created from the optically measured data and then compared to the design data base to identify out-of-tolerance areas. The system is completely off-line programmed, developing its scan path from the design data base. Furthermore, it operates in real-time, thereby providing data soon after scanning a surface. A multiprocessor architecture provides wideband processing which off-loads the computer. Global accuracy is +/- 0.025 inch (0.6mm). Local accuracy is better than +/- 0.0225 inch (0.06mm) over a 6 inch (152mm) radius. A photograph of the system is shown in Figure 7.

Propeller Optical Finishing System

As a follow-on to the APOMS project, RVSI is developing for the U.S. Navy a Propeller Optical Finishing System (PROFS). This system will use the inspection data from the APOMS to off line generate paths to be executed by two Cincinnati Milacron 776 robots to automatically finish grind the propellers.

Next Generation 3-D Vision

RVSI has developed a laboratory model of a true 3-D real time measurement system. The system is based on a film based 3-D copying system developed by RVSI (then known as Solid Photography) in the mid 1970's. That system was able to make high quality portrait sculptures as shown in Figure 8. The laboratory model was demonstrated during the summer of 1984. It is able to acquire and process in real time (less than 3 seconds) a complete 3-D data base of every resolvable element in a cubic field of view of 2 inches (50.8mm) on side. Since the system uses optical triangulation, only those surfaces which are viewable to both the camera and the projector can be measured. A photograph of the accuracy module mounted on a Cybotech H80 robot is shown in Figure 9.

The 3-D sensor has an accuracy of +/- 0.005 in. (0.13mm) for the 2 inch (50.8mm) on side field of view. The field of view can be increased or decreased with the corresponding scaling in accuracy. RVSI is continuing work on this product and a production model will be available. The speed of data acquisition can be decreased to approximately 0.3 sec. in production units depending on improvements in a key component of the sensor.

CONCLUSION

It is generally acknowledged that 3-D vision is a key to the use of intelligent industrial robots in automated manufacturing operations. The successful use of RVSI's 3-D vision system in a number of robotic applications underscores the present availability of a proven and versatile 3-D vision sensor to meet the present needs of robotic manufacturing systems. The development of the true 3-D volumetric sensor will ensure that 3-D vision will meet the future needs of manufacturers.

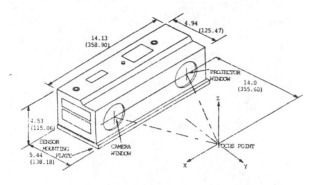

Fig. 1. Line Drawing of Robo Sensor Vision Module.

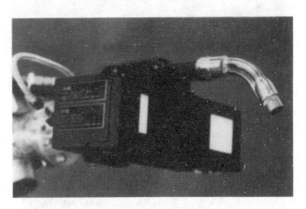

Fig. 2. Robo Sensor Vision Module.

Fig. 3. Robo Sensor Processor Unit.

Fig. 4. Automated Component Optical Measurement System.

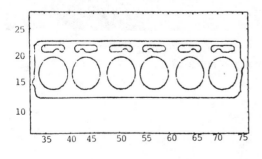

Fig. 5. Graphic Printout of Automated Component Optical Measurement System.

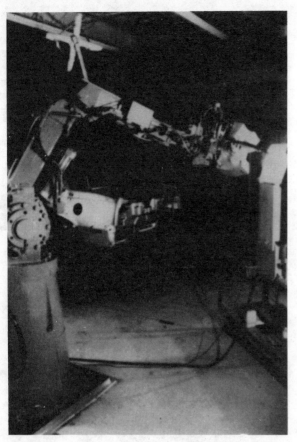

Fig. 6. Seam Automated Welding System for Silicon-Bronze welding.

Fig. 7. Automated Propeller Inspection and Welding System.

Fig. 8. Photograph of 3-D Replicator.

Fig. 9. Photograph of Accuracy Module.

THREE-DIMENSIONAL SENSING

A New 3D Sensor for Teaching Robot Paths and Environments

M. Ishii, S. Sakane, M. Kakikura
Electrotechnical Laboratory, MITI
and
Y. Mikami
Tokyo Metropolitan Technical College, Japan

ABSTRACT

A new 3-D sensor which can simultaneously measure 3-D position and orientation of robot environments is proposed. The 3-D sensor consists of a camera using a position sensitive device and multiple LEDs attached to robotic devices such as a robot's hand and a teaching device. Their 3-D position and orientation are calculated in real time based on both 2-D coordinate values of the LEDs in an image plane of the camera and geometrical relationships among 3-D positions of the LEDs. The 3-D sensor has advantages of simplicity and practical usefulness for getting 3-D information of robotic devices at high speed. The experimental results of applications to visual detection of robot's hand position and orientation, teaching of robot paths, and modeling of robot environments are described.

1. INTRODUCTION

Recently, techniques of acquiring 3-D information at high speed and with accuracy have been pursued with increasing demands for advanced and sophisticated tasks that a robot has to carry out. Several methods of obtaining range information in robot vision systems have been developed. These methods can be categorized into two groups: (1) range finders for acquisition of range image data to recognize objects, (2) visual sensors for obtaining 3-D position and orientation of robotic devices.

The former are noncontact range sensors for measuring objects without special arrangements such as marking of robot environments. Active and passive sensors have been the primary techniques used for robot vision systems in this group. Actitve range finders project a laser spot [1] or slit beam [2] onto objects and observe an image of the reflected light by a TV camera. Then, the 3-D position is calculated based on a triangulation technique. These methods, however, cannot always detect the reflected light because of the object's surface conditions. Passive stereo vision systems using a couple of cameras consume much time in stereo matching to find corresponding primitives of two images [3].

On the other hand, the methods of the second group utilize special arrangements of robot environments for sensing 3-D information. For example, bright points are attached to robotic devices as marks to define unique positions [4].

This paper is concerned with the second type of range finder which can simultaneously detect the 3-D position and orientation of robotic devices.

The 3-D sensor proposed in this paper is composed of a position sensitive device (PSD) camera, and multiple LEDs attached to robotic devices such as a robot's hand and a teaching device. Since the sensor directly receives emitted lights of the LEDs, it can avoid the problem of receiving reflected lights. Moreover, it has an advantage of fast operation using a simple algorithm and a non-scanning analog image sensor such as a PSD camera.

Experimental results of three applications using the 3-D sensor are presented. They are as follows: (1) visual detection of robot's hand position and orientation, (2) teaching of robot paths by a 3-D teaching device, (3) modeling of robot environments by a 3-D tablet device. Finally, future subjects and extensions of the 3-D sensor are discussed.

2. THE 3-D SENSOR SYSTEM

Overall structure of the 3-D sensor system is shown in Fig. 1. Multiple infrared LEDs attached to a rectangular plate on a robotic device are turned on sequentially with intensity control by a microprocessor. As soon as a PSD camera receives the light of these LEDs, it detects the brightest spot in the image. Analog output voltages are obtained proportional to 2-D coordinate values of the spot with respect to the image plane[5]. The analog outputs are converted to digital values and then input to the microprocessor. The 3-D positions of the multiple LEDs are determined by using both 2-D coordinate values of multiple LEDs with respect to the image plane and geometrical relationships among 3-D positions of the LEDs. An equation of a plane which represents the plate on the robotic device is determined based on the 3-D positions of at least three LEDs. Consequently, the 3-D position and orientation of the robotic device are obtained. This 3-D information is transferred to a host computer and is used for various robotic applications.

Calculation of camera parameters

To obtain accurate 3-D information in the 3-D sensor system, estimated PSD camera parameters must be calculated in advance. The camera parameters which must be estimated are as follows: (1) 3-D position Os (Xos, Yos, Zos) of the PSD camera in the robot coordinate system, (2) rotation angles of pan α, tilt β and swing γ of the image plane in the camera coordinate system, and (3) focal length F of the lens used for the PSD camera.

As shown in Fig. 2, the robot coordinate system is represented by (Xr, Yr, Zr). Or is the origin of the robot coordinate system. (Xs, Ys, Zs) represents the camera coordinate system whose Zs axis coincides with the principal line of the lens. Os, the origin of the camera coordinate system, is located at a focal point of the lens.

The relation between these two coordinate systems is given by the following vector-matrix equation.

$$\begin{pmatrix} Xs \\ Ys \\ Zs \end{pmatrix} = R \cdot \begin{pmatrix} Xr - Xos \\ Yr - Yos \\ Zr - Zos \end{pmatrix} \qquad (1)$$

where, rotation matrix R is given as follows:

$$R = \begin{pmatrix} \cos\gamma & \sin\gamma & 0 \\ -\sin\gamma & \cos\gamma & 0 \\ 0 & 0 & 1 \end{pmatrix} \cdot \begin{pmatrix} \cos\beta & 0 & -\sin\beta \\ 0 & 1 & 0 \\ \sin\beta & 0 & \cos\beta \end{pmatrix} \cdot \begin{pmatrix} 1 & 0 & 0 \\ 0 & \cos\alpha & \sin\alpha \\ 0 & -\sin\alpha & \cos\alpha \end{pmatrix}$$

Xc and Yc denote the 2-D coordinate values of the image plane of the PSD camera. Sx and Sy represent the 2-D position of the principal line in the image plane.

$$Xc = -F \cdot Xs / Zs + Sx \qquad (2)$$
$$Yc = -F \cdot Ys / Zs + Sy \qquad (3)$$

From Equations (1), (2), and (3), camera parameters are estimated using a Newton-Raphson method based on more than four known 3-D positions of the LEDs and corresponding 2-D coordinate values with respect to the camera [6] (See Appendix A).

The method used to calculate camera parameters in this system has the advantage that a camera can be set up freely at an optimal position for a robot.

Algorithm of extracting 3-D position and orientation

Here, we describe an algorithm to extract 3-D position and orientation of a robotic device. When the geometrical relationship among the four positions of the LEDs on the plate of the robotic device is given, the 3-D positions of the LEDs are uniquely determined based on the 2-D positions of the LEDs with respect to a camera under perspective transformation [7]. As shown in Fig. 2, $\vec{qi}$ (i=1,..,4) represents a vector from the origin Os of the camera coordinate system to each 3-D position of the LEDs and $\vec{pi}$ (i=1,..,4) represents a vector from Os to the corresponding position in the image plane of the camera.

Assuming a perspective transformation, the following equations are obtained.

$$\vec{qi} = s_i \cdot \vec{pi}, \quad s_i > 0 \quad (i=1,..,4) \qquad (4)$$
where, $\vec{qi} = (X_s, Y_s, Z_s)^T$, $\vec{pi} = (X_c, Y_c, -F)^T$.

Since four LEDs are mounted at the four corners of a rectangle, the following five equations are obtained.

$$\vec{q1} + \vec{q3} = \vec{q2} + \vec{q4} \qquad (5)$$
$$|\vec{q2} - \vec{q1}| = D \qquad (6)$$
$$|\vec{q3} - \vec{q1}| = \sqrt{D^2 + S^2} \qquad (7)$$
$$|\vec{q4} - \vec{q2}| = \sqrt{D^2 + S^2} \qquad (8)$$
$$|\vec{q1} - \vec{q4}| = S \qquad (9)$$

Here, $|\vec{qi} - \vec{qj}|$ represents a distance between two vectors $\vec{qi}$ and $\vec{qj}$. Equation (5) implies that the center of the line L1-L3 coincides with the center of the line L2-L4. Equations (6),..,(9) indicate that two neighbouring sides of the rectangle are orthogonal to each other and their lengths are D and S respectively.

By solving the above equations, s_i (i=1,..,4) and $\vec{qi}$ (i=1,..,4) in Equation (4) are uniquely determined. Then, the 3-D positions of the four LEDs are obtained by Equation (1) as shown in Appendix B.

Figure 3 shows the geometrical relationship between the position of a specified point P and the four LEDs, L1, L2, L3, and L4. The specified point P of the robotic device is located at distances of a on the x axis, D/2 on the y axis, and b on the z axis from the LEDs. The orientation at the specified point P is represented by a matrix $M=(\vec{xp}, \vec{yp}, \vec{zp})$ where $\vec{xp}$, $\vec{yp}$, and $\vec{zp}$ are unit vectors. Since the position and orientation at the specified point are determined using three of the LEDs, the following linear equations are obtained.

$$\vec{P} + a \cdot \vec{xp} + D/2 \cdot \vec{yp} + \qquad b \cdot \vec{zp} = \vec{L1} \qquad (10)$$
$$\vec{P} + a \cdot \vec{xp} - D/2 \cdot \vec{yp} + \qquad b \cdot \vec{zp} = \vec{L2} \qquad (11)$$
$$\vec{P} + a \cdot \vec{xp} - D/2 \cdot \vec{yp} + (S+b) \cdot \vec{zp} = \vec{L3} \qquad (12)$$

Then, the solution is obtained as follows:

$$\vec{xp} = \vec{yp} \times \vec{zp} \qquad (13)$$
$$\vec{yp} = (\vec{L1} - \vec{L2}) / D \qquad (14)$$
$$\vec{zp} = (\vec{L3} - \vec{L2}) / S \qquad (15)$$

From Equation (10) and the orientation matix $M=(\vec{xp}, \vec{yp}, \vec{zp})$, the 3-D position of the point P is derived as follows:

$$\vec{P} = \vec{L1} - a \cdot \vec{xp} - D/2 \cdot \vec{yp} - b \cdot \vec{zp} \qquad (16)$$

Consequently, Equations (13) through (16) uniquely determine the 3-D position and orientation of the specified point P of the robotic device.

3. VISUAL DETECTION OF ROBOT'S HAND POSITION AND ORIENTATION

To accomplish a variety of tasks, a robot's hand must be located accurately in 3-D space. A multi-joint-manipulator requires much computation to determine pose (position and orientation) of the hand since 3-D geometrical transformations must be calculated using angles of the joints. Another problem is that bending components of the manipulator cause the postion error to accumulate at the hand. In addition, when a new robot's arm with any kind of flexible structure is developed in the future, a direct sensing method will be required to determine the pose of the hand for its control in 3-D space.

We describe below an application of the 3-D sensor system used to determine the pose of a robot's hand. Since the proposed 3-D sensor system permits direct detection of the pose of a hand, it works very fast and can be applied even to the arms with flexible structures.

System structure of sensing robot's hand pose

Figure 4 shows an overview of a system for sensing the robot's hand pose. The system consists of LEDs attached to a plate on a robot's hand and a PSD camera fixed over the hand to see the field of the movements. Totally, sixteen LEDs are attached to the hand. Eight of them are attached on top of the hand and the rest on the sides. In general, four LEDs in a single plane are enough to determine the pose of the hand. Therefore, a suitable set of four LEDs can be selected adaptively depending on the pose. In the experiments, however, only a set of four LEDs near the four corners of the plate is used. The specifications of the LED and the PSD camera used in the experiments are described in Appendix C.

Experimental results

In the calculation of the camera parameters, the hand moved to three positions within the field of view. Since the eight LEDs were used in the calcutation, twenty-four sets of 2-D positions of the image plane and 3-D position of the robot coordinate system were obtained to estimate the camera parameters. The calcutated parameters are as follows:

$$Xos = 1154, \quad Yos = 7, \quad Zos = 480 \ (mm), \quad F = 5365,$$
$$\alpha = 0.060, \quad \beta = -0.534, \quad \gamma = 0.006 \ (rad.)$$

Figure 5 shows the structure of the robot's hand and the four LEDs. The actual values of D, S, a and b are as follows:

$$D = 150, \quad S = 50, \quad a = 40, \quad b = 30 \quad (mm)$$

The position of the hand is obtained by Equation (15). The orientation angles (A1, A2, A3) of the hand are calcutated using the matrix M as follows:

$$A1 = atan(-zp(2)/zp(3)) \tag{17}$$
$$A2 = asin(-zp(1)) \tag{18}$$
$$A3 = atan(yp(1)/xp(1)) \tag{19}.$$

The robot's hand was moved to a certain pose. Then the positions of the LEDs were calculated. Table 1 compares calculated and actual values. Table 2 compares the calculated hand pose and the actual one. The average error is within 3 mm in position and 2 degrees in orientation. As shown in Tabel 1, the X-coordinate data has an error which is 5 mm at the maximum. This is because the depth from the camera to the hand roughly coincides with the X-coordinate and the depth data can be less acculate compared with other coordinates. Since the depth is about 600 mm in this case, the relative error is less than 1 percent.

4. TEACHING ROBOT PATHS USING A 3-D TEACHING DEVICE

One application of the 3-D sensor is teaching robot paths. In general, teaching robot how to move its hand is a crucial task for operators. A teaching box is widely used in many industries. Since robots are active during the teaching, the operator sometimes exposes himself to danger. Moreover, such a teaching task requires much time. We developed a new teaching method which is a remedy for such problems since it does not require robots to be in an active mode during the teaching. The teaching system uses a 3-D teaching device based on the 3-D sensor described in Section 2.

System configuration of teaching robot paths

Figure 6 shows a system configuration of teaching robot paths. An operator grasps a 3-D teaching device which consists of a grip, a square plate with four LEDs which form a rectangle, and a needle in front of the plate for a pointer. Three buttons to give instructions to a microprocessor are attached on the surface of the grip. When a button on the grip is pushed, a measurement action starts. The four LEDs are turned on sequentially. A PSD camera fixed over the work field detects the 2-D position of the bright spot. A set of data in terms of the four LEDs is then transferred to a host computer to calculate the 3-D position and orientation at the pointer. The data of the pointer is stored as a database to generate a robot path. In the operating mode, the robot's hand moves according to the database. As currently implementated, the database is combined with a robot language on PETL which is a LISP-like language developed at Electrotechnical Laboratory [8].

Experimental results

(a) Teaching a robot path

Figure 7 (a) shows a robot path which has four turning points. The actual robot path in the operating mode is shown in Fig. 7 (b). In Fig. 7 (b), a LED was attached to the center of the hand gripper to make the path visible.

(b) Teaching orientation of robot's hand

Figure 8 (a) shows how this device is used to teach orietation angles. In this case, the object has a slant surface for grasping. Figure 8 (b) shows the results in the operating mode. The results show that the orientation given by the teaching system is reproduced successfully.

5. MODELING ROBOT ENVIRONMENTS BY A 3-D TABLET DEVICE

Recently, much attention has been paid to modeling robot environmets, since it is certain that model-based vision and manipulation will play a very important role in realizing advanced tasks. There are various kinds of methods for modeling objects. Though an engineering drawing may be a source of modeling, it does not satisfy the needs for modeling objects directly in real environmets, especially when no drawing exists to give its actual size. In contrast, a laser-spot range finder can be used as a measuring tool for the modeling [9]. The system, however, is still too expensive for practical use. In addition, poor reflectance conditions on the object's surface may disturb the measurement.

Since the proposed 3-D sensor has the advantage of the simpler system configuration and reliability in light reception, we applied the 3-D sensor to this kind of modeling task.

System configuration of modeling robot environments

The system configuration is the same as that of teaching robot paths described in Section 4 except for a 3-D tablet used to create the third dimension in modeling. There are 16 (4 x 4) LEDs attached to a horizontal plate which slides vertically along an iron bar. A PSD camera, attached near the top of the bar, is calibrated beforehand using lights of the LEDs on the plate. The operator grasps the teaching device and places it on the specified positions of the objects depending on the geometrical shape of the models. The 3-D data of the objects are transferred to a solid modeling system GEOMAP [10] to generate a primitive volume. These primitive volumes can be used to generate more complex models using geometric operations such as union, intersection, subtracton, etc. Then, the generated models are stored in files as a database of model-based tasks for robots.

Experimental results

A fundamental experiment was performed on modeling a power unit which has a rectangular parallelepiped shape. Figure 9 (a) shows the modeling experiments in progress. Four vertices, which give three edges of the rectangular parallelepiped, are measured to generate the model. In order to visually confirm that the model is generated properly, a superimposed display is implemented. Wireframe lines of the generated model are superimposed on an image taken by a monitor TV camera over the field of the modeling task. Since the primitive volume is generated at the origin of the world coordinate system in GEOMAP, a geometric transformation of translation and rotation is determined so that the transformed vertices have minimum error between the model and the actual data in 3-D space. A least-square technique is applied to the determination. The actual objects dimensions were 100 x 50 x 95 (mm). The dimensions of the generated model were 100.3 x 52.4 x 97.6 (mm) which is a mean value of three trials. In each trial, 3-D position of a vertex is measured three times and the average value is used for the calculation. The experimental results, though few in numbers, demonstrate that the proposed 3-D sensor can be used in the modeling as a measuring tool which has a simpler system configulation compared with other methods.

6. CONCLUSIONS

A new approach to obtain 3-D information of robot environments is proposed. The 3-D sensor described in this paper has the advantages of simple system configuration and real-time response to reliably acquiring 3-D information.

The speed and accuracy of the sensor system in the current implementation can be summarized as below: (1) Overall time to collect 3-D information is about 113 msecs (30 msecs to acquire data of four LEDs, 80 msecs to transfer data from the microprocessor to the host computer, and 3 msecs to calculete 3-D position and orientation in the host computer). Since the PSD has an optimal intensity of input spot light, the intensity of the LEDs are controlled by a program in the current implementation. However, the feedback control can be done easily in hardware. Therefore, the time to collect the data of four LEDs can be improved up to about 4 msecs. Besides, when all the processing is implemented in the microprocessor, the data transfer time can be reduced. (2) The average error is within 3 mm in position and 2 degrees in orientation. A major error source, possibly as much as 2.5 percent, is the PSD's inherent position detecting error. The other factors are position error in mounting the LEDs on the plate and error in the camera parameters.

As applications of the 3-D sensor system, three basic experiments, detecting pose of robot's hand, teaching robot paths, and modeling robot environments, are performed. They show the feasibility of applying the 3-D sensor to these kinds of robotic tasks. However, the following subjects are left for improvements and extensions: (1) enlargement of the view angle of the PSD camera using high power LEDs and wide-angle lens; (2) improvements of accuracy and reliability in acquiring 3-D information by combining the sensing system and stereo vision method; and (3) link between the teaching system and robot software system such as an off-line programming system using robot languages.

[Acknowledgements]

The authors greatly appreciate Dr. Seiji Wakamatsu, the director of the Automatic Control Division, for his continuous encouragements and Dr. Tomomasa Sato, Dr. Takashi Suehiro, and other members of the Robotics Research Group at Electrotechnical Laboratory for their assistance and suggestions on the experimental systems.

[References]

[1] Ishii,M and Nagata,T, "Feature extraction of three-dimensional objects and visual processing in a hand-eye system using laser tracker," Pattern Recognition, vol.8, pp.229-237, 1976.

[2] Shirai,Y and Suwa,M, "Recognition of polyhedron with a range finder," 2nd International Joint Conference on Artificial Intelligence, p.80, 1971.

[3] Barnard,S.T and Fischler,M.A, "Computational Stereo," ACM Computing Surveys, pp.553-572, vol.14, no.4, Dec.1982.

[4] Kasai,T, Asahi,T, Yoshimori,T and Tsuji,S, "Measurement system of 3-D motion using a pair of position sensing detector camera (in Japanese)," Trans. of the Society of Instrumental and Control Engineers, vol.19, no.20, pp.61-67, Dec. 1983.

[5] Kanade,T and Sommer,T.M, "An optical proximity sensor for measuring suface position and orientation for robot manipulation," Proc. of Third Int. Conf. on Robot Vision and Sensory Controls, pp.301-309, Nov. 1983.

[6] Shimasaki,M, "Some discussions on inverse of projection (in Japanese)", SIGIE Records, The Institute of Electronics and Communication Engineers, of JAPAN, IE79-15, pp.83-92, May 1979.

[7] Lowe,D.G, "Solving for parameter of object models from image descriptions," Proc. of ARPA Image Understanding Workshop, pp.121-127, Maryland, April 1980.

[8] Tsukamoto,M, "PETL system reference manual (in Japanese)," Electrotechnical Laboratory, 1982.

[9] Hasegawa,T, "An interactive system for modeling and monitoring a manipulation environment," IEEE Trans. on System, Man and Sybernetics, vol.SMC-12, no.3, May/June 1982.

[10] Hosaka,M and Kimura,F, "An interactive geometric design system with handwriting input," Information Processing 77, Amsterdam, North-Holand, 1977.

[Appendix A]

Unknown parameters in Equations (2) and (3) are (Xt, Yt, Zt) of translation, (α, β, γ) of rotation and F of focal length, where $(Xt, Yt, Zt)^T = R (Xos, Yos, Zos)^T$. Since the original equations produce a non-linear system, a Newton-Raphson method is employed where the estimates of the parameters are corrected iteratively. The errors of a point between a model and an actual image can be expressed as the sum of the products of partial derivatives with respect to each camera parameter:

$$Dx(\partial Xi/\partial Xt) + Dy(\partial Xi/\partial Yt) + Dz(\partial Xi/\partial Zt)$$
$$+ D\alpha(\partial Xi/\partial\alpha) + D\beta(\partial Xi/\partial\beta) + D\gamma(\partial Xi/\partial\gamma) + DF(\partial Xi/\partial F) = Ex \quad (A-1)$$
$$Dx(\partial Yi/\partial Xt) + Dy(\partial Yi/\partial Yt) + Dz(\partial Yi/\partial Zt)$$
$$+ D\alpha(\partial Yi/\partial\alpha) + D\beta(\partial Yi/\partial\beta) + D\gamma(\partial Yi/\partial\gamma) + DF(\partial Yi/\partial F) = Ey \quad (A-2)$$

Dx, Dy, Dz, Dα, Dβ, Dγ, and DF represents errors of each parameters. Ex and Ey are errors in x and y components between estimated values of a model and an actual point. The partial derivatives of Xi and Yi with respect to each of the seven parameters are as follows:

$(\partial Xi/\partial Xt) = F/(Zr-Zt),$ $\qquad (\partial Yi/\partial Xt) = 0,$

$(\partial Xi/\partial Yt) = 0,$ $\qquad (\partial Yi/\partial Yt) = F/(Zr-Zt),$

$(\partial Xi/\partial Zt) = -F(Xr-Xt)/(Zr-Zt)^2,$ $\qquad (\partial Yi/\partial Zt) = -F(Yr-Yt)/(Zr-Zt)^2,$

$(\partial Xi/\partial\gamma) = F\, Yr/(Zr-Zt),$ $\qquad (\partial Yi/\partial\gamma) = -F\, Xr/(Zr-Zt),$

$(\partial Xi/\partial\beta) = -F\, Zr/(Zr-Zt) - F\, Xr(Xr-Xt)/(Zr-Zt)^2,$

$(\partial Yi/\partial\beta) = -F\, Xr(Yr-Yt)/(Zr-Zt)^2,$

$(\partial Xi/\partial\alpha) = F\, Yr(Xr-Xt)/(Zr-Zt)^2,$

$(\partial Yi/\partial\alpha) = F\, Zr/(Zr-Zt) + F\, Yr(Yr-Yt)/(Zr-Zt)^2,$

$(\partial Xi/\partial F) = -(Xr-Xt)/(Zr-Zt),$ $\qquad (\partial Yi/\partial F) = -(Yr-Yt)/(Zr-Zt)$

Simultaneous linear equations are solved by initially estimating the values of the parameters. When the number of corresponding pairs of points between a model and an actual image are greater than the number of the unknown parameters, a least-squares method can be applied. In the experiments, several iterations are enough to obtain convergence of the parameters.

[Appendix B]

From (4) and (5),

$$s1\, \vec{p1} + s3\, \vec{p3} = s2\, \vec{p2} + s4\, \vec{p4} \quad (B-1)$$

Outer vector products of (B-1) and p4 become

$$s1(\vec{p1} \times \vec{p4}) + s3(\vec{p3} \times \vec{p4}) = s2(\vec{p2} \times \vec{p4}) \quad (B-2)$$

where $\vec{p4} \times \vec{p4} = 0$.

Also, inner vector products of (B-2) and p3 are

$$s1\vec{p3} \cdot (\vec{p1} \times \vec{p4}) = s2\vec{p3} \cdot (\vec{p2} \times \vec{p4}) \quad (B-3)$$

where $\vec{p3} \cdot (\vec{p3} \times \vec{p4}) = 0$.

From (B-3), the following equation is obtained.

$$s2 = s1\vec{p3} \cdot (\vec{p1} \times \vec{p4}) / \vec{p3} \cdot (\vec{p2} \times \vec{p4})$$
$$= c21 \; s1 \tag{B-4}$$

Likewise, c31 and c41 are obtained as follows:
$$s3 = c31s1 \tag{B-5}$$
$$s4 = c41s1 \tag{B-6}$$

From (6) and (B-4),
$$D^2 = s1^2 \vec{p1}^2 - 2s1 \; s2 \; (\vec{p1} \cdot \vec{p2}) + s2^2 \vec{p2}^2$$
$$= s1^2 \vec{p1}^2 - 2c21 \; s1^2 \; (\vec{p1} \cdot \vec{p2}) + c21^2 \; s1^2 \; \vec{p2}^2 \tag{B-7}$$

Therefore,

$$s1 = D / \sqrt{\vec{p1}^2 - 2c21 \; (\vec{p1} \cdot \vec{p2}) + c21^2 \; \vec{p2}^2}. \tag{B-8}$$

Similarly, from (7) through (9), each value for s1 is also calculated. The parameter s1 is obtained as an average value of these four values in terms of s1. By substituting s1 into (B-4), (B-5), and (B-6); s2, s3 and s4 are determined, respectively.

Then, from Equation (4), $\vec{qi}$ = (Xs,Ys,Zs) (i=1,...4) are calculated. Consequently, by means of inverse transformation of Equation (1), 3-D positions of four LEDs, that is (Xr,Yr,Zr), are uniquely determined.

[Appendix C]
The devices used in the experiments are as follows:
(1) The PSD camera is C1454-05 (Hamamatsu Photonics Inc.). Sampling speed: 300 Hz. Accuracy: 1 to 2.5 %.
(2) The infrared LED is NJL1120-L (Shin-Nihon Musen Inc.). Peak wave length: 900 nm. Power: 15mW.
(3) The robot arm is SMART200 (Sumitomo Juuki Inc.). Degrees of freedom: 5. Accuracy: +- 0.02 mm.
(4) The host computer and the microprocessor are VAX11/780 and LSI11/23 (DEC.), respectively.

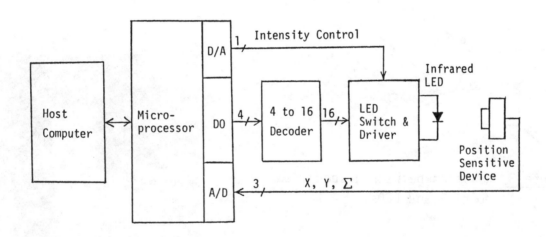

Figure 1   Block diagram of a 3-D sensor system.

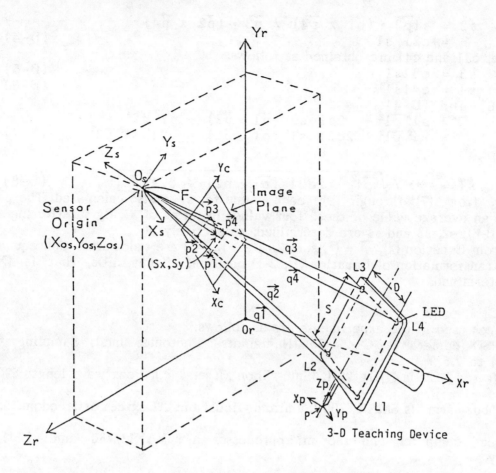

Figure 2 Relationships for the coordinate systems.

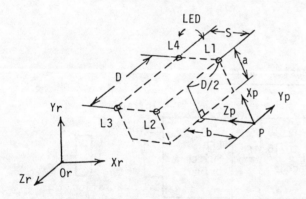

Figure 3 Relation between a specified point P and LEDs.

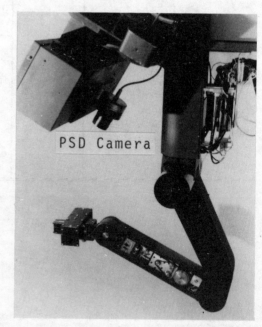

Figure 4 Overview of a hand pose sensor.

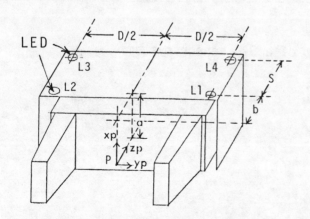

Figure 5 Robot hand with four LEDs.

162

Table 1 3-D values of four LEDs.

N O	X r	X hps	Y r	Y hps	Z r	Z hps
L 1	6 7 7	6 7 6	7 5	7 5	1 9 2	1 9 1
L 2	6 7 7	6 7 9	- 7 5	- 7 5	1 9 2	1 9 4
L 3	6 6 0	6 5 9	- 7 5	- 7 5	2 3 9	2 4 0
L 4	6 6 0	6 5 5	7 5	7 6	2 3 9	2 3 7

Table 2 Robot hand pose data.

P O S E	X r	Y r	Z r	A 1	A 2	A 3
R O B O T	6 5 0 . 0	0 . 0	1 5 0 . 0	0 . 0	2 0 . 0	0 . 0
H P S	6 5 2 . 5	- 1 . 1	1 4 9 . 6	- 0 . 4	2 3 . 2	- 1 . 3

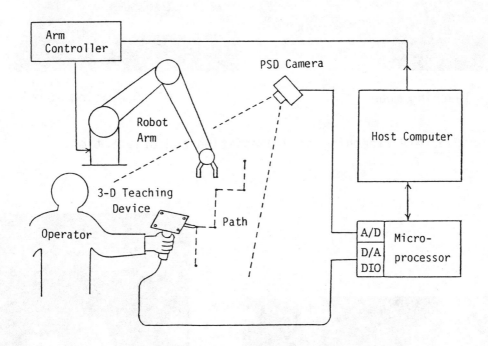

Figure 6 A robot teaching system.

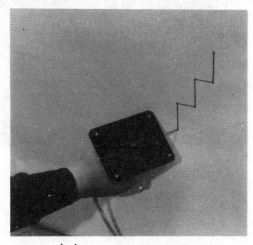

(a) Teaching mode.

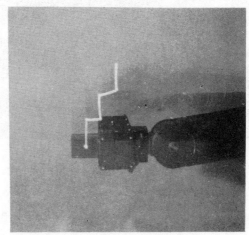

(b) Operating mode.

Figure 7 Teaching robot path.

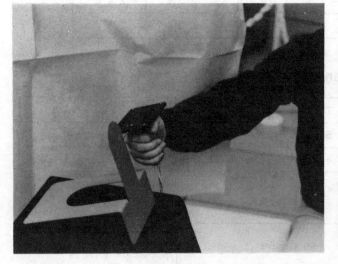

(a) Teaching mode.   (b) Operating mode.

Figure 8 Teaching of 3-D position and orientation.

(a) Pointing to an object.   (b) Superimposed display.

Figure 9 Modeling of robot environments.

Range from Brightness for Robotic Vision

R. A. Jarvis
Australian National University, Australia

ABSTRACT

Range recovery has become an important component in sensory-based robotics research. Robust, semantic-free, low level 3D scene analysis can be effectively supported by good quality registered intensity/range data. This paper explores the use of brightness variations in colour imagery as a means of recovering range data that is, by construction, exactly registered with intensity data. Some preliminary results of experimentation carried out in the Computer Vision and Robotics Laboratory at the Australian National University are presented and future developments discussed.

1. INTRODUCTION

It is clear from the recent literature [1,2] that range recovery from 3D scene analysis, particularly in support of robotic navigation or manipulation, has become one of the important preoccupations of the Computer Vision research community. The recognition of the significance of range data to robust 'semantic-free' analysis at the low level end of scene analysis is not misplaced, since position, shape and orientation are recoverable from such data and these in turn are, together, the basis of recognition, task planning and trajectory control (See Figure 1).

Two basic approaches to range recovery have dominated the literature. The first involves multiple views or triangulation calculations of some variety and can be either passive or active in terms of energy source. This category includes displacement stereo [3-6], temporal stereo [7,8] (time sequence analysis with moving camera), light striping, [9] simple triangulation [10,11] and Moire fringe analysis [12]. Both forms of stereo analysis tend to be computationally complex and both suffer from the 'missing parts' problem where not all portions of an image viewable from one camera position are viewable from another. Light striping and simple triangulation usually involve opto-mechanical instrumentation and tend to be slow in recovering a medium to high resolution 'range picture'. The Moire fringe approach shows promise but requires fairly specialised (and expensive) equipment.

The second major approach to range recovery involves direct determination of transit times of an energy beam, (either ultrasonic or light) to and back from a target on the scene. Since a single viewing position is involved, the 'missing parts' problem which frustrates triagulation based methods can be avoided by adopting a design where outward bound and return beams are essentially coaxial. Since only a single range value is obtained at a time, equipment for scanning the beam over the surfaces of the scene viewable from the instrument position must be provided. The ultrasonic approach has the advantage that the transit times involved are fairly easily measured since sound velocity in air is not high, but it is both difficult to narrow the beam for high resolution scanning and to avoid non-return of the beam, or worse, false targets which result from specular reflection which can occur even for relatively rough surfaces (surface undulations of the scale of the wavelength of the ultrasonic energy source). Two varieties of laser range finders have been reported. For the first [13] phase shift of a modulated continuous laser beam between outgoing and return apertures is measured using sensitive instrumentation. Energy return attenuations adjusted for range can be used to estimate the albedo of the target surface. For the apparatus described in [13], several hours were required to obtain 128 x 128 resolution range pictures of relatively high range accuracy. The second approach uses a pulse laser source and measures pulse transit times directly. With low power lasers, integrations over a large number of shots for each point is necessary to improve the signal/noise ratio so as to obtain acceptably accurate range data. For the instrument reported in [14] it took in the order of 3 min (ideally) to acquire a 128 x 128 range picture (2cm accuracy over a 1-3m range). With the instrument reported in [15], sub-centimeter accuracy for 64 x 64 range pictures could be obtained in 40 seconds. However, despite the possibility of improvement, such instruments are not likely to be widely used in the near future because of their expense.

Another category of range recovery methodology worth mentioning despite the fairly small exposure in the literature is also basically monocular but is usually passive. This category includes range from focus [16,17] and the use of refletance maps to recover 'shape from brightness' [18]. The first of these depend on being able to determine which parts of an image are in focus for each of a range of lens positions. The second is applicable to analysis of scenes containing objects with the one surface type for which the reflectance map has been previously determined.

The steropsis and focus related methods mentioned above are, in addition to computational complexity limitations, also restricted to recovery of range data at parts of scene which are visually distinct (e.g. at edges) but can not be used on uniform surfaces.

The method described in this paper can recover range over uniform, textured, or coloured (or any combination) surfaces by using image data acquired separately for each of two point light source positions . The range data is, by construction, registered with the image data, involves a single viewing positions and is computationally trivial to calculate.

The next section describes the basis of the method, the following gives implementation details. Then follows a section presenting some preliminary results. A final section discusses future developments and presents the conclusions of the work.

2. RANGE FROM BRIGHTNESS

Consider Figure 2. which shows the optical geometry for illuminating the scene by a point light source at two positions (S_1 and S_2) along the optical axis of the single viewing camera. The surface normal of a patch on the scene is at angle θ to the line from the point light sources to the reference point of the surface. Assuming Lambertian scattering surfaces, the brightness values (I_1 and I_2) as measured by the camera for the reference point on the surface for the two point light source positions are related as follows:

$$\frac{I_1}{I_2} = \frac{\ell^2}{(\ell + d)^2} \qquad\qquad (1)$$

simple algebra shows that the solution to the quadratic in ℓ

$$(I_2-I_1)\ell^2 - 2I_1 d\ell - I_1 d^2 = 0 \qquad \text{is given by}$$

$$\ell = \frac{d(I_1 \pm \sqrt{I_1 I_2})}{(I_2-I_1)} \qquad\qquad (2)$$

, the only realistic solution being given by the + ve alternative in the numerator, the denominator being always positive.

Letting $R = I_2/I_1$ which is always ≥ 1.0,

$$\ell = \frac{d(1+\sqrt{R})}{(R-1)} \qquad\qquad (3)$$

This relationship is graphed in Figure 3.

The sensitivity of the method improves with the size of d which is the distance between the light source positions. The closeness of S_1 to the scene and the value of d are restricted by practical considerations of the sensitivity and linearity range of response of the camera and the 'cosine effect' as illustrated in Figure 4. In Figure 4 it is clear that, should the closer light position and distance between the light positions be too small and too large, respectively, the different angles the incident beam makes to the surface normal of an off optical axis surface patch for the two positions would allow the cosine factor to dominate the brightness ratio calculation and confound range recovery using the simple formula (3). Also, the approximation $\ell_1 \approx \ell_2$ + d becomes less acceptable as the surface patch is further off centre of the camera axis.

3. EXPERIMENTAL SET-UP

It becomes obvious, when considering the shape of the graph of Figure 2 and the fact that the acceptable signal noise ratios of most video cameras are to some extent dictated by the integration over time and space on a monitor screen and the retinal integration effect in the eye of the human viewer, that a great deal of care should be exercised in obtaining accurate intensity data in digitised form from the camera. Fortunately, a video stream processor in the Computer Vision and Robotics Laboratory at the Australian National University permitted, amongst other things, real time video frame integration. Full colour data was gathered for each of three light conditions (background lighting, far point light source + background, close point light source + background). A Kodak 35mm slide projector was used to simulate the point light source, the effective distance being considerably further than the physical distance. The background image components were subtracted pixel by pixel from each of the close-lit and far-lit image components to isolate the lighting effects due solely to the point-light source. For each of the red, green and blue channels, for each of these lighting conditions, 256 video frames were integrated, (eight bits for each colour of each pixel over a 128 x 128 spatial resolution). Some 45 seconds were thus involved with image aquisition itself. The image data was tranferred via a parallel interface to a VAX 11/750 which was used to compute the range, using the red, green and blue image data separately. Some 7 seconds of processor time were taken for the three sets of range calculations. The results were tranferred in colour intensity form back to the video stream processor for viewing. Combining the signals for the three range intensities permitted a good appreciation of the quality of the resulting range data. The image transfers to and from the VAX consumed about 12 seconds, the transfer rate being quite low. Of the total time of approximately one minute taken for one complete experiment (excluding the time for shifting the light source which could be quite small) 75% was involved with image acquisition. Further experiments may indicate that this can be reduced under better lighting conditions or with a higher quality camera.

4. PRELIMINARY RESULTS

Consider the simple scene shown in Figure 5(a). The objects depicted are uniformly white and well separated in the depth direction with respect to the single camera viewing position. Figure 5(b) shows the intensity histogram for the far lit point source illumination of the scene and Figure 5(c) the intensity histogram for the close lit point source illumination situation. The peaks marked in Figure 5(b), which indicate the intensity/range relationships for the three objects, have spread out in Figure 5(c), showing the effect of changing the illumination conditions from far lit to close lit. In particular, the peaks not only have all moved to the right (brighter) but the spacings between them indicate that the variation is related to the closeness of the objects to the light source (i.e. range), the gap between the right most pair of peaks being larger than the gap between left most pair of peaks. Figure 5(d) shows the range picture for this scene with intensity proportional to the inverse of range (close components brighter). This study captures the fundamental aspects of the range from brightness methodology described in this paper.

Figure 6(a) is a general view of a more complex scene example which includes a block with a slanted face. Figure 6(b) shows the background (mostly black) illumination, far lit and close lit intensity data collected and the corresponding range picture, the last being shown in more detail in Figure 6(c). It is clear that the range recovery from brightness calculation has correctly range tagged the surface elements visable from the camera position.

Although the method is applicable to curved/coloured/textured objects as well as flat faced, uniformly coloured ones, the calculations seem prone to greater error for multicoloured scenes with curved objects as exemplified in Figure 7. This is probably due to less accurate data collection for objects that are not white and the complex secondary reflection effects caused by curved surfaces. More research and better camera equipment might well accomodate some of these problems. The examples shown are only preliminary and detailed analysis of the results has not yet been carried out.

5. DISCUSSION AND CONCLUSIONS

The range recovery method described in this paper has the following properties:-

    (a) region, not edge based (can range to visually uniform surfaces)
    (b) monocular (one camera position, no 'missing parts')
    (c) range registration with image data is intrinsic to the method.
    (d) shadow effects minimal with approximate camera optical axis positioned
        lighting.
    (e) computationally trivial.
    (f) independent of camera 2-D geometry (insensitive to spatial distortion)
    (g) can be applied to monochrome or colour imagery of uniform or textured
        surfaces.
    (h) critical w/r to camera-sensitivity, linearity, dynamic range and
        signal/noise ratio.
    (i) confounded by secondary reflection lighting effects.
    (j) restricted to Lambertian scattering surfaces.

Of these only (h),(i) and (j) can be considered detremental. Better quality cameras are becoming very much cheaper so attribute (h) of the method should pose no threat to the practical future of this approach. The difficulties associated with (i) and (j) above, are however, inherently more difficult to resolve.

In section 2 the 'cosine effect' was presented as a restricting influence on the placement of the point light sources such that surface normal direction with respect to incident beam does not become the dominant factor in brightness variation for off axis surfaces. However, if range is first determined using light source positions which avoid the 'cosine effect', the image for close point lit situations can perhaps be analysed to determine surface slopes, possibly more accurately than can be evaluated from the raw range data.

The method presented should be applicable to 'camera in the hand' analysis where the robotic manipulator carries the camera and points it at various parts of the scene for initial scene analysis or to support visual servoing at the object approachment stage of a grasping or fitting task. In this case, the point lighting set up would be part of the 'on board' apparatus.

Overall, this simple approach shows promise as a monocular image based ranging method for robotic vision support where the required form of structured lighting is acceptable and only Lambertian scattering surfaces are involved.

1. Kak, A.C., "Depth Perception for Robots", School of E.E., Purdue University W. Lafayette, Indiana, Technical Report TR-EE83-44 October 1983.

2. Jarvis, R.A., "A Perspective on Rangefinding Techniques for Computer Vision", IEEE Trans. on Pattern Analysis and Machine Intelligence, Vol. PAMI-5, No. 2, March 1983, pp 122-139.

3. Yakimovsky, Y. & Cunningham, R., "A System for Extracting Three-Dimensional Measurements from a Stereo Pair of TV Cameras", Comput. Graphics Image Processing, Vol. 7, pp 195-210, 1978.

4. Moravec, H.P., "Visual Mapping by a Robot Rover", Prov. 6th Int'l Joint Conf., A.I., 1979, pp 598-620.

5. Baker, H.H., "Edge Based Stereo Correlation", Proc. ARPA Image Understanding Workshop, University Maryland, April 1980.

6. Marr, D. & Poggio, T., "Cooperative Computation of Stereo Disparity", Artificial Intelligence Memo 364, MIT A.I. Lab., Cambridge, Mass., June 1976.

7. Prazdny, K., "Motion and Structure from Optical Flow", Proc. 6th Int. Joint Conf. Artificial Intell., Tokyo, Japan, 1979, pp 702-704.

8. Williams, T.D., "Depth from camera motion in a real world scene", IEEE Trans. Pattern Anal. Machine Intell., Vol. PAMI-2, pp 511-516, November 1980.

9. Popplestone, R.J., Brown, C.M., Ambler, A.P. & Crawford, G.F., "Forming Models of Plane-and-Cylinder Faceted Bodies from Light Stripes", Proc. 4th Int'l Joint Conf. A.I., 1975, pp 664-668.

10. Connah, D.M., & Fishbourne, C.A., "Using a Laser for Scene Analysis", Proc. 2nd. Int. Conf. on Robot Vision and Sensory Controls, November 2-4, Stuttgart, Germany, pp. 233-240.

11. Nimrod, N., Margalith, A., & Mergler, "A Laser-based Scanning Range Finder for Robotic Applications", Proc. 2nd. Int. Conf. on Robot Vision and Sensory Controls, November 2-4, Stuttgart, Germany, pp 241-252.

12. Idesawa, M., Yatagai. T., & Soma, T. "A method for automatic measurement of three-dimensional shape by new type of Moire fringe topography", in Proc. 3rd Int. Joint Conf. Artifical Intell., Coronada, CA, November 8-11, 1976, pp 708-712.

13. Nitzan, D., Brain, A.E. & Duda, R.O., "The Measurement and Use of Registered Reflectance and Range Data in Scene Analysis", <u>Proc. IEEE</u>, Vol. 65, No. 2, February 1977, pp 206-220.

14. Lewis, R.A., & Johnston, A.R., "A Scanning Laser Rangefinder for a Robotic Vehicle", <u>Proc. 5th Int'l Joint Conf. A.I.</u>, 1977, pp 762-768.

15. Jarvis, R.A., "A Laser Time-of-Flight Range Scanner for Robotic Vision", Dept. of Computer Science, Australian National University, Technical Report TR-CS-81-10. Also in <u>Trans IEEE on PAMI</u>, Vol. PAMI-5, No. 5, September 1983, pp 505-512.

16. Horn, B.K.P., "Focussing", M.I.T., Project MAC, AI Memo. 160, May 1968.

17. Jarvis, R.A., "Focus optimisation criteria for computer image processing", <u>Microscope</u>, Vol. 24, pp 163-180, 2nd quarter, 1976.

18. Horn, B.K.P., "Shape from shading: A method for obtaining the shape of a smooth opaque object from one view", M.I.T. Project MAC, MAC TR-79, November 1970.

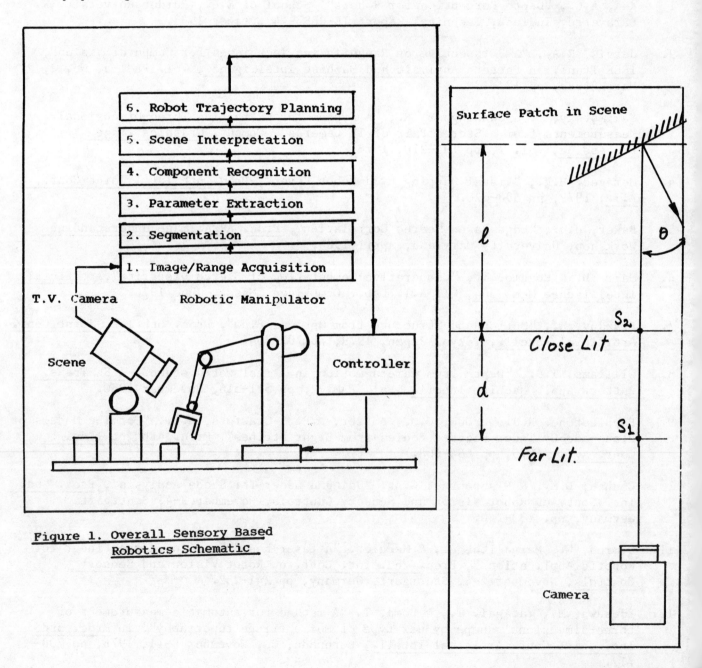

Figure 1. Overall Sensory Based Robotics Schematic

Figure 2. Range from Brigh Set-Up

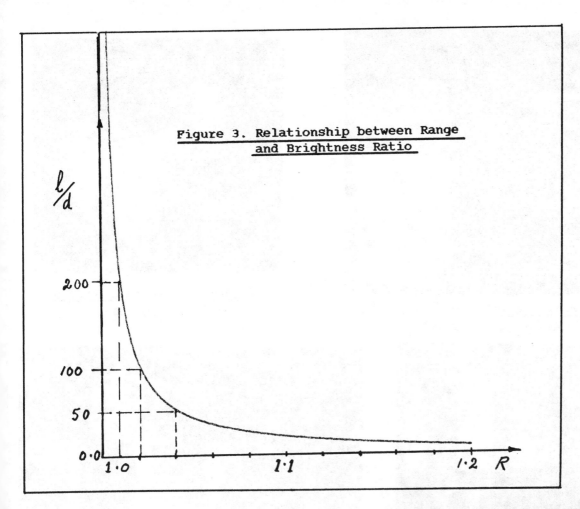

Figure 3. Relationship between Range and Brightness Ratio

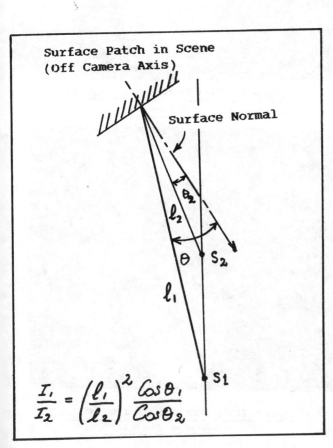

$$\frac{I_1}{I_2} = \left(\frac{\ell_1}{\ell_2}\right)^2 \frac{\cos\theta_1}{\cos\theta_2}$$

Figure 4. 'Cosine Effect' Geometry

Figure 5.

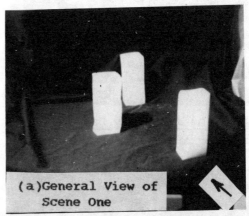

(a)General View of Scene One

(b)Far Lit Histogram

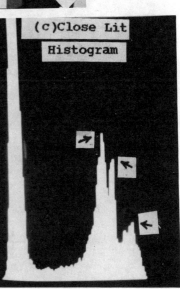

(c)Close Lit Histogram

171

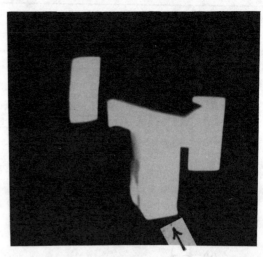

Figure 5(d) Range Picture for Scene One
(Brighter is Closer)

Figure 6(a) General View of
Scene Two

Figure 6(b) Intensity Data and
Range Picture for Scene Two

Figure 6(c) Detail for Range
Picture for Scene Two

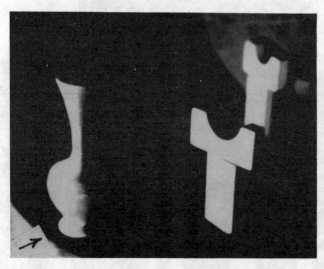

Figure 7(a) General View for
Scene Three (Coloured)

Figure 7(b) Range Picture for
Scene Three

172

3D Reconstruction of Silhouettes

G. Sandini
DIST-University
M. Straforini, V. Torre
and
A. Verri
University of Genoa, Italy

ABSTRACT

A stereo algorithm to compute the location in space of flat objects (silhouettes) is presented. The algorithm is simplified by solving the correspondence problem only for the points of zero curvature of the shape contour. The points of zero curvature also show the property of being good perspective invariants allowing the stereo images to be acquired from quite different viewpoints (to increase accuracy). Moreover, because the camera parameters and the viewing geometry are known, the matching between points of zero curvature was reduced from a two-dimensional to a one-dimensional problem by searching for matching along conjugate epipolar lines. The equation of the plane on which the silhouette lies was determined with a precision of at least 99%.

1. INTRODUCTION

Measurements of depth is a critical problem in robotic vision. Depth information can be obtained from motion[1], from shading[2,3] or through stereographic vision, that is from the analysis of two different images of the same scene. The key problem of stereo is to match points in the two images that correspond to the same physical location (correspondence problem). Solution of the correspondence problem requires the answer to two basic questions: what are the image primitives on which matching is based and what are the rules guiding the matching.

The image primitives on which matching is based could be edges [4,5,6,7,8], patches of the convolution sign[9] or surfaces[10].

The above procedures obtain good matches when the two images are similar, that is are taken from two close points of view. The drawback of this approach is a large sensitivity of the absolute depth evaluation on the accuracy of measurements involved. It is easy to show that in order to reduce the error in depth evaluation is necessary to have images taken from two very different points of view. Therefore it is necessary to look to features of the image that are "good" perspective invariants. It will be shown that the points of zero curvature (PZC) of planar curves are "good" perspective invariants. Using this property we have developed a stereo algorithm able to find the location in space of a flat object (silhouette) with a relative error of less than 1%.

2. THE STEREO GEOMETRY

Let (x_L, y_L, z_L) and (x_R, y_R, z_R) be two systems of coordinates, the former fixed in the left camera, the latter in the right one, so that their origins coincide with the intersection of the optical axis with the camera target and the plane $Z_L = 0$, $Z_R = 0$ coincide with the targets. The stereo geometry is now described by the focal length f of the camera optics (the foci obviously lie on the z_L and z_R axis), by the displacement vector $D = (k_L, l_L, m_L)$ between the centers of the targets (or the origins of the systems of coordinates) and by ϕ, ϑ, ψ Euler angles determined by the two systems of coordinates. Note that in this notation y_L and y_R are the abscissae axis.

We will suppose that the two optical axis are coplanar and that the two cameras have been rotated around the optical axis so that we have $\phi = 0$, $\psi = 0$ and $k_L = 0$. Assuming perspective projection the point $P = (x_L, y_L, z_L)$ in the left system of coordinates is projected into

$$x_{Lp} = \frac{f \, x_L}{f + z_L} \tag{1}$$

$$y_{Lp} = \frac{f \, y_L}{f + z_L} \tag{2}$$

and in the second screen(right system of coordinates)

$$x_{Rp} = \frac{f \, x_R}{f + z_R} \tag{3}$$

$$y_{Rp} = \frac{f \, y_R}{f + z_R} \tag{4}$$

The relation between (x_L, y_L, z_L) and (x_R, y_R, z_R) is

$$x_R = r_{11}(x_L - k_L) + r_{12}(y_L - l_L) + r_{13}(z_L - m_L)$$

$$y_R = r_{21}(x_L - k_L) + r_{22}(y_L - l_L) + r_{23}(z_L - m_L) \tag{5}$$

$$z_R = r_{31}(x_L - k_L) + r_{32}(y_L - l_L) + r_{33}(z_L - m_L)$$

where r_{ij} are elements of the well known ortogonal rotation matrix[11]. If the point (x_{Lp}, y_{Lp}) on the first screen is matched with (x_{Rp}, y_{Rp}) on the second screen the absolute depth Z_L is

$$Z_L = f + \frac{(y_{Rp}\sin\vartheta + f\cos\vartheta)(l_L - y_{Lp}) + (f\sin\vartheta - y_{Rp}\cos\vartheta)m_L + f\,y_{Rp}}{(y_{Rp}y_{Lp} + f^2)\sin\vartheta + f(y_{Lp} - y_{Rp})\cos\vartheta} \tag{6}$$

The absolute error Δz_L is

$$\Delta z = \left|\frac{\partial z_L}{\partial y_{Lp}}\right|\Delta y_{Lp} + \left|\frac{\partial z_L}{\partial y_{Rp}}\right|\Delta y_{Rp} + \left|\frac{\partial z}{\partial x_{Lp}}\right|\Delta x_{Lp} + \left|\frac{\partial z}{\partial x_{Rp}}\right|\Delta x_{Rp} +$$

$$+ \left|\frac{\partial z}{\partial \phi}\right|\Delta\phi + \left|\frac{\partial z}{\partial \psi}\right|\Delta\psi + \left|\frac{\partial z}{\partial \vartheta}\right|\Delta\vartheta + \left|\frac{\partial z}{\partial k_L}\right|\Delta k_L + \left|\frac{\partial z}{\partial m_L}\right|\Delta m_L \quad + \tag{7}$$

$$+ \left|\frac{\partial z}{\partial l_L}\right|\Delta l_L + \left|\frac{\partial z}{\partial f}\right|\Delta f$$

The explicit expression for Δz is easily obtained when $f \rightarrow \infty$, that is in the case of orthographic projection:

$$\Delta z = \frac{1}{|\sin\vartheta|}\Delta y_{Rp} + \frac{1}{|tg\,\vartheta|}\Delta y_{Lp} + \frac{1}{|tg\,\vartheta|}\Delta l_L + \Delta m_L + \frac{|x_{Lp}|}{|tg\,\vartheta|}\Delta\phi +$$

$$+ \frac{|y_{Lp} - \cos\vartheta\,y_{Rp} - l_L|}{\sin^2\vartheta}\Delta\vartheta + \frac{|x_{Lp}|}{|\sin\vartheta|}\Delta\psi \tag{8}$$

When $l_L, m_L \gg x_{Lp}, y_{Lp}, x_{Rp}, y_{Rp}$ (that is when l_L, m_L are much larger than the diameter of the vidicon tube)

$$\Delta z = \frac{|f \sin\vartheta + l_L \cos\vartheta + m_L \sin\vartheta|}{|f\ tg\ \vartheta\ \sin\vartheta|}\Delta y_{Lp} + \qquad (9)$$

$$+ \frac{|l_L + f \sin\vartheta|}{|f \sin^2\vartheta|}\Delta y_{Rp} + \frac{1}{|tg\vartheta|}\Delta l_L + \Delta m_L + \frac{m_L l_L}{f \sin^2\vartheta}\Delta\vartheta$$

Now in both cases the only relevant terms to the absolute error are:

$$\Delta z = \frac{1}{|tg\vartheta|}\Delta l_L + \Delta m_L + \frac{|y_{Lp} - \cos\vartheta\ y_{Rp} - l_L|}{\sin^2\vartheta}\Delta\vartheta \qquad (10)$$

for the former case (orthographic) and

$$\Delta z = \frac{1}{|tg\vartheta|}\Delta l_L + \Delta m_L + \frac{m_L l_L}{f \sin^2\vartheta}\Delta\vartheta \qquad (11)$$

for the latter case.
As can be easily shown, therefore, in both cases, Δz strictly decreases when ϑ increases and is minimum for $\vartheta = \pi/2$.

   Expression (10) can be effectively used when the focal lenght f is much larger than the size of the vidicon tube or the CCD matrix. For instance with common camera lenses with f = 7.5 cm. equation (10) can be satisfactorily used. In the general case the explicit expression for Δz is rather complex but the same qualitative conclusion can be reached[12] : to minimize Δz is necessary to observe the scene from two very different points of view.

3. GOOD PERSPECTIVE INVARIANTS
   To minimize Δz we have to solve the correspondence problem with rather different images of the same scene. Therefore it is necessary to look to image primitives or features that are "preserved" under perspective projections. Using standa rd techniques of differential geometry (Verri, Yuille in preparation) it is possible to show that:
Let be $\gamma(t) = (x(t), y(t), z(t))$ a planar regular curve with curvature $K(t)$ and $\gamma_p(t) = (x_p(t), y_p(t))$ the projected curve, where $x_p(t)$ and $y_p(t)$ are as in equations (1)-(2), with curvature $K_p(t)$, if

t^* is such that $K(t^*)=0$ than $K_p(t^*)=0$ and if t^* is such that $K_p(t^*)=0$ than $K(t^*)=0$.

In other words zeros of the curvature of planar curves are "good" perspective invariants. The discovery and use of other perspective invariants is likely to be a major step to obtain efficient stereo algorithms.

4. A STEREO ALGORITHM FOR PLANAR SILHOUETTES

On the ground of previous results we have implemented on a PDP computer connected to a VDS 701 image processing system (512x512 pixels - 8 bit) a stereo algorithm for planar objects able to find the equation of the plane on which the silhouette lies, with a precision of at least 99%.

The algorithm is composed of five steps, namely:
a) - Extraction of shape contour;
b) - Detection of points of zero curvature of shape contour;
c) - Epipolar transformation of shape contour;
d) - Matching of points of zero curvature on conjugated epipolar lines;
e) - Geometrical recosnstruction of the 3-D shape from matched points and camera parameters.

We now illustrate the algorithm providing the 3-D reconstruction of the black silhouette.
The image acquired by the left camera is shown in Fig. 1A, while Fig. 1B represents the right stereo image. The origin of the coordinates system was fixed in the center of the vidicon tube of the left camera, with the x and z axis lying in the optical plane determined by the optical axis of the two cameras. The focal lenght of the lenses was 12.5 mm. and the parameters of the right camera were:

$$m_L = 18.5 \text{ cm.} \qquad l_L = 45.7 \text{ cm.} \qquad \vartheta = 0.785 \text{ rad} (\sim 45°)$$

The black silhouette of Fig. 1 was lying in a plane π , whose equation in our coordinate system was:

$$ax+by+cz = 1 \quad \text{where} \quad a=1.11 \ 10^{-2} \quad b = -6.56 \ 10^{-3} \quad c = 1.92 \ 10^{-2}$$

The goal of the algorithm is to determine the equation of the plane and consequently the 3-D reconstruction of the silhouette of Fig.1.

4.a. Extraction of shape contour

The two images of Fig. 1. were first binarized to facilitate the extrection of shape contour . Subsequently the binarized images were filtered with a two-dimensional symmetrical gaussian (with σ = 3 pixels) and the zero-crossings of the Laplacian of the filtered images[13,14] were extracted. The obtained contours are shown in Fig.2A (left image) and Fig. 2B (right image).

4.b. Detection of points of zero curvature of shape contours

To extract points of zero curvarure we used a contour following algorithm computing the curvature according to the algorithm proposed by Freeman and Davis [15]. The identified PZC are indicated in Fig.2 by white dots.

4.c. Epipolar transformation

When the viewing geometry is known it is possible to reduce the matching task from a two-dimensional to a much simpler one-dimensional problem. This simplification is obtained by searching for matching along conjugated epipolar lines . The epipolar transformation maps conjugated epipolar lines into horizontal lines with the same height. The amount of image distorsion introduced by the epipolar transformation increases with the distance of the eipole from the camera. Large distorsions are present whenever the epipole is on the vidicon target itself or just outside. Negligible distorsions occour when the epipole is farer from the camera at least 10 times the size of the vidicon tube. In this case the two images are already "registered" and matching can be perormed on horizontal lines with the same height. For instance when f=7.5 cm. and ϑ =30° the distance of the epipole from the vidicon tube is about 30 cm., that is about 15 times the size of the tube.

In our case the epipole was at 3.0 cm. from the vidicon tube center. The results of the epipolar transformations are shown in Fig. 2C for the left and 2D for the right shape contours. In Fig. 3 we have superimposed the contours before and after the epipolar transformation for the left and right image. We immediately see that appreciable disto rsion can be observed above and below the principal epipolar line. The correspondence between left and right PZC is graphically shown in Fig. 4. In this figure horzontal lines were drawn from the PZC of the left image for both the original and the transformed images. The one-dimensional correspondence between PZC of the transformed images is evident, particularly for PZC lying on epipolar lines distant from the principal epipolar line.

4.d. Matching of points of zero curvature

When there are few points of zero curvature in the shape contour it is rather unlikely that more than a single PZC lies on the same epipolar line. If two or more than two PZC lie on the same epipolar line, matching of corresponding PZC could be guided by the ordering constraint or by other rules. In robotic vision the use of the ordering constraint can be fallacious. Since in order to reduce the depth error z the two cameras must be far apart, the ordering con- straint may be often violated. In fact if a planar object lies on a plane intersecting the line joining the camera foci, it will be in the "ambiguous zone". For lines, curves, segments within the ambiguous zone the ordering constraint is not a good procedure for matching.

When two or more than two PZC were found on the same epipolar line matching was guided by procedures based on the qualitative nature of PZC. With these procedures it is possible to determine in many cases if the planar silhouette is in the ambiguous zone or not. However it is possible to construct patterns that remain ambiguous and whose 3-D structure is not uniquely determined.

4.e. Geometrical reconstruction of the 3-D shape

Using eqs. (1), (2) and (6) we can compute the 3-D location of matched PZC. For the silhouette shown in Fig. 1., 14 PZC were detected and in Table 1 the (x,y,z) values of the PZC are reported along with their distance from the original plane.

TABLE 1

PZC	X_L	Y_L	Z_L	DIST
1	-5.7	2.3	49.5	.1
2	-1,5	2.1	51.8	.1
3	2.3	.6	53.6	-
4	5.8	3.0	56.1	.3
5	5.3	7.2	57.4	.2
6	3.3	11.2	57.6	.2
7	.2	12.9	56.4	.2
8	-2.7	12.3	54.8	-
9	-4.5	12.5	53.9	.1
10	-6.1	13.7	53.2	-
11	-9.1	12.8	51.2	-
12	-12.1	9.9	48.6	.1
13	-11.5	6.5	47.8	.1
14	-10.1	3.0	47.5	.2

Table 1: The computed position in space (x_L, y_L, z_L) of the 14 PZC of the silhouette of Fig. 1, is reported. The values are expressed in cm. with a precision of 1. mm.
DIST represents the absolute value of the difference between z_L and the real depth values directly measured from the experimental set-up.

From Table 1 we see that the computed points are always within 2 or 3 mm. from the original plan. In our case m_L = 18.5 cm. l_L = 45.7 cm., that is $l_L, m_L \gg x_{Lp}, y_{Lp}, x_{Rp}, y_{Rp}$.
Therefore we can use equation (11) with $\Delta \vartheta$ = 2/1000 ϑ = 45° f = 12.5 mm, obtaining Δz = 0.5 cm.
Observe that if ϑ = 10° Δz is about 5 cm. and the precision on z would drop from 99% to 90%
The extension of the proposed algorithm to arbitrary non planar curves is currently under investigation and , from some preliminar results we can foresee its usefulness in more general situations.

REFERENCES

1 Ullman,S. "The interpretation of Structure from Motion". Proc.R.Soc. London B,203 405,426 (1979)

2 Horn,B.K.P. "Obtaining Shape from Shading Information" in the: in the: Psychology of Computer Vision, P.H.Winston ed., Mc Graw-Hill Publ. New York 1975,pp.115,155 (1975)

3 Icheuchi,K. and Horn,B.K.P. "Numerical Shape from Shading and Occluding Boundaries". Artificial Intelligence 17 141,184 (1981)

4 Marr,D. and Poggio,T. "A Computational Theory of Human Stereo Vision" Proc.R.Soc. London B,204 301,328 (1979)

5 Grimson,W.E.L. "A Computational Theory of Visual Surface interpolation". Phil. Trans.R.Soc. London B,298 395,427 (1982)

6 Arnold,R.D. and Binford,T.O. "Geometric Constraints in Stereo Vision" Proc. S.P.I.E., San Diego 238 281,292 (1980)

7 Baker,H.H. and Binford,T.O. "Depth from Edge and Intensity Based Stereo". Proc. 6th. Int. Conf. Al. Tokyo (1981)

8 Mayhew,J.E.W. and Frisby,J.P. "Psychophysical and Computational Studies towards a Theory of Human Stereopsis". Artif. Intell. 16 349-385 (1981)

9 Nishihara,K. "PRISM: A Practical Realtime Imaging Stereo Matcher" in Intelligent Robots: 3th. Int. Conf. on Robot Vision and Sensory Controls. Society of Photo-Optical Instrumentation Engineers, Proc. 449, Cambridge MA. (1983)

10 Faugeras,O. and Herbert,U. "A 3-D Recognition and Positions Algorithm Using Geometrical Matching Between Primitive Surfaces". Proceedings of the Conference on Artificial Intelligence, Karlsrhue (1983)

11 Goldstein,H. Classical Mechanics. Addison Wesley Publ. 1950

12 Verri,A. Metodi Matematici per la Visione Stereografica. Ph.D. Thesis Dept. of Physics – University of Genoa, Genoa, Italy (1984)

13 Marr,D. and Hildreth,E.C. "Theory of Edge Detection". Proc.R.Soc. London B,207 187,217 (1980)

14 Hildreth,E.C. "The measurement of Visual Motion". Ph.D. Thesis, Electrical Engineering and Computer Science Department, M.I.T., Cambridge, MA (1980)

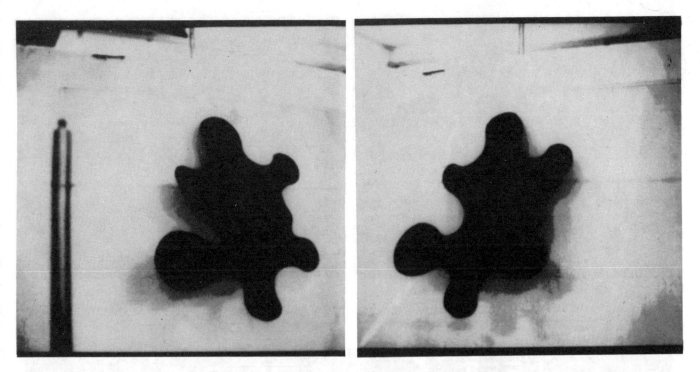

Fig. 1: Original images acquired by the two cameras.

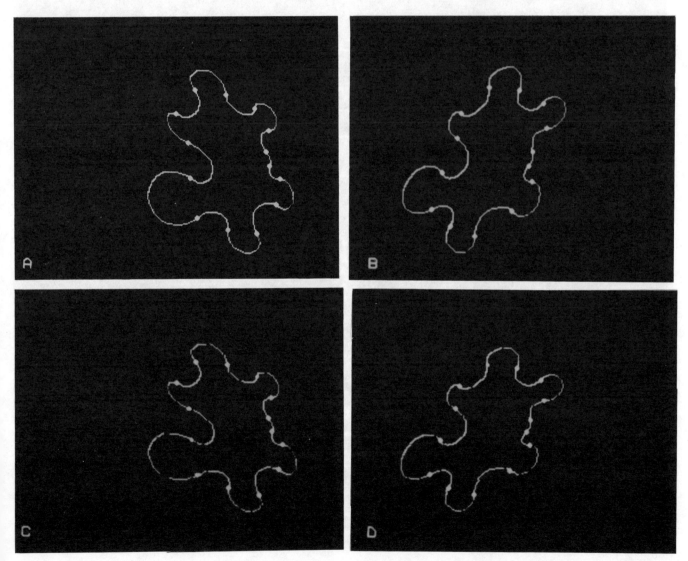

F ig. 2: Contours and points of zero curvature (dots). A) left image before transformation; B) right image before transformation; C) and D) epipolar transformations.

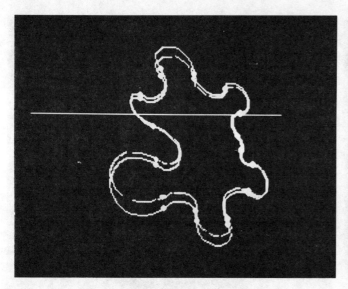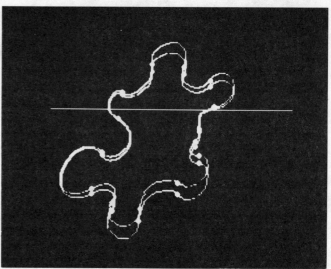

Fig. 3: left: superposition of Fig.s 2A and 2C; right: superposition of Fig.s 2B and 2D.

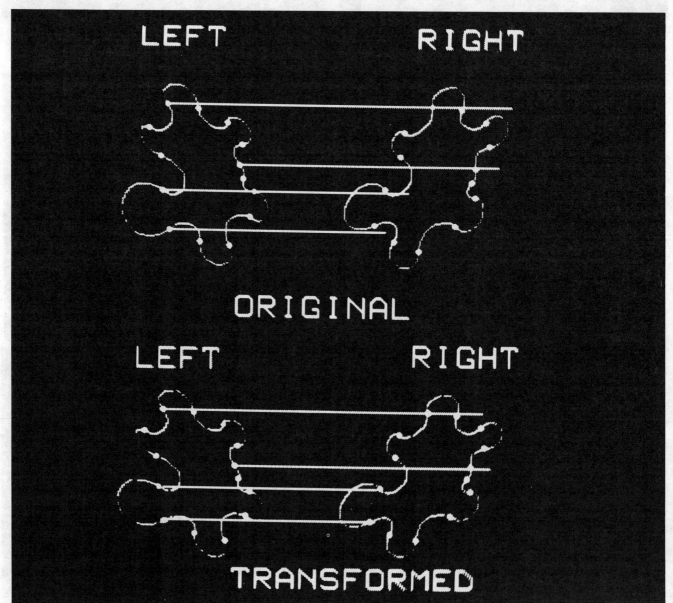

Fig. 4: Graphical representation of the correspondence between PZC. Upper: before transfomation; lower: after transformation.

Object Recognition with Combined Tactile and Visual Information

R.-C. Luo

Robotics and Automation Laboratory, University of Illinois at Chicago, USA

W.-H. Tsai

National Chiao Tung University, Taiwan

and

J. C. Lin

Robotics and Automation Laboratory, University of Illinois at Chicago, USA

Abstract

This paper describes a system which combines 2-D vision and taction for 3-D object recognition. A conventional video camera is used to obtain the top view of an object and two tactile sensing arrays mounted on a robot gripper are used to measure information about the lateral surfaces of the object. 3-D reference object models are established as a decision tree, and recognition of unknown objects is accomplished through measuring and comparing input object features hierarchically with those of the reference objects associated with the decision tree. Advantages of the proposed approach over the conventional ones (with visual or tactile information alone) are also identified.

I. Introduction

In industrial applications of robotics, it is often found necessary to recognize industrial objects and manipulate them with robot arms for various purposes. Examples include object sorting followed by loading (into classified compartments, e.g.), assembly with mixed parts on conveyor lines, bin picking followed by sorting, etc. In these applications, usually objects are first brought to and placed on a flat work platform for recognition. Once recognized, possibly with their positions and orientations computed, objects are manipulated (picked, rotated, etc.) and transferred by robot arms to appropriate positions for next application tasks.

Conventionally, object recognition is performed using visual information. With the advent of tactile array sensors [1-3] also useful for object shape measurement, we propose in this paper a new approach to object recognition which combines the use of visual and tactile information. Non-contact visual information of object shapes is obtained from a video camera mounted right above the work platform, and tactile information of lateral object shapes is measured by two array sensors included in the robot grippers. Both types of 2-D shape information are utilized for recognition. The recognition scheme is defined in a hierarchical manner so that an object is recognized first with the 2-D visual information along, followed by, if necessary, the 2-D tactile information measured when the object is grasped by the robot gripper. At least, the following advantages can be found in this approach:

(1) Part of the object recognition work can be performed during robot arm manipulation. This means that the total cycle time for object recognition plus object handling can be reduced.
(2) A lot of industrial objects are recognizable by their top-view 2-D shapes. This type of recognition can be done using 2-D visual information alone. Only when an object cannot be recognized by its top-view alone will lateral tactile shape information be used next. This hierarchical approach also improves recognition speed because it avoids direct 3-D measurement and recognition [4-6] which usually are time-consuming and unnecessary for most industrial objects.
(3) Since only 2-D information is processed, well-developed 2-D shape analysis techniques can be utilized for recognition without involving the more difficult 3-D scene analysis problem which is still not fully solved yet [7].
(4) In case that recognition using the above two types of information is still inadequate, then with the help from the robot arm, the object can be rotated so that both types of sensors can obtain more discriminant information from different object views.

The proposed approach will use moment invariants [8, 9] of object silhouette shapes as the features for object recognition. The reason for this is twofold. First, the lateral object shapes measured with the tactile array sensor are in low resolution because of the small array size available. Also, the shapes result from touching or taction of object surfaces on array sensor elements, so the number of points included in a shape ranges from one point (e.g., when the object is a sphere), a line (e.g., when the object is a cylinder), a surface patch (e.g., when the object is a polyhedron), to possibly a combination of the former three cases. The boundaries of such shapes, especially of the first two cases (a point and a line), are not meaningful enough for most boundary-based shape descriptions (such as Fourier descriptors, chain codes, syntactic string representations, etc.) to be applicable here. Instead it is better to base the shape analysis on shape regions, and moment invariants are appropriate for this purpose. Next, since robot manipulation on the objects is necessary, it is often required to find out the position and the orientation of a given object so that proper grasp of the object with the robot gripper can be accomplished. For this, moments turn out to be the best choice. In particular, object centroids and principal axes, which defines object positions and orienations, can be easily derived as functions of low-order moments.

The hierarchical recognition scheme is based on a decision tree [10] which can be conducted automatically in the learning phase from a set of given objects. Object shape ambiguity is resolved further as more tree levels are expanded until all shapes are discriminated or until further resolution is impossible. In the recognition phase, the decision tree is traversed when input object features are compared with reference object features until a decisive tree node is reached or until the input object is determined indiscriminable in its current stable state. For the latter case, the gripper is operated to pick up and rotate the object so that a new object state can be obtained. Another phase of object recognition is then started.

In the remainder of this paper, the proposed system configuration is first described in Sec. II with emphasis on tactile information measurement. An overview on object learning and recognition schemes is then presented in Sec. III. Moment invariants are reviewed and related visual and tactile features proposed in Sec. IV.

II. SYSTEM CONFIGURATION AND TACTILE INFORMATION MEASUREMENT

The system configuration and tactile information measurement are described in this section.

A. System Configuration

The system we use for object recognition and manipulation is as shown in Fig. 1. Both the TV camera and the array sensors are controlled by a microcomputer. The camera is mounted right on the top of a work platform so that perspective effects on object images can be reduced to a minimum. The camera optical axis (going through the camera lens center) is made perpendicular to the platform plane. The object to be recognized and manipulated is always positioned directly under the TV camera.

The gripper we use includes two square-shaped array sensors. The sensing elements on the arrays are arranged in a form of grid patterns. The elements are attached close enough to the array edges so that slant contact of an object surface with any array edge can also be detected. The array sensors are made sensitive enough so that contact of the sensing elements with object surfaces can be detected easily before the gripper movement torque forces the object to possibly change its orientation. This ensures that the tactile information measured by touching an identical object from a fixed lateral direction (fixed with respect to the principal axis of inertia of the object, as will be discussed) will always be identical or <u>stable</u>. Such stable tactile information is relied upon by the proposed approach for object recognition. On the other hand, because of the requirement imposed by the proposed recognition scheme, the two arrays on the gripper are so constructed that they are always operated <u>parallelly</u>.

Before the tactile information measurement process can be described, we define some notations to facilitate geometric descriptions of the system configuration. Let A_R denote the right-hand side array sensor as viewed from the gripper wrist, and A_L denote the left-hand side one. In some cases, A_R and A_L will also be used to specify the planes going through the sensing-element surfaces. Each array, after touching any object surface, will provide an array image of tactile information. Let I_R denote the image provided by A_R and I_L provided by A_L. Let O_R, the center of A_R, be the origin of the image coordinate system for I_R. The origin O_L for I_L is similarly chosen. The line going through O_R and O_L will be called the <u>gripper lateral axis</u> and denoted by L_L. Also, when A_R and A_L are maximally extended, the middle point of the line segment joining O_R and O_L will be called the <u>gripper center</u> and denoted by G. The plane defined by the surface of the platform will be denoted by P_P. And the line going through G and perpendicular to P_P will be called the <u>gripper vertical axis</u> and denoted by L_V. The direction defined by the vector from G to O_R will be called the positive direction of L_L. Another useful structure is the plane going through G and L_V, and perpendicular to L_L (and so parallel to A_R and A_L), which will be called the <u>gripper central plane</u> and denoted by P_G. Fig. 2 shows the spatial relation among all the geometric structures defined above.

B. Tactile Information Measurement

The first requirement is that the bottom edges of A_R and A_L align with the platform plane P_P. When this is satisfied as in Fig. 2, P_G, L_V, A_R and A_L will all be perpendicular to P_P, and L_L parallel to P_P. A lateral direction from which the measurement will be made is selected next. This direction is determined as an angle (in the counterclockwise sense as viewed from the top) measured with respect to the inherent orientation of the object. Specifically, after the top-view silhouette of the object is obtained (after appropriate image processing operations are applied on the raw visual image of the object), we compute the centroid and the principal axis of the object and denote them by C and L_C, respectively. Computation of C and L_C will be described in Sec. IV. The direction defined by the vector from C to one of the end points of L_C which is closer to the gripper wrist will be called the <u>positive direction</u> of L_C.

To measure tactile information of the object from a <u>lateral direction</u> θ, A_R and A_L are first opened to the maximum extent and made vertical to P_P. They are next brought, with their bottom edges touching P_P, close to the object in such a way that C aligns with L_V (i.e., G and C are co-linear on L_V) and the positive direction of L_L remains at angle θ from the positive direction of L_C. The arrays are then moved toward each other in parallel to sandwich the object without changing the alignment of C with L_V and the angle θ of L_L with respect to L_C until contact with the object is detected by one of the arrays. At this time, array movement is stopped immediately and the tactile information on the touched array is stored as image I_L or I_R. Depending on which side, A_L or A_R, is touched first (say A_L), the robot arm is then manipulated such that the gripper (and the two arrays) is moved horizontally with the untouched side (A_R now) toward the object. During this movement, the lateral direction θ of L_L with respect to L_C is still kept unchanged though the alignment of C with L_V is destroyed. Also, the arrays are now moved in the same direction instead of toward each other. This movement is continued slowly until contact of the other array (A_R) is detected. After the corresponding tactile information is stored (as I_R), the tactile measurement is completed, resulting in two tactile images I_L and I_R. For example, the object shown in Fig. 2 is being measured from the lateral direction of 90°. Fig. 3 shows another example which illustrates the gripper movement sequence.

III.   PROPOSED OBJECT LEARNING AND RECOGNITION SCHEMES

We will first describe the basic idea we use for object recognition, followed by a brief introduction to proposed object learning and recognition procedures.

A.   Recognition Principle

As mentioned in the introduction (Sec. I), objects are recognized hierarchically, first with their top-view visual images and then with their lateral tactile images, if necessary. Visual object images, taken with the TV camera are processed to obtain object silhouette shapes. Shape boundaries are then extracted for visual recognition (i.e., recognition based on visual information). Holes and other object characteristics within the boundaries, if any, are ignored in this study because most objects are discriminable from each other without using within-boundary information. On the contrary, because of the low tactile image resolution, all object points in tactile images are utilized for lateral recognition (i.e., recognition based on tactile information).

One approach for lateral recognition is to measure a sequence of tactile images around the top-view silhouette boundary and then compare such image sequences of an input object with those of reference objects. However, because of the nature of tactile measurement, a single measurement often results in images that correspond to only two top-view silhouette boundary points, such as in the case of measuring a thick circular disk. This means that it will possibly take a long time for the gripper to measure an adequate sequence of tactile images to cover the whole lateral object surface. Actually from a mathematical point of view, if the lateral object surface is all convex-curved (such as that of a disk), it requires an infinite number of points to compose the top-view object boundary. Thus, an infinite number of tactile measurements is required to completely cover the lateral object surface. Although image digitization relaxes the requirement of infiniteness, sequential measurement around the whole boundary by rather slow gripper operations is certainly impractical for real applications, especially for measuring curved object surfaces.

Fortunately, most objects, if indiscriminable from their top views, can still be discriminated without using complete lateral object surface information. If two objects look the same from their top views, there must exist some specific lateral direction in which the two objects differ in their lateral surface structure. Otherwise, the two objects must be totally identical unless their invisible bottom surfaces have different indentations. Therefore, it is feasible to find out such discriminable lateral directions in the learning phase and measure tactile information in these directions for discrimination in the recognition phase. And this is the basic principle of the proposed object recognition scheme.

B.   Object Shape Learning

Accordingly, a block diagram showing the major steps of the

187

learning phase is included in Fig. 4. After all the reference objects are discriminated according to their top-view silhouette boundaries, there may still exist some groups of objects, each containing several objects which are visually indiscriminable. Then, for each such group, the gripper is operated to measure tactile information for further discrimination. This precedes by first selecting a set of preselected lateral directions and then measuring a pair of tactile images for each of these directions. The lateral direction most effective for discriminating each group of visually-ambiguous objects is selected. Stop if all groups of ambiguous objects are discriminable at this stage; otherwise, repeat the process by selecting another most effective lateral direction from the remaining directions for each subgroup of still ambiguous objects until all lateral directions are tried. The effectiveness of a lateral direction for lateral discrimination is defined as the number of subgroups which result from the discrimination based on the lateral direction. The above learning procedure also involves selection of shape features from visual and tactile object images for object discrimination. Feature selection will be discussed in Sec. IV. The result of learning will be a hierarchical decision tree with each tree node representing a group of ambiguous or indiscriminable objects and each tree link (below the first-level nodes) being associated with a most effective lateral direction . The emphasis here is that the whole learning procedure can be made totally automatic.

Unfortunately, the above learning procedure does not guarantee that all objects are completely discriminated. That is, some nodes of the decision tree may include more than one object which is indiscriminable based on all preselected lateral directions. This problem will be solved in the recognition phase discussed next.

C. Object Shape Recognition

As shown in Fig. 5, the recognition procedure begins with taking the top-view visual image of a given unknown object after it is brought right under the TV camera on the platform. The object is then discriminated according to features extracted from the visual image. If the object is not discriminable with its visual image alone, a pair of tactile images are then measured from the alteral direction specified in the decision tree. After object features are extracted, the object is discriminated further. This step may be repeated more than once if the number of tree levels is more than two. Most objects should be recognizable after this step.

But as mentioned previously, there may still exist some objects which are indiscriminable if they are in certain stable states on the platform. For example, the two objects shown in Fig. 6(a) and (b) are indiscriminable according to their top views and tactile measurement from any lateral direction because the only difference between them, the cubic on the top of the first object, is actually "invisible" to the system (because only the top-view silhouette boundary is utilized for recognition). It is also "untouchable" from any lateral direction with the measurement method described in Sec. II. One way to solve the problem is to change the stable state of the input object so that the originally "invisible" and "untouchable" portion of the object can become "visible" and "touchable." Thus, if object 1 in Fig. 6 (a) is rotated 90° with respect to its top-view principal axis (L_c) and laid down again, it looks like Fig. 6(c) which now is discriminable from Fig 6(b). This change of object states can be performed by the robot gripper. Subsequently, the new object state is subjected to the same recognition proceudre as described above, i.e., discrimination first with its visual image and next with its tactile image pairs, if necessary.

D. An Illustrative Example

Some objects and their stable states are shown in Fig. 7 (only states with higher stability are included). The decision tree constructed by the proposed learning procedure is shown in Fig. 8. We draw measured shapes or shape pairs right above each ambiguous shape group. Note that all objects have their top-view principal axes coincident with their most elongated direction. Since most object top-view silhouette shapes are rectangular, they are visually indiscriminable, as is partitioned into the first-level node S_1. With the measurements from lateral direction 0°, all the shapes in S_1 are discriminated except those included in tree node $S_{1.2}$. With another measurement from direction 90°, shape S_{91} (in $S_{1.2.2}$) is discriminated From S_{21} and S_{71} (in $S_{1.2.1}$). However, because of the untouchable rectangular prism on top of S_{71}, S_{71} is indiscriminable from S_{21} in its current state. Therefore $S_{1.2.1}$ is non-partitionable. Note that although S_{91} is discriminable from S_{21} and S_{71} directly from direction 90° (without measuring from direction 0_o first), direction 0° is the most effective lateral direction to partition S_1 and so is selected to partition S_1 first.

In the recognition stage, given an input object which looks like S_{71} shown in Fig. 7(g). Then based on the decision tree shown in Fig. 8, the input will be found indiscriminable from S_{21} when the non-partionable node $S_{1.2.1}$ is reached during the recognition steps. After the object is rotated and dropped onto the platform by the gripper, its state may be changed into that of S_{72} as shown in Fig. 7(g). In that case, after another phase of recognition is started, the first-level node S_3 will be reached. Since $S_{1.2.1}$ includes object 2 and object 7 and S_3 includes object 7 and object 8, we get an intersection as $S^* = S_{1.2.1} \cap S_3 = \{7\}$ which includes only object 7. So, the input can be correctly recognized as object 7, and the second recognition phase can be stopped.

IV. MOMENT INVARIANTS AS SHAPE FEATURES

In this section, we first give the definitions of moments of binary digital images, and then describe various shape features for object recognition in terms of various moment invariants.

A. Definitions of Moment Invariants

Let B denote a set of N discrete points used to describe a shape in a binary image. Each point in B has a value 1. The (p, q) moment of B is defined as:

$$m_{pq} = \sum_{i=1}^{N} x_i^p \ y_i^p,$$

and the (p, q) central moments of B are defined as:

$$\bar{m}_{pq} = \sum_{i=1}^{N} (x_i - \bar{x})^P (y_i - \bar{y})^q,$$

where

$$\bar{x} = m_{10} / m_{00},$$

$$\bar{y} = m_{01} / m_{00},$$

and $(\bar{x}, \bar{y})$ is the centroid of B.

Moments can be given a physical interpretation if we regard each point of the binary image as possessing an identical quantity of mass. Thus $m_{00} = \bar{m}_{00}$ is the total mass of the image which actually is the image area in terms of the number of points. A lot of invariant shape features with respect to translation and rotation can be derived from the central moments as shall be discussed subsequently.

On the other hand, moments are also useful for computing the principal axis of shape B which is defined as the line L_C going through the point (α, β) with slope $\tan\theta$, i.e.,

$$(y - \beta) \cos\theta = (x - \alpha) \sin\theta,$$

about which the moment of inertia of B, defined as

$$I = \sum_{i=1}^{N} [(x_i - \alpha) \sin\theta - (y_i - \beta) \cos\theta]^2,$$

is minimum. It can be shown [11] that $(\alpha, \beta) = (\bar{x}, \bar{y})$, i.e., L_C goes through the centroid of B, and that θ is just the direction of the eigenvector corresponding to the larger eigenvalue of the following moment matrix:

$$M = \begin{bmatrix} \bar{m}_{20} & \bar{m}_{11} \\ \bar{m}_{11} & \bar{m}_{02} \end{bmatrix}.$$

It can be shown [8] that θ is such that:

$$\tan 2\theta = \frac{2\bar{m}_{11}}{\bar{m}_{20} - \bar{m}_{02}}.$$

In practice, there are always two solutions for θ :

$$\theta_1 = \tfrac{1}{2} \tan^{-1} \frac{2\bar{m}_{11}}{\bar{m}_{20} - \bar{m}_{02}}$$

and

$$\theta_2 = 180^\circ + \theta_1.$$

The one pointing to the end point of L_C which is closer to the gripper wrist has been defined as the positive direction of L_C in Sec. II. Note that the principal axis direction of an object is often relied upon in robot arm manipulation as the best direction to grasp the object stably.

B. Shape Features for Visual Recognition

We first discuss the features we use for recognizing visual images. Since the TV camera can take images with a rather high resolution and since object silhouette boundaries reveal, in most cases, enough shape information for object recognition, the point set used for defining moments is chosen to include just the shape boundary points instead of all silhouette points. This set will be denoted as B_v. This also makes the features, to be defined next, more informative about minute details of the shape boundaries. Furthermore, moment computation can be significantly enhanced because fewer points are involved.

The first feature v_1 we use is $m_{00} = \bar{m}_{00}$ which is the area of B_v. All features for recognizing visual images will be denoted as v_i. Since B_v is the silhouette boundary, v_1 actually is the perimeter of the boundary [11]. Another useful shape feature is shape elongation or eccentricity. An eccentricity measure in terms of central moments is described in [12] as:

$$e = [(\bar{m}_{02} - \bar{m}_{20})^2 + 4\bar{m}_{11}^2]/\bar{m}_{00}.$$

To avoid time-consuming division, we choose v_2 to be the numerator of e above, i.e., $v_2 = (\bar{m}_{02} - \bar{m}_{20})^2 + 4\bar{m}_{11}^2$. The third feature v_3 is the moment of inertia around the centroid which is:

$$v_3 = \sum_{i=1}^{N}[(x-\bar{x})^2 + (y-\bar{y})^2]$$
$$= \bar{m}_{20} + \bar{m}_{02}.$$

Two other pertinent features are selected in terms of higher-order central moments and are defined as follows [8]:

$$v_4 = (\bar{m}_{30} - 3\bar{m}_{12})^2 + (3\bar{m}_{21} - \bar{m}_{03})^2,$$
$$v_5 = (\bar{m}_{30} + \bar{m}_{12})^2 + (\bar{m}_{21} + \bar{m}_{03})^2.$$

In the following sections, we use V to denote the feature vector composed of the five features v_1 through v_5 described above, i.e.,

$$V = [v_1, v_2, v_3, v_4, v_5]'.$$

V will be called the visual feature vector. All visual features as defined above are invariant with respect to rotation and translation [8].

C. Shape Features for Tactile Recognition

191

Since tactile images I_L and I_R have much lower resolutions, all non-zero points in I_L and I_R (resulting from contact of array sensing elements on the object surfaces) will be used to compute the moment values. In the following, the non-zero points in I_L and I_R will be denoted as B_L and B_R, respectively. Features extracted from B_L and B_R will be denoted as t_i^L and t_j^R, respectively, and called <u>tactile features</u>. Also, the moments m_{pq} (or $\bar{m}_{pq}$) computed from B_R and B_L will be superscripted as m_{pq}^R and m_{pq}^L (or $\bar{m}_{pq}^R$ and $\bar{m}_{pq}^L$), respectively.

Again, the areas of B_L and B_R can be used as tactile features. We select t_1^L as $m_{00}^L = \bar{m}_{00}^L$ and t_1^R as $\bar{m}_{00}^R = \bar{m}_{00}^R$. Next, since B_L and B_R are measured with array sensors A_L and A_R fixed spatially with respect to the platform plane P_P, the centroids and the directions of the principal axes of B_L and B_R reveal a certain amount of 3-D structural information about the object. Therefore, we choose t_2^L and t_3^L to be $\bar{x}_L$ and $\bar{y}_L$, t_2^R and t_3^R to be $\bar{x}_R$ and $\bar{y}_R$, respectively, and t_4^L and t_4^R to be:

$$t_4^L = \tfrac{1}{2} \tan^{-1} \left(\frac{2\bar{m}_{11}^L}{\bar{m}_{20}^L - \bar{m}_{02}^L} \right),$$

$$t_4^R = \tfrac{1}{2} \tan^{-1} \left(\frac{2\bar{m}_{11}^R}{\bar{m}_{20}^R - \bar{m}_{02}^R} \right).$$

Finally, t^L and t^R are selected similarly as v_2 such that:

$$t_5^L = (\bar{m}_{02}^L - \bar{m}_{20}^L)^2 + 4(\bar{m}_{11}^L)^2,$$

$$t_5^R = (\bar{m}_{02}^R - \bar{m}_{20}^R)^2 + 4(\bar{m}_{11}^R)^2.$$

The two vectors $T^L = [t_1^L, t_2^L, t_3^L, t_4^L, t_5^L]'$ and $T^R = [t_1^R, t_2^R, t_3^R, t_4^R, t_5^R]'$ together are called the <u>tactile feature vector pair</u> and sometimes will be collected as a single vector $T = [T^L, T^R]'$. Tactile features similar to v_3, v_4, and v_5 are not used because the above 10 tactile features are found adequate for tactile recognition.

D. <u>Feature Matching for Object Recognition</u>

Assuming that the robot is working in a controllable environment with minimal noise, we adopt a very simple way to compare features for object recognition, with possible extension to more complicated pattern recognition theory. Two features u_1 and u_2 of the same type are said to be similar if their absolute difference $|u_1 - u_2|$ is less than a predetermined threshold for that type. Two feature vecotrs U_1 and U_2 are said to be similar if every two corresponding features u_{i1} in U_1 and u_{i2} in U_2 are similar.

On the other hand, when comparing one tactile feature pair T_1^L, T_1^R with another T_2^L, T_2^R, since there are two possible principal axis directions for each object, it is necessary to perform two comparisons before drawing any conclusion. That is, we first compare T_1^L with T_2^L and T_1^R with T_2^R, and then compare T_1^L with T_1^R and T_1^R with T_2^L. If either comparison is successful, then we say that the two tactile feature pairs are similar. Otherwise, they are declared dissimilar.

References

[1]   M. H. Raibort and J. E. Janner, "A VLSI Tactile Array Sensor," <u>Proceedings of the 12th International Symposium on Industrial Robots</u>, Paris, France, 1982.

[2]   M. Briot, "The Utilization of an 'Artificial Skin' Sensor for the Identification of Solid Objects," <u>Proceedings of the 9th International Symposium on Industrial Robots</u>, Washington, D.C. 1979.

[3]   P. Dario, R. Bardelli, D. de Rossi, L. R. Wang, P. C. Pinotti, "Touch-Sensitive Polymer Skin Uses Piezoelectric Properties to Recognize Orientation of Objects," <u>Sensor Review</u>, Vol. 2, 1982.

[4]   M. Oshima and Y. Shirai, "Object REcognition Using Three-Dimensional Information," <u>IEEE Trans. Patt. Anal. Mach. Intell.</u>, Vol. PAMI-5, No. 4, pp. 353-361, July 1983.

[5]   Y. Sato and I. Honda, "Pseudodistance Measurement for Recognition of Curved Objects," <u>Ibid.</u>, pp. 362-372, July 1983.

[6]   R. Nevatia, "Prescription and Recognition of Curved Objects," <u>Artificial Intelligence</u>, Vol. 8, 1977.

[7]   R. Bajcsy, "Three-Dimensional Object Representation," in <u>Patt. Recgo. Theory and Appl.</u> (J. Kittler, K. S. Fu, and L. F. Pau, Eds.), D. Reidel Publishing Co., U. K., 1982.

[8]   M. K. Hu, "Visual Pattern Recognition by Moment Invariants," <u>IRE Trans. Inform. Theory</u>, Vol. IT-8, pp. 178-187, Feb. 1962.

[9]   S. A. Dudani, K. J. Breeding and R. B. McGhee," Aircraft Identification by Moment Invariants," <u>IEEE Trans. Computers</u>, Vol. C-26, No. 1, pp. 39-45, Jan. 1977.

[10] J. K. Mui and K. S. Fu, "Automated Classification of Nucleated Blood Cells Using a Binary Tree Classifier," <u>IEEE Trans. Patt. Anal. Mach. Intell.</u>, Vol. PAMI-2, No. 5, Sept. 1980.

[11] A. Rosenfeld and A. C. Kak, <u>Digital Picture Processing</u>, Academic Press, New York, 1982.

[12] D. H. Ballard and C. M. Brown, <u>Computer Vision</u>, Prentice-Hall, New Jersey, 1982.

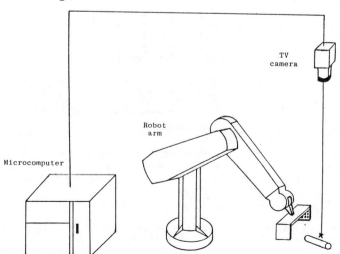

Fig. 1 System configuration for object recognition and manipulation.

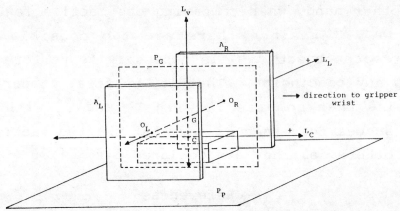

A_R : right array sensor;	L_L : gripper lateral axis;
A_L : left array sensor;	L_V : gripper vertical axis;
O_R : origin of A_R;	P_G : gripper central plane;
G : gripper center;	P_P : platform plane;
C : center of object top-view image;	L_C : principal axis of top-view object image.

Fig. 2   Spatial relation among various system planes, axes, and centers. "+" denotes positive directions of axes L_L and L_C. The object is being measured for its tactile information from C direction 90°.

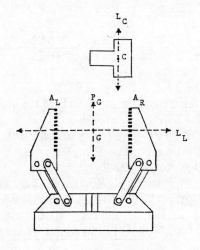

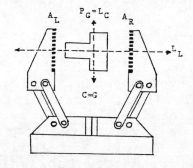

(a)   Before measurement. The gripper is being moved to enclose the object.

(b) P_G and L_C, C and G, and L_L and measure direction 90° are all aligned.

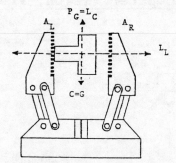

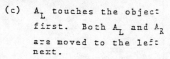

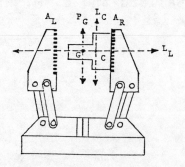

(c) A_L touches the object first. Both A_L and A_R are moved to the left next.

(d) A_R touches the object now and the measurement is completed.

Fig. 3   Gripper movement sequence for 90° tactile measurement. Arrows in each figure show moving directions of the array sensors.

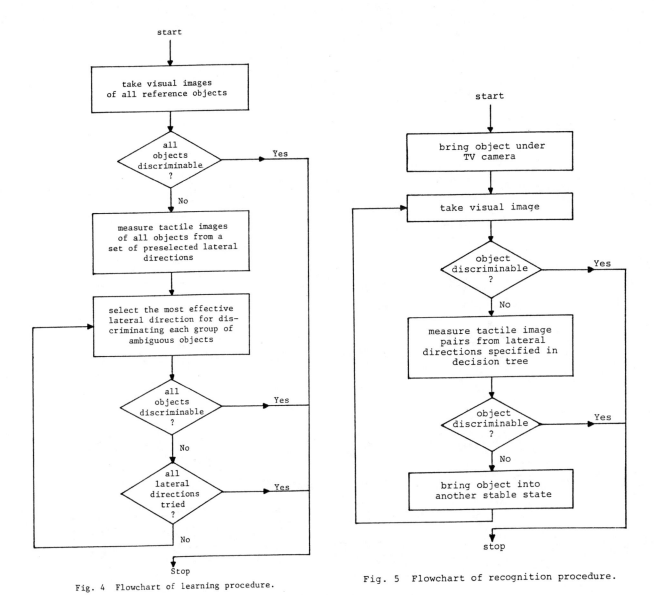

start

take visual images
of all reference objects

all objects discriminable ? — Yes

No

measure tactile images
of all objects from a
set of preselected lateral
directions

select the most effective
lateral direction for discriminating each group of
ambiguous objects

all objects discriminable ? — Yes

No

all lateral directions tried ? — Yes

No

Stop

Fig. 4 Flowchart of learning procedure.

start

bring object under
TV camera

take visual image

object discriminable ? — Yes

No

measure tactile image
pairs from lateral
directions specified in
decision tree

object discriminable ? — Yes

No

bring object into
another stable state

stop

Fig. 5 Flowchart of recognition procedure.

(a) Object 1.

(b) Object 2.

(c) Another stable
state of object
1.

Fig. 6   Two indiscriminable objects ((a) and (b)). Object 1
in (a) is discriminable from object 2 in (b) if its
stable state is changed into the one as shown in (c).

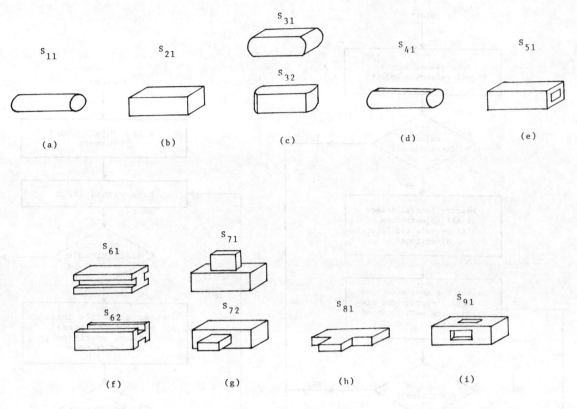

Fig. 7 Objects for learning to construct decision tree. Two stable states are shown for object 3 in (c), object 6 in (f), and object 7 in (g).

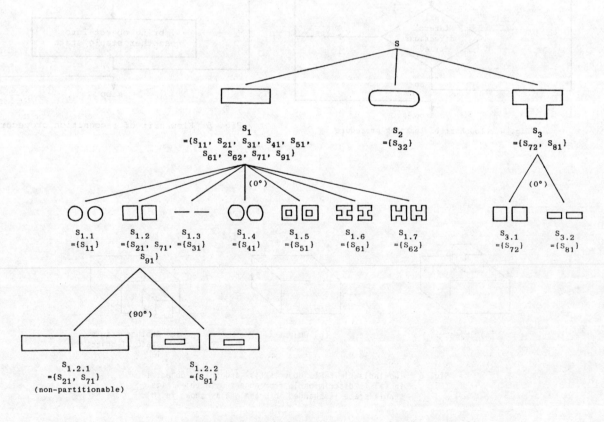

Fig. 8 Constructed decision tree by proposed learning procedure for objects shown in Fig. 7.

ARCHIE – An Experimental 3D Vision System

R. J. Fryer
University of Strathclyde, UK

ABSTRACT

The Strathclyde ARCHIE (Artificial Retina and Cortex Heuristic Imaging Experiment) project is an investigation of a pipelined pseudo-parallel architecture for real-time derivation of the three-dimensional structure of a viewed scene. This will have an interface to an intelligent database which will use the perceived structure, together with other clues such as colour or texture, to interpret the scene and recognise objects. The overall configuration bears strong resemblance to some schematics of vertebrate visual systems, hence the acronym.

INTRODUCTION

Electromagnetic radiation, in the form of visible light and adjacent parts of the spectrum, provides the richest source of detailed environmental information available to mankind. Using it one can gain knowledge, often to a high degree of precision, about the shape, size, distribution, temperature, composition and microstructure of surrounding visible surfaces. All this information may be derived to a greater or lesser degree of precision without direct physical contact with the surfaces observed and in many cases making use of a free resource - ambient light.

The evolutionary history of mankind has been such as to make vision pre-eminent among the senses and to cause the civilisations of all continents to structure themselves around visually directed activities. All technological activity from the earliest agrarian developments on has, except in the most unusual and special circumstances, been based on the visual monitoring of processes as part of a feedback loop. Scientific advance has in large measure been based on the making visible of the hitherto invisible, whether directly as with microscopes and telescopes or indirectly via the translation of non-visual environmental characteristics into visual ones such as meter readings or charts.

As long as the rate and scope of technological activity is such as to permit humans to form part of the control loop such manual/visual translation and monitoring is very successful.   It was implicit in the origins of the industrial revolution and has served us well up to the recent past.   Current pressures of cost, volume and rate of manufacture, and also of quality assurance standards, are placing severe strains on this type of control loop, however, leading to the various automation procedures in use today.

For high volume long runs of standard parts dedicated automated production systems are cost effective.   For relatively low volume short runs the high investment required for dedicated automated machinery will often not be justifiable, and yet for many such situations the 'traditional' methods remain inappropriate.   This has led to the concept of the 'flexible manufacturing system' (FMS) which is able to adapt its activity to the needs of the moment.   Such flexibility of activity implies flexibility of feedback which in turn implies an expansion in the number, sophistication and variety of sensors which are required.   The potential contribution of image acquisition and analysis schemes to FMSs has long been recognised and a great deal of work has been performed on both the theory and practice of such techniques.

Unfortunately the analysis of single images has, except in isolated special cases, proven very difficult.   This is for two principal reasons - (a) the sheer amount of data associated with an image, and (b) the intrinsic complexity of 'real' scenes in terms of intensity and colour distributions.   To illustrate the first point consider the case of a low resolution image of 128x128 pixels; this is a matrix of 16,384 individual points each recorded to some specific precision and having some degree of error due to noise. It is not only the individual pixel values but also the relationship of each to its neighbours which defines the image contents.   As illustration of the second point I refer the reader to any photograph of a real scene.

It is perhaps worth noting in passing the implications for serial processing machines of the data volumes in images.   Image handling is necessarily a lengthy process on such machines.   Consider a simple operation applied to the low resolution image mentioned above : using reasonable if optimistic figures such as a 1MHz processor clock, five clock cycles per machine instruction, and a loop of twenty such instructions applied successively to each pixel, a total processing time of 1.6 seconds is obtained.   Such a loop might be capable of simple image enhancement, it would certainly be quite insufficient by orders of magnitude for any detailed image analysis and recognition procedure.   Serial machines are thus elminated as the principal processing agents in any real-time image analysis/recognition system.

A widely perceived response to the unsuitability of serial machines for image processing is to postulate the fully parallel array processor as the necessary base architecture.   Certainly array processors are a powerful means of reducing processing time for many useful operations applied to single images.   For the cross correlation of multiple images, as is needed for some purposes, their utility is less obvious except as fast preprocessors.   A more serious drawback from the point of the industrial user is their expense.   Array processors for image handling are currently complex and expensive devices.   It is, I believe, true to say that no suitable off-the-shelf array processor exists which would be cost effective for use on an FMS workstation.   Such devices will certainly become available.   Current work into the theory of parallel processing coupled with forecastable advances in very large scale integration (VLSI) will make them practicable in the foreseeable future.   The key question is 'when?'.   Image array processors may well be on the horizon, it is much less likely that they are round the corner.

3D VISION AND THE IMAGE SEGMENTATION PROBLEM

One of the principal problems facing the developers of any image recognition system is that of image segmentation.   For current purposes this maybe defined as the identification and isolation of significant portions of the scene for higher level interpretation.   Examples of portions to be isolated would be major edges or surfaces of objects and significant surface markings.   Once identified such scene components may be correlated and used to deduce the nature and/or orientation of the objects of which they form part.

The problem of achieving meaningful segmentation of individual images is vastly complicated by lighting variations. Variations of surface albedo and photometric properties, non-parallel (and possibly diffuse or mutiple) lighting, light scattered from adjacent surfaces, and shadowing all add to the overall complexity of an image. The use of silhouettes from binary images can be useful for some classes of industrial artefact but is subject to severe limitations in the general case. Current work on the use of grey-scale data for such objects will ease some of these limitations, but for many situations of interest the increased capacity will still be insufficient.

It is noteworthy that for many or most of these situations of interest, from automated assembly to self-guided vehicles, the object of the image segmentation is to enable the building of three-dimensional models of the viewed scene as a basis for activity planning. In these circumstances image segmentation is simply one means (and not necessarily the best) to an end. Hence it is reasonable to ask whether the sequence of operations can be reversed? In other words, could a three-dimensional representation of the scene be created and used as a key to the way individual scene components can be distinguished from their surroundings?

OBTAINING THE 3D REPRESENTATION (THE 2.5D SKETCH)

A true three dimensional representation is almost impossible to generate since it implies a complete knowledge of the volume occupancy of every scene component whether hidden or not. An actual observer, human or machine, is necessarily limited to a knowledge of the distribution of those surfaces actually visible; all further knowledge must either be extrapolation based on experience of similar situations, or memory of the same objects from another viewpoint. The intermediate, limited, physically realisable representation has come to be known as the 2.5D sketch. It is this limited, observer centred representation which is considered throughout the remainder of this paper.

The search for efficient methods of generating the 2.5D sketch has been in progress for more than a decade. Many approaches are possible and have been investigated, as surveyed by Jarvis in 1983[1]. Roughly speaking the available techniques may be divided into two classes - active and passive systems. Active systems include such techniques as contrived lighting, laser or acoustic time-of-flight, and the use of phase information from coherent illumination. Such active systems can produce good results in appropriate circumstances but usually involve a high off-line processing load. They are in general less well adapted than are passive systems to working in environments suitable for and including humans.

Passive systems take advantage of such scene components as texture gradients, image brightness, relative occlusion of identified objects, focus optimisation, scene changes resulting from camera motion, and comparison of images from multiple viewpoints (including stereoscopic disparity). Not all these techniques produce absolute ranges (e.g. use of occlusion); use of texture or of image brightness requires strong assumptions to be made concerning the nature of the surfaces viewed. Only range from camera motion and use of simultaneous images from multiple viewpoints are both independent in principle of such assumptions and capable of high precision measurement. Both require the identification of equivalent features in each of two or more images and then derive absolute positions by the application of geometric principles plus a knowledge of camera positions and characteristics.

The minimum number of images for such analysis is two (the stereoscopic case) though larger numbers have been used, for instance by Morasso and Tagliasco[2] and by Tsai[3]. The analysis of such views is straightforward in principle but requires the prior localisation of corresponding features in each image. Often such localisations and correlations have been made manually though automatic procedures have been used with some success, for instance by Henderson et al[4], by Yakimovsky and Cunningham[5] and by Baker[6].

Such systems as exist for the use of multiple viewpoints are non real-time, seemingly requiring minutes (at least) of processing on powerful computers for their action. This is partly due to the algorithms used for feature extraction (implemented on serial machines with the consequences discussed above), and partly because the camera orientations used complicate the subsequent analysis.

It is the contention of this paper that both of these constraints can be eased or adjusted sufficiently to permit real-time (of the order of fractions of a second) determination of a 2.5D sketch from industrially interesting scenes using current state-of-the-art components in a stereoscopic system. In order to see how this may be done it is necessary to look more closely at the theory of stereoscopy.

STEREOSCOPY

The General Case

A thoughtful survey of stereoscopic systems research up to 1982 has been published by Barnard and Fischler[7].

Assuming flat image planes the general stereo geometry is as illustrated in Figure 1. Images are formed on the two 'retinas' by lenses at F and F'. The orientations of the two cameras are such that the optic axes converge on a point of interest P which therefore forms images on the principal points C and C'. A point object, O, not in the plane PCC' forms images at X and X'. The plane defined by OFF' interesects the images along 'epipolar lines'; as O moves to other points in this same plane its projections X and X' move along their respective epipolar lines.

In this general case the epipolar lines in a given image are not parallel to each other even when the lens and camera characteristics are identical. In other words the 'stereoscopic disparity' or 'parallax' of the images of O has components in both the horizontal and vertical directions relative to the plane PCC'. The recovery of the absolute position of O in space, given X and X', is therefore non-trivial. This is true not only because it is necessary to include the relative camera orientations in the computation but also because given, say, X the search for X' must be made along a line which in general will not correspond to the sampling pattern of the image.

The Special Case

Fortunately there exists a special camera orientation which makes the epipolar lines parallel. This is illustrated in Figure 2. The trick, if such it may be called, is simply to move the point P in Figure 1 to infinity causing the optic axes of the cameras to be parallel, and to ensure that the line FF' joining the lens centres is perpendicular to the optic axes, i.e. parallel to the image planes. As before the plane defined by the optic axes intersects the image planes along lines through C and C'. Again any point object O defines a plane OFF' which intersects the image planes along epipolar lines which include its images at X and X'. Now however, since the line FF' is parallel to the image planes, the epipolar lines are each parallel to FF', and hence to each other.

This situation can be put to good use for rectangularly sampled (and particularly raster scanned) images where the eipipolar lines can be caused to be parallel to the sampling direction/raster line. In particular, by causing the focal lengths and other camera/sampling characteristics to be identical, so that the images are coplanar, equivalent scene points fall on the same sample/raster line greatly simplifying the analysis. In this case the cameras are said to be "in correspondence".

THE ARCHIE PROJECT - CAMERAS IN CORRESPONDENCE

Broad Theoretical Basis

The ARCHIE project is based on the concept that industrially useful 3D imaging systems must be real-time (response < 1 second), reliable (hence as theoretically and constructionally simple as possible) and soon. The latter requirement implies that currently available hardware should be used wherever possible with dependence on technical advances reduced to the minimum.

Most image acquisition systems available off-the-shelf and which provide an electronically encoded signal are raster scanned. Video cameras are widely available and relatively cheap but suffer from unreliable image sampling geometry. Charge coupled diode (CCD) devices provide good geometric fidelity but are relatively expensive. Both types of device are capable of delivering single images, but they have the specific property of repeating or refreshing images at a high rate. This property should be exploited wherever possible.

Figure 3 illustrates the imaging geometry in plan. By similar triangles

$$d/f = e/CX = (e + s)/C'X' \qquad\qquad \ldots\ldots[1]$$

i.e. $\qquad\qquad d = s.f/\beta \qquad\qquad\qquad\qquad\qquad \ldots\ldots[2]$

where $\qquad\qquad \beta = C'X' - CX \qquad\qquad\qquad \ldots\ldots[3]$

Note that, since s and f are constants, for a given value of d the disparity β is single valued. Further, since s, f and d are all positive then β must be positive for all d 'in front of' the lenses. Finally, the value of β is independent of the height of O above or below the plane of the optic axes.

These facts are taken advantage of in the ARCHIE system by arranging the cameras in correspondence with scan lines parallel to the epipolar lines. The input images are scanned in synchronism. Since equivalent features lie on the same scan line in each image and appear consistently earlier in one image than in the other the measurement of stereoscopic disparity is reduced to one of simple timing provided that the necessary identification can be made.

Intrinsic Properties

Two classes of system property are of interest - timing, and depth/disparity relationships.

Pixel Dwell Time is the fundamental timing constraint for the electronics. A video image is scanned 50 times per second and contains, allowing for interlace, some 312 lines. Allowing for an image aspect ratio of width:height = 4:3 the dwell time for a square sample pattern is approximately 150 nanoseconds. The pixel sampling clock must therefore be of order 6.5 MHz.

Depth Resolution. Equations [1] to [3] embody the basic depth/disparity relationship for the system. As may be seen from [2] range, in terms of the perpendicular distance from the lens plane, is inversely proportional to the measured disparity. The relation is independent of position in the image enabling a conversion to be made by simple table lookup.

In the real case, however, the measured disparity can only take discrete values set by the sampling interval. This governs the precision with which depth determinations can be made. If the sample frequency is n per unit distance then an integer disparity may be defined by $\delta = \mathrm{trunc}\ (\beta.n)$. Substituting in [2] one obtains

$$d = x.f.n/\delta \qquad\qquad\qquad\qquad\qquad \ldots\ldots[4]$$

The range resolution, Δ, for any given distance d is then given by the decrease in the estimate of d as δ increases by one.

i.e. $\qquad\qquad \Delta = s.f.n/\delta - s.f.n/(\delta+1) = s.f.n/\delta(\delta+1) \qquad \ldots\ldots[5]$

Figure 4 illustrates the relationship between d and Δ for the particular case : s = 10cm; f = 6cm; n = 330 pixels/cm (a typical value for a CCD device). As can be seen from the figure, at a distance of 1m such a system is theoretically capable of depth resolutions of 5mm; at 2m this has deteriorated to 2cm, and at 4m to 8cm. To the right of the figure two alternative vertical scales (corresponding to the same curve) are presented. The leftmost gives the value of δ corresponding to a given depth, d; as d decreases δ increases rapidly. The rightmost gives the ratio Δ/d as a percentage; as a fraction of actual depth the intrinsic resolution is small for normal working distances and improves for increasing proximity. Similar curves are readily computable for other system dimensions and have the same form.

Working Volume. The curves of Figure 4 are based on assumed use of a CCD device for image acquisition. This device is assumed to have 400 pixels within a breadth of 1.2cm. Thus the acceptance angle of a camera using such a device is, for a focal length of 6cm, limited to $\pm5.7°$. The acceptance cones of two such cameras separated by 10cm first intersect at 50cm range and at one metre overlap by 10cm laterally. Thus the 'common area' between the two images at this range covers 200 pixels (corresponding to 0.5mm/pixel lateral resolution) each of which could, in principle, have its range determined to within 2.5mm. Thus any derived range representation for an object at this distance must in general consist of a series of contours widely spaced relative to the sampling frequency. In order to derive inclinations of faces considerable smoothing could be required.

THE ARCHIE SYSTEM - AN EXPERIMENTAL PROTOTYPE

Conceptual Model

A prototype is currently under construction in the Department of Computer Science at Strathclyde University to try to implement the above principles in a real system. The conceptual model is shown in Figure 5 though not all aspects are currently scheduled for construction.

A pair of cameras in correspondence are synchronised and their outputs sampled and digitised under the control of a system clock. The resulting serial bit stream may, by judicious use of shift registers and/or memories, have local operators applied to groups of adjacent pixels to pick out specific low-level scene characteristics. These characteristics may be correlated on a line-by-line basis to produce disparity estimates which may be 'tied' to the position of the generating characteristic in that image in which it appears later. A number of such characteristics may be extracted in parallel for a single frame and/or various characteristics may be extracted on specific frames of a short cycle. Such measures may be accumulated in a frame buffer and interpolated in synchronism with the image scanning, validity tags ensuring that measurements are not altered by the interpolation process. Further, since the tags may be assigned values according to the confidence which a specific measure holds, and since individual measurements be assigned a finite lifetime in terms of frames (longer for higher confidence) the contents of the disparity memory may be made to evolve in response to changes in the viewed scene. In this scenario static scene components will frequently be refreshed, evolving components or noise will not.

The interpolated disparity representation may be converted to a range representation (2.5D sketch) for subsequent use. Alongside the sketch other representations may be made available. For instance the differential of the range representation gives information on the magnitude and orientation of inclinations of surfaces and the location of breaks of slope (major edges); colour or texture representations may be extracted from the input data, etc.

In the conceptual model all these representations are simultaneously available to a microprocessor for higher level operations such as cross correlations, area measurements and so on. The microprocessor communicates with an intelligent knowledge based system which contains world knowledge in a form suitable for interpreting the various representations in terms of scene components.

The overall conceptual model coincidentally resembles simple schematics of a vertebrate visual sysem. In the biological system the retinae of each eye perform simple preprocessing of the input data; this data is further processed and sorted in its passage to the visual cortex where it appears as an array of high-level representations which are in turn worked on by the 'higher faculties' for scene recognition. For this reason, and because the system is experimental in nature, the project acronym was chosen to be ARCHIE - for Artificial Retina and Cortex Heuristic Imaging Experiment.

Current Status of the Hardware

An early ARCHIE prototype is in the final stages of construction and so far fully meets its anticipated performance. The cameras used are a pair of synchronised video cameras, chosen in part for their cheapness. The video signal is digitised to six bits of which only the four most significant are currently used. The features extracted for correlation are so far based solely on the variation in intensity along a scan line. This is one of the simplest and least robust schemes possible but it is easy to set up and performs well with simple contrasty scenes. Work has been performed on the application of 3x3 local operators to feature extraction with some success and will be reported elsewhere[8].

Simple correlation hardware has been built and tested, and demonstrated to be capable of generating disparity measures as required. The procedure relies upon the absence of confusing signals in the form of multiple features. This is easy to arrange for simple scenes; as the scenes are made more complex it will be necessary to utilise more and more specific and restrictive local operator feature identifiers.

An interpolating memory system is in the final stages of construction. This will permit the accumulation of disparity measures and the investigation of optimum interpolation schemes. The allowance for occluding edges provides an interesting challenge here.

No alternative representations or connection to an IKBS are yet scheduled.

CONCLUSION

Thus, at the time of writing a conceptual model of a complete 3D vision system exists as the end objective of an ongoing project. A start has been made on the construction of a simple demonstrator prototype which shows every sign of living up to expectations. Due to the modular, pipelined architecture stepwise refinement of the hardware should be possible by systematically enhancing weaker sections.

ACKNOWLEDGEMENTS

The author gratefully acknowledges support from the Scientific Equipment Fund of the University of Strathclyde for a pump-priming grant. Also the assistance of R.Watson and G.Haran during design and implementation of the hardware.

REFERENCES

[1]   Jarvis, R.A., "A Perspective on Range Finding Techniques", IEEE Trans. on Pattern Analysis and Machine Intelligence, Vol. 5, No. 4, pp 122-139 (1983)

[2]   Morasso, P., and Tagliasco, V., "Analysis of Human Movements : Spatial Localisation with Multiple Perspective Views", Med. and Biol.Eng. & Comput., Vol.21, pp 74-82 (1983).

[3]   Tsai, R.Y., "Multiframe Image Point Matching and 3-D Surface Reconstruction", IEEE Trans. on Pattern Analysis and Machine Intelligence, Vol.5, No. 2, pp 159-174 (1983)

[4]   Henderson, R.C., Miller, W.J., and Grosch, C.B., "Automatic Stereo Reconstruction of Man-Made Targets", Proc.Soc. Photo-Opt. Intrument Eng., Vol.186, pp 240-248 (1979)

[5]   Yakimovsky, Y., and Cunningham, R., "A System for Extracting Three-Dimensional Measurements from a Stereo Pair of TV Cameras" Comput.Graphics Image Proc. Vol.7, pp 195-210 (1978)

[6]   Baker, H.H., "Depth from Edge and Intensity Based Stereo" Ph.D. Dissertation, University of Illinois at Urbana-Champlain (1981)

[7]   Barnard, S.T., and Fischler, M.A., "Computational Stereo" Computing Surveys, Vol.14, No.4, pp 553-572 (December 1982)

[8]   Fryer, R.J., and Davidson, A., "Local Image Operators Applied to Binary Coded Video Data Streams", in preparation.

FIGURES

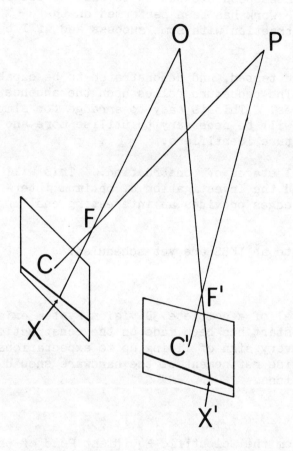

Figure 1
Stereoscopic Geometry - General case

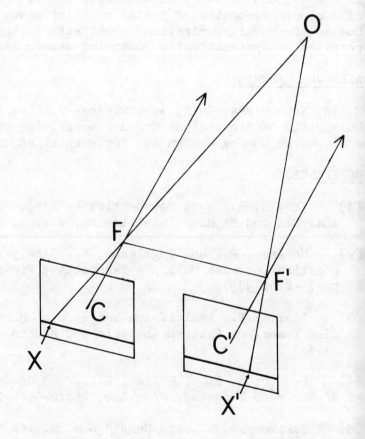

Figure 2
Stereoscopic Geometry - Special case

204

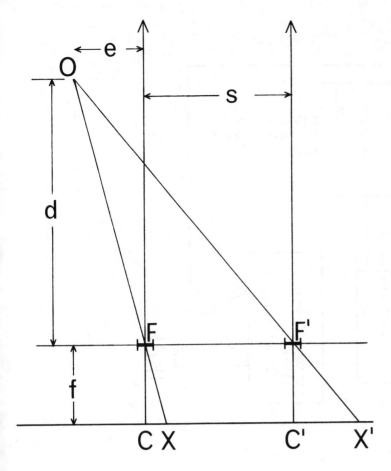

Figure 3
Projection of the 3D special case
geometry onto the plane of the
optic axes

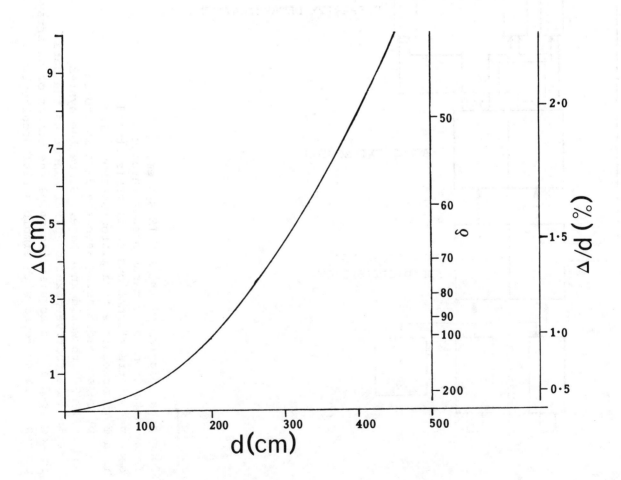

Figure 4   Depth/disparity relationships

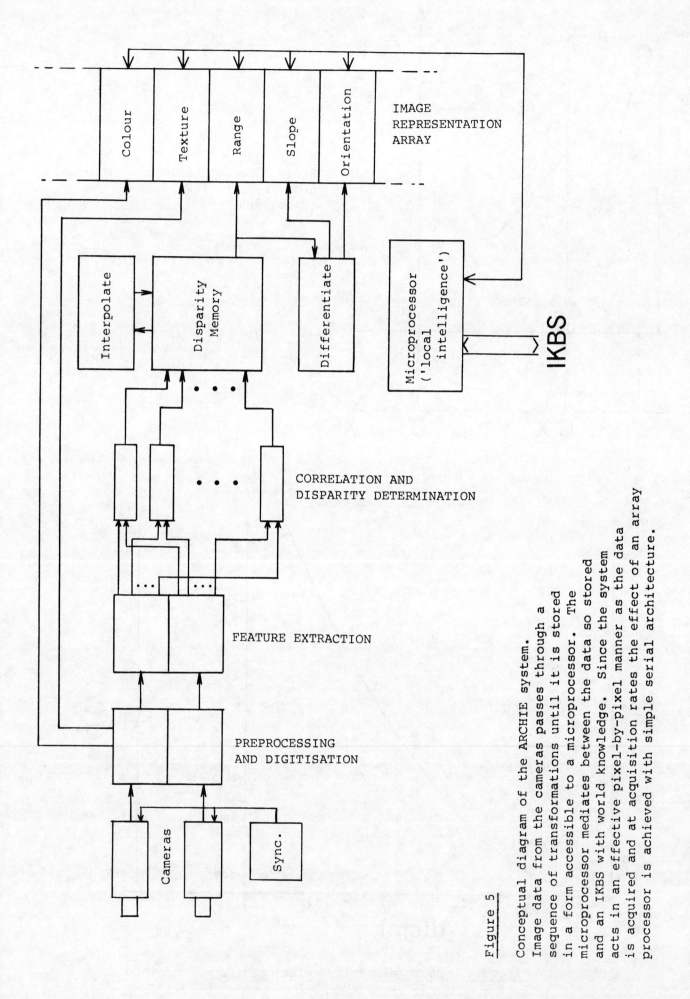

IMAGE
REPRESENTATION
ARRAY

Colour Texture Range Slope Orientation

Interpolate Disparity Memory Differentiate

Microprocessor ('local intelligence')

IKBS

CORRELATION AND
DISPARITY DETERMINATION

FEATURE EXTRACTION

PREPROCESSING
AND DIGITISATION

Cameras

Sync.

Figure 5

Conceptual diagram of the ARCHIE system.
Image data from the cameras passes through a
sequence of transformations until it is stored
in a form accessible to a microprocessor. The
microprocessor mediates between the data so stored
and an IKBS with world knowledge. Since the system
acts in an effective pixel-by-pixel manner as the data
is acquired and at acquisition rates the effect of an array
processor is achieved with simple serial architecture.

206

ADDENDUM

Since the time of writing the preceding paper the interpolating memory mentioned therein has been completed. This addendum presents the first results in their raw state. The memory allows the capture and interpolation of data in a subset of the entire image some 128 x 128 pixels in extent. Point measures of disparity are made and deposited into an appropriate memory element; thereafter these data points are not permitted to evolve. Values at intervening points are estimated by the repeated application of a running-mean computation based on 3 x 3 pixel sets.

This procedure has the advantage that it is simple to apply, but it does not produce satisfactory interpolation except for very dense arrays of measures. The effect of the algorithm as applied here is to propagate high measures over their surroundings except where they are locally pulled down by fixed points of lower value.

Figure 6 shows a view of the test object as viewed by one of the video cameras. The image quality is not high but is adequate. Figure 7 shows corresponding edges derived from each view. The geometric modification of the images due to the differing viewpoints is readily visible. Also on Figure 7 are streaks joining equivalent points on the two images. These streaks are generated by the setting and resetting of a flip-flop by the feature extraction circuitry; where they do not terminate in the right-hand image the extraction circuitry has failed to 'fire' on that path. The pixel clock is counted during a streak and its value clocked into the interpolating memory at its termination point.

Figure 8, which is a monochrome copy of an RGB monitor display, illustrates the results of a full data-collection/interpolation cycle. Eight grey values (colours) are displayed generated by the top three bits of the disparity at each point. The oblique bands are due to edge effects since in this early version the top and left edges are effectively held to zero whereas the right and lower edges are high; the strong propagation of the high values across the field, and even into the interior of the frog, is readily visible. The differing disparity values of points within the data are also visible however.

Figure 9 illustrates this variation in accumulated/estimated disparity and also allows the variation to be related to the position of the object. Only the topmost bit of the disparity data is displayed; overlaid on this is the position of every pixel at which, for this particular run, a disparity measure was made. The point to note is that the near and far rear feet of the frog are clearly distinguished in this data set.

Evidently there is much room for improvement in these results. A better interpolation scheme is required and this is under active development. Thereafter the emphasis will be on the reliable acquisition of point measure via more powerful feature extraction and correlation schemes. Even in its current state, however, ARCHIE shows promise of growing into a useful system.

Figure 6

207

Figure 7

Figure 8

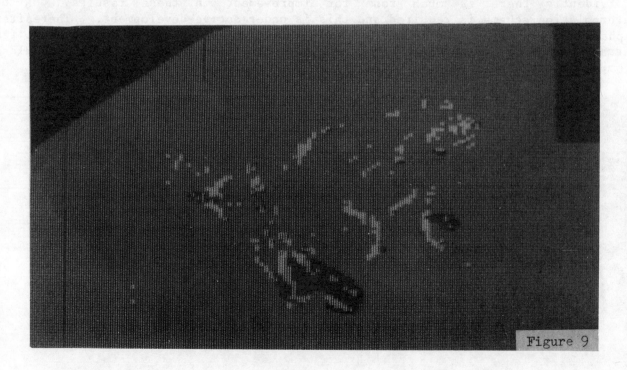

Figure 9

208

ROBOT GUIDANCE AND SENSORY CONTROL

A Practical Solution to Real Time Path Control of a Robot

P. Sholl

Unimation (Europe), UK

and

C. Loughlin

Electronic Automation, UK

ABSTRACT

This paper describes the real time path control capability, ALTER, of the UNIMATION robot control language VAL II. It covers the language, communications and implimentation (equations) aspects of ALTER and includes examples of internal and external path control. A specific example of the use of ALTER with the I-SIGHT 32 vision system for real time path control is included.

A PRACTICAL SOLUTION TO REAL TIME PATH CONTROL.

1.0   THE CONCEPT OF REAL TIME PATH CONTROL

1.1   Introduction

     The Unimation robot control language VAL II offers a wide range of
advanced capabilities to the robot programmer. Amongst these is a
comprehensive set of communications facilities covering four levels of
communications. These are:

     1. Supervisory communications
     2. Real time path control
     3. I/O ports
     4. Binary I/O

1.1.1 Supervisory Link To Host Computer. - A high quality serial link
featuring the rigorous DDCMP protocol. This may be used for off line
programming, networking, supervision etc.

1.1.2 ALTER Real Time Path Control. - An EIA RS-423 serial link
compatible with the EIA RS-232C standard. To allow the necessary
immediacy of data transfer and use by the robot this link has a
considerably less rigorous protocol than the supervisory link. This
level will be considered in detail below.

1.1.3 Input/output Via IOPUT/DAC Instructions And IOGET/ADC Function. -
This allows user defined input and output on

     1. Standard serial ports
     2. Standard parallel ports
     3. Standard analog ports
     4. Any user non-standard ports

1.1.4 Binary Input/output On Standard I/O Lines. -

     The advanced capability considered in this paper is real time path
control. This feature is known in VAL II as ALTER and it allows a range
of ways of exchanging robot motion and other data.

1.2 Alter Capabilities

     There are two distinct types of ALTER in VAL II.

A PRACTICAL SOLUTION TO REAL TIME PATH CONTROL.

1.2.1 External Alter - This is invoked by the ALTER instruction in VAL II. The instruction causes communication to begin over the dedicated serial link.

First a startup message is transmitted to inform the external computer that a path control communication is being initiated. It may also relay user specified information to the external computer. A response is required from the external device within a certain time limit or else an error condition arises.

If a satisfactory reply is received then normal ALTER messages are transmitted until path control is terminated by:

1. A NOALTER instruction is executed.
2. A DETACH instruction is executed. (This removes the arm from computer control).
3. A non straight line motion is attempted by the robot.
4. Robot control program execution ceases for any reason.

The output message consists of some control bytes, followed by an optional number of data bytes (which may be used to inform the external computer of the robot trajectory or status) or anything else, then another control byte and a checksum.

The response message to the robot contains the following information:

First a code to enable (when it has a negative value) a subroutine specified in the ALTER instruction to be called from the robot control program. The code value may be read by this subroutine as a means of error trapping, or used as a method of transfering non motion information from the external computer to the robot using the EXCEPTION function of VAL II.

Then a selection byte, followed by dx, dy, and dz and optional rx, ry, rz (decided by the selection byte) components are sent to cause the real time path control as outlined below.

Sample programs are given in section 4.

1.2.2 Internal ALTER - This ALTER facility requires no transmission of messages to an external computer. Instead VAL II expects an ALTOUT instruction to be executed every major clock cycle (currently 28ms) to internally pass control data to the robot control program. VAL II features a Process Control (P.C.) program which appears to the user as running in parallel with the robot control program; typically used for cell control. It may however also be used to calculate correction data to transfer to the robot control program via the ALTOUT instruction. Unlike the ALTER instruction no robot motion information or data list is output. ALTOUT is used in conjunction with the ALTER instruction which specifies the mode (world or tool) in which the data transfered is to be interpreted.

The Process Control program may therefore be used for instance to read data input on an analog input port (using the ADC function in VAL II), process the information according to a control algorithm, and transfer the correction data to the robot control program. An example of

211

A PRACTICAL SOLUTION TO REAL TIME PATH CONTROL.

the use of ALTOUT is given in the appendix.

2.0 IMPLIMENTATION OF ALTER IN VAL II.

The data sent from the external computer to the robot can be in 8 different modes. These are the combinations of the following:

1. World or Tool mode.
2. Absolute or incremental mode.
3. Short or long form.

1) determines whether the changes sent are interpreted as changes in the robot world coordinate frame or the robot tool coordinate frame.

2) determines whether the changes sent are absolute or incremental ie. cumulative or non-cumulative. See fig.

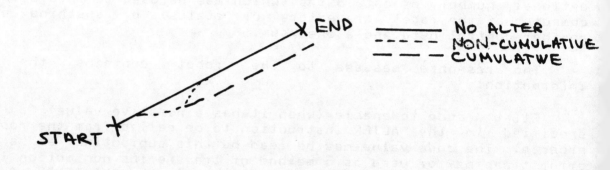

3) determines whether a rotational change is sent or not. All data is assumed to be short form unless it is long form, ie. a rotation correction is sent. This has implications to the amount of processing necessary in the robot computer. DEFINITIONS:

XYZ represents the position vector [x,y,z] ROT represents the rotation matrix dXdYdZ represents the ALTER position change data rXrYrZ represents the ALTER rotation change data Sum () represents the accumulated change data specified

2.1 World Absolute Short

In the case of World Absolute Short mode there is only one equation to be evaluated, and hence it is the least computationally expensive mode. The assignment equation is as follows:

Final XYZ := dXdYdZ + Unaltered XYZ   (1)

A PRACTICAL SOLUTION TO REAL TIME PATH CONTROL.

2.2   World Absolute Long

    In this mode the above assignment equation is used, but additionally
the rotation component is sent as well and has to be evaluated as
follows:

        Final ROT := rXrYrZ * Unaltered ROT   (2)

2.3   World Incremental Short

    In this case more computation is necessary because the changes have
to be accumulated. Even though it is a short mode there may be
accumulated rotation so the final rotation must be evaluated.

        New Sum(dXdYdZ) := Old Sum(dXdYdZ) + dXdYdZ   (3)

        Final XYZ := New Sum(dXdYdZ) + Unaltered XYZ   (4)

        Final ROT := Sum(rXrYrZ) * Unaltered ROT   (5)

2.4   World Incremental Long

    In addition to the assignment equations in World Incremental Short
mode, before the Final ROT evaluation it is necessary to find the new
accumulated rotation as follows:

        New Sum(rXrYrZ) := Old Sum(rXrYrZ) * rXrYrZ   (6)

2.5   Tool Absolute Short

    The ALTER information is transmitted in tool mode so it must be
converted to world mode by multiplying it by the unaltered rotation.

2.6   Tool Absolute Long

    As in the case of World Absolute Long above it is necessary to
evaluate the final rotation, after conversion from tool mode to world
mode by multiplying by the unaltered rotation matrix.

2.7   Tool Incremental Short

    This is similar to the World Incremental Mode except that the ALTER
dXdYdZ data is multiplied by the unaltered rotation to set it back to
world mode.

2.8   Tool Incremental Long

For the long mode, the ALTER rotational corrections are also converted from tool to world mode. This is done because it is easier for the user to send data corresponding to the original tool co-ordinate system rather than dynamically changing the tool co-ordinate system.

3.0   REAL TIME PATH CONTROL WITH I-SIGHT 32 VISION SYSTEM

3.1   System Overview

The basic operation of the system is that the camera takes in visual information, processes this picture to determine either position, distance or orientation information and then instructs the robot to move such that, for example, either the position is centralised, a constant height is maintained, or orientation is corrected.

The most important feature of the system is that steps are taken to maintain a stable reference situation. For example, if a hole is located in the top right-hand corner of the camera's field of view, then the vision system will instruct the robot to move right a bit and up a bit in an iterative procedure in such a way as to centre the hole within the field of view. Once the hole is centred, the robot then knows where it is. Until it is in the centre the system does not consider any further actions other than the operation of centring. Thus the system never considers the action "I want to insert a rivet at X location 15, Y location 8", but instead first centres and then undertakes the action "I want to insert a rivet in the hole above which I am now centred". The distinction between these two processes is not obvious but it is important as a means of obtaining high accuracy and good performance.

3.2   Real Time Path Control

Visual feedback can be used in two modes which might appear to be different in operation but which are, in fact, identical. To illustrate these modes, we will take two examples. In the first, which might be edge or contour following, the robot has basically been told to move from A to B. It uses visual feedback to maintain a constant distance and compensate for variations in the expected path. As the robot is moving from A to B cumulative alter data (Dx,Dy) that it received will cause an immediate displacement of the robot arm by Dx,Dy and also modify the position of the target location B by the same amount.

In the second example which is used in applications such as hole location, the points A and B are at the same place i.e. the robot is stationary. Cumulative alter data that is received will therefore cause an immediate displacement of both the robot's position and the location of point B (which are the same thing) by the amount Dx,Dy. Thus the processes of path and position modification are identical in operation.

3.3   The Camera System

Two types of optical sensors are to be described. Both use lasers and both only detect the physical and not optical characteristics of the part in view. Indeed, the camera does not see the object at all but simply the laser light reflected off the object's surface. This has a number of benefits; the first is that the colour, light intensity or visual contrast of the picture becomes completely irrelevant and the same information can be derived off black, white or grey surfaces. The second advantage is that we are measuring the physical location of the edge/surface and not the rather incidental occurrence of a change in light intensity. As it is almost always the physical boundaries that we wish to detect, it is obviously better that they be obtained 'first hand'.

3.4   Single Spot Laser Probe

The use of the miniature camera in conjunction with a collimated solid state laser enables the vision system to act as a ranging device. The laser is a solid state continuous wave 2mW laser (MULLARD CQL13A) operating in the near infra-red at a wave-length of 820nm which coincides approximately with the peak sensitivity of the camera (Fig.1). An infra-red filter (KODAK 88A) is used within the camera in order to attenuate the effects of illumination in the visible spectrum.

The laser is mounted vertically with the camera adjacent to it but at an angle to the laser axis which is set according to the range and depth of field that is required (Fig.2).

P=camera/laser offset

$$r = \frac{P}{\tan \alpha}$$

r=range
d=depth of field
α=camera/laser angle

$$d = \frac{P}{\tan (\alpha - \theta)} - \frac{P}{\tan (\alpha + \theta)}$$

θ= camera viewing angle
Example: P = 50mm θ = 15 α = 30
            r =   86mm
            d = 136mm

The laser projects a dot of light onto the surface and it is this dot that is picked up by the camera. The picture seen by the camera will contain the single blob and it is the vertical position of the blob that defines the distance of the laser from the surface. If the blob is in the vertical centre of the picture then the surface will be at a distance of 'r'. It should be noted that the vertical position of the blob is not linearly proportional to distance but this does not materially affect the operation of the system as the feedback loop will always act to keep the robot at a distance 'r' from the surface.

The parallel beam projected by the laser is 5.4mm in diameter which is large by collimated laser standards. This is important, however, in view of of the increased accuracy that is made possible in the determination of the beam (Fig.2.).

The low resolution of the camera also requires that a broader beam be used as it would otherwise be possible for the laser light to fall not on a photosite but rather on the warp and weft of the silicon structure and thereby not show up on the camera image.

3.5 Applications Of The Single Spot Laser Probe

3.5.1 Contour Following/Height Sensing - If we take an example where the robot is required to lay a track of glue in a straight line across a curved surface, without height sensing it would be necessary to teach the robot a large number of points across the surface so that a constant height could be maintained. Typically, this might take in excess of an hour to programme, would be prone to operator error and would have to be repeated for all other glue paths across the surface.

Using the height-sensing feedback from the vision system, it is only necessary to teach the start and finish points to the robot as all height variations would automatically be detected by the height sensor and the robot path would be modified accordingly. This would cut programming time down to about 5 minutes, eliminate operator error and also compensate for variations in the position of the surface caused by jig tolerances.

3.5.2 Object Location/Dimension Checking - In this application example, we wish to check certain dimensions of a welded frame structure. The vision system is low resolution and therefore cannot be used by itself to obtain precise measurements. However, a robot such as the Unimation PUMA 560 has a resolution of 0.1mm (4 thou) and can therefore be used effectively as a measuring instrument provided it can sense what it is measuring.

Assume that the welded frame is a simple rectangular construction and that we wish to take a number of key measurements in order to determine its quality. The process of obtaining each measurement is that the robot starts with the single spot scanner pointing at free air and then moves in towards the expected location of the edge and keeps moving until the vision system detects the edge. The robot will then servo under visual feedback until the laser dot is in the centre of the camera's picture. Once this has been achieved, the robot then knows the precise location in three dimensions of the edge of the frame. Further measurements can be undertaken in the same way until all relative dimensions of the frame are known and can therefore be used to determine whether the position, size and squareness of the frame is to within manufacturing tolerances. These calculations can all be undertaken by the robot control system and an accept/reject decision made.

The key to this application is that the measurement system is the robot itself with the vision system acting solely as a sensor or guide that tells the robot when to take a measurement.

216

A PRACTICAL SOLUTION TO REAL TIME PATH CONTROL.

3.6   Line Of Light Laser Probe

A lot of work (e.g. reference 4) has already been done on the detection of edges using a projected line of light and a camera system that analyses the angles and curvature of the projected line and detects discontinuities in it which represent the edge or border in question (Fig.3). This work has primarily concentrated on the automatic following of edges and seams during welding operations. This paper will aim to identify new application areas for the line of light laser techniques.

The line of light is generated by a laser and can be produced either by the use of a cylindrical lens (e.g. a glass rod) in front of a collinated laser beam or more conveniently by using the direct output of a rather special solid state continuous wave laser whose output beam diverges at 50 in one axis but only 6 in the other so that effectively projects a diverging line of laser light onto a surface. The particular device used in our research (STC LC06-03) is a 6mW laser with peak emission in the near infra-red at 850 nm. One major advantage of this line of light laser probe is its small size (about the same as a cigarette lighter) and its low weight which mean it can be used in very confined or restricted areas.

The single spot laser probe enabled surface measurement in one axis to be made (i.e. distance). With the line of light laser it is possible to determine surface angle from the tilt of the perceived line and also detect the position of edges more rapidly than is possible with single point inspection.

3.6.1   Edge Following - If the line of light is projected so that it runs orthogonally to the edge (Fig.3) then the camera will only see part of the line that hits the top surface with the end of the line marking the edge. The vision system processes this picture and then sends offset information (normally in just one axis) to the robot in order to keep the end of the line (the edge) in the centre of the picture. If the surface also undulates up and down this will show up as a vertical displacement of the line. This height information can be fed back to the robot in a similar manner to the contour following example.

In this way it is therefore possible to accurately follow an edge in three dimensions using very simple, cost-effective hardware.

3.6.2   Object Location - In another example let us assume that cylindrical objects are presented to the robot in pallets but there is a variation in the absolute location of the cylinders caused by jigging and packing tolerances. The cylinders are very delicate and must be accurately located ($\pm$0.1mm) before they can be picked up by the robot's gripper.

A high resolution overhead vision system might well be able to identify the position of the cylinders but would not be able to offer sufficient accuracy of location due to parallax errors, optical distortion, scaling errors and mismatch between the camera's and the robot's co-ordinate systems.

A PRACTICAL SOLUTION TO REAL TIME PATH CONTROL.

By using a line of light laser in a similar manner to the previous edge following application it is possible to locate precisely four positions (two in X axis and two in Y) that will enable the centre of the cylinder to be accurately determined (Fig.4). The principle limiting factor on accuracy in this application will be set by the robot rather than the vision system for the same reasons as discussed in dimension checking.

4.0   SAMPLE PROGRAMS

4.1   Simple External Alter Program

```
PROGRAM alter
;basic alter test
        CALL initialise ;subroutine to initialise variables
        MOVES #safe ;go to safe precision point
        DO
        ALTER (channel, mode, sub, level) ;start altering
        ;ALTER servos the tool about #safe
        DELAY 0.5 ;move nowhere
        NOALTER ;stop altering
        UNTIL FALSE ;loop forever
END


PROGRAM frame.change
;redefines robot frame according to data received from alter processor
        CALL initialise ;subroutine to set up variables
        ;loop to move to two approximate points
        FOR point.count = 0 TO 1
            MOVES frame[point.count]
            BREAK ;cease motion
            ALTER (channel, mode, sub, level) ;start altering
            ; when sensor indicates point found ALTER calls subroutine
            ; "sub" which notes current arm position and sets found
            ; flag
            DELAY 0.5 ;move nowhere
            DO
                ;nothing
            UNTIL found == TRUE ;end of loop
            NOALTER ;stop altering
        END ;next point.count
        CALL redefine ;call subroutine to redefine robot frame
END
```

4.2   Simple Internal Alter Programs.


```
PROGRAM internal
; moves arm between two locations altering according to P.C. program
        CALL initialise; subroutine to set up variables
        mode = world.cumulative
        PCEXECUTE alt,-1,0; start P.C. program looping
        DO
            FOR count =0 TO 1
```

A PRACTICAL SOLUTION TO REAL TIME PATH CONTROL.

```
                    ALTER (internal,mode,sub,1); start internal ALTER
                    MOVES loc[count]
                    NOALTER
              END
          UNTIL FALSE; loop forever
END
PROGRAM alt
; accepts alter information from the teach pendant
        except = 0
        dx = 0
        CASE PENDANT(1) OF
            VALUE rec.button:
              except = -1; set exception code to end programs
            VALUE jt1.minus:
              dx = -PENDANT(3)*factor; dx magnitude from speed control
            VALUE jt1.plus:
              dx = PENDANT(3)*factor; dx magnitude from speed control
            ANY
              dx = 0
        END
        ALTOUT except,dx,0,0,0,0,0; ALTER output to main program
END

PROGRAM sub
; specified alter exception subroutine
        PCEND; kills P.C. program
        HALT; kills robot control program
        RETURN
END
```

5.0   CONCLUSION

      The examples given bear witness to the power, versatility and simple
language instructions of VAL II's ALTER. The implications for real time
path control which includes robot trajectory information as in ALTER are
many as it stands, and yet ALTER provides user programmable structures to
allow expansion to meet future needs.

      It has been shown that, through its ALTER facility, VAL II offers a
versatile, user friendly and a practical solution to real time path
control of a robot.

6.0   REFERENCES

1. Loughlin, C. and Hudson, E. 'Eye in hand Robot vision' 2nd Rovisec,
Stuttgart, 1982 pp 264-270.
2. Loughlin, C. and Morris, J. 'Applications of eye in hand vision' 7
th B.R.A., Cambridge, England, May 1984.
3. Hill, Rosen et al.   'Machine intelligence research applied to
industrial automation' Eighth report N.S.F. Grant April 1975 - L3074.
SRI project 4391. SRI International, Menlo Park, California, August
1978.
4. Clocksin, W.F. and Davey P.G. 'Progress in visual feedback for
robot arc-welding of thin sheet steel' 2nd Rovisec, Stuttgart 1982.

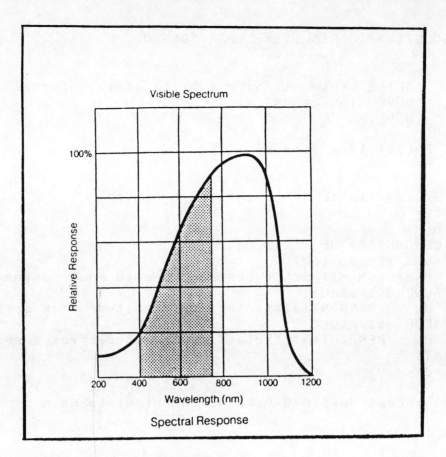

FIG 1

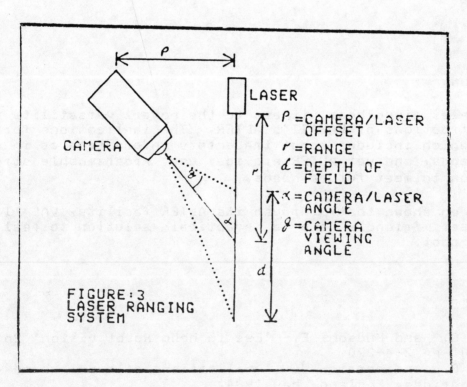

FIG 2

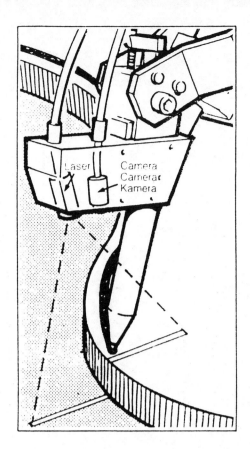

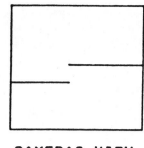

CAMERAS VIEW

FIGURE 3

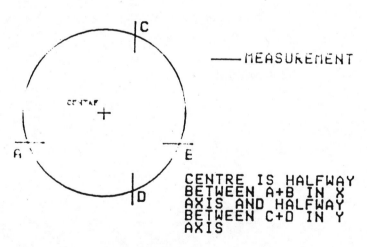

CENTRE IS HALFWAY
BETWEEN A+B IN X
AXIS AND HALFWAY
BETWEEN C+D IN Y
AXIS

FIGURE:4 CYLINDER LOCATION

Location of Work-pieces and Guidance of Industrial Robots with a Vision System

J. Amat, A. Casals
and
V. Llario
Facultat d'Informatica de Barcelona, Spain

ABSTRACT

Location of work-pieces in industrial environments by means of vi
sual perception needs the use of pattern recognition techniques that
rarely allow their use in real time. The paper which is presented describes a specialized image processor that detects an object's appearance in complex scenes and obtains the object's coordinates and orientation in a very short time (100 ms). A low cost microcomputer processes
the information about the object obtained by isolating it from its background and elaborates the needed data for guiding the robot. The image
processor detects object's appearance on the scene by the comparison of
the obtained image, 128x128 pixels, with a scene reference image. This
image is continuosly actualized to adapt itself to lighting fluctuations
or environment condition variations. The obtained information consisting
in diferences between images is filtered to avoid false detections.

1.- INTRODUCTION.

A wide number of actual industrial Robot application are "pick and
place" operations. The task to be performed by the robot in such applica
tion is very simple but the environment must be conveniently arranged.
If the system is provided with some kind of perception a great improvement is obtained [1,2].

One of the applications of visual perception is location of objects. One of the most common problems we run into in industrial applica
tions is that of picking up randomly oriented parts on a conveyor belt
in a predeterminated way. If visual perception is added to the system it
becomes more flexible and, as a result, the number of possible applica-

tions grows enormously.

The trouble with vision systems arises from the big amount of data that has to be processed if it has to be used in a real-time general purpose system. That's the reason why so many studies have been done in this area imposing restrictions which reduce generality but also cost and processing time [1,2,3].

The vision system which is exposed in this paper has the aim of bringing a solution to some robot applications that need real time control [4] at a low cost.

The goal of the work which is presented is to locate a unique known part appearing in the scene with unknown location and impredictable orientation.

The detection of a piece is accomplished by comparing a usual scene image with the present image at the moment of appearance of the part. The substraction of both images separates the part from the background. During the following step a window is opened around the part in order to delimit the zone of the image to be considered for further processing [5]. In order to work in industrial environments, usually with adverse conditions, and to be able to assure good results in the guidance of the robot the whole image inside the window is reconstructed and then the orientation of the part is determined.

As processing time is an important aspect in the system, the orientation of the part is determined by means of a looking up table built up during the training phase. The grasping points are also obtained in the same way.

The low cost system tried out allows guidance of the robot to the part appearing in the scene in a time which usually is lower than 0.6 seconds.

2.- PART EXTRACTION.

The block diagram of the vision system which is developed to guide the robot is shown in figure 1.

The goal of the extraction function is to distinguish beetween the part and the background in order to eliminate irrelevant information as well as noise in the video signal.

There are many images that can have varying background gray-levels which might interfase with the assigned task. In general, if no special lighting conditions are considered, the task of detection may be strongly disturbed.

Good results have been obtained trying to solve this problem in some applications by adapting the image of background to a periodic image actualisation using the algorithm.

$$M_j(x,y) = I_j(x,y) - M_{j-1}(x,y)$$

where $M_j(x,y)$ es the matrix of the image in memory, $M_{j-1}(x,y)$ is the one corresponding to the previous image, and $I_j(x,y)$ the present image obtained from the TV camera.

This filtering process allows continuos absortions of background variations and giving the convenient threshold it's possible to detect the appearance of a new part.

The prototype tried out in the laboratory performs this filtering and detection by means of a special purpose circuit. Fig. 2.

This structure allows the obtention of the part in 20 ms by substrac tion of the background.

The results are shown in fig. 3, 4 and 5.

Some problems arise from this method because sometimes it is not pos sible to extract all the points of the part due to the coincidence of gray levels of the part and background.

3.- IMAGE RECONSTRUCTION.

As it has been mentioned above, during the background substraction phase some problems may arise from gray level coincidence. The method tried to reconstruct the image with good results consists of a specific lighting system based on three light sources electronically controlled which are projected on the scene, figure 7.

Every time a source falls upon the part, there is a shadow projection from the object. Now, substraction of the background is performed with M_j and the three new images. Then a gradient operator is applied to the image which has been obtained from the substraction between background and the three images that contain the generated shadows. Among the changes detected by the gradient operator only those corresponding to the inner contour are considered the resulting image corresponds to the contour of the part.

Fig. 8 shows a situation with gray level coincidence as mentioned be fore. As a consequence, the substraction process obviously gives the bad results shown in figure 9, although the whole object can be reconstructed from its boundary. Fig. 10.

4.- LOCATION OF THE WORKPIECE.

Once the images have been substracted and the part is detected it is necessary to locate it. Thus it is possible to give the grasping points to the control unit in order to generate the correct commands to control the arm motions. In order to determine the grasping points both location and orientation of the part have to be calculated. The location of the part is achieved by generating its boundary rectangle. The generation of this window is performed by the μ C in a very short time. Moreover, the boundary rectangle defines the needed contrains on the image and thus re duces the process time. This location doesn't give enough information to guide the robot. It is also necessary to calculate its orientation and to define the points of prehension.

The table used to determine the orientation of the part is built up during the train-in phase by calculating two descriptors for each orien tation. Thus the process of determination of the orientation will con sist of the comparison between all these descriptors with those corre sponding to the part which is detected. Then the nearest neighbour rule is applied. Once the orientation is determined automatically the points of prehension are read.

5.- LEARNING PHASE.

The shape descriptors are obtained by showing the model to the vision system. This task is performed by putting the object on a rotating base which rotates under the control of the μC. Each time a picture is taken the μC reads the position of the base and calculates the shape descriptors.

Once the μC has calculated the boundary rectangle, the external hardware deals with the task of calculating, during next frame, the distances P and Q.

P_n is the number of 0's accumulated during scanning of line n inside the boundary rectangle till the first pixel corresponding to the object is detected. Q_n is the number of 1's corresponding to the object accumulated during scanning of line n.

These two parameters are read by the μC at the end of each line (every 64 μs) in order to calculate the two employed descriptors, which are.

$$D_1(\theta) = \sum_{n=0}^{N} n.P_n$$

$$D_2(\theta) = \sum_{n=0}^{N} n.Q_n$$

The look-up table is built up from this pair of descriptors $D_1(\theta)$ and $D_2(\theta)$, and the output gives the information about the grasping points.

The grasping points depending on the value of the orientation θ and also on the boundary rectangle, are calculated by the μC at the end of the training phase. The points are shown in a reference orientarions by the user.

6.- EXECUTION PHASE.

Every time a known part is detected on the scene the μC calculates the boundary rectangle in 20 ms. During the next frame $D_1(\theta)$ and $D_2(\theta)$ are calculated simultaneously to the scanning. This operation needs 20 ms. Finally the μC needs approximatively 50 ms to look for the nearest neighbour on the table, getting the values of the grasping points coordinates. Thus the system is able to get the position and orientation of the part in a time lower than 100 ms if it isn't necessary to use the special lighting system to detect the whole object. When this is not possible and the special lighting is required, three consecutive image acquisitions are needed to reconstruct the part. In this case the operation time is about 400 ms.

7.- CONCLUSIONS.

The vision system which is described in this paper was designed to determine the orientation and grasping points of randomly oriented parts on heterogeneous backgrounds. It's a high speed lowcost system suitable for a lot industrial applications.

Parts without irrelevants contour features are inadequate to the system. It's a versatil system and the time of learning new parts is quite short

Acknowledgements:

This system has been developped upon research support of the CAICYT. We also gratefully acknowledge the contribution of Miguel Las Heras.

BIBLIOGRAPHY

1 - E. Lundquist. "Robotic vision systems eye factory applications".
    Mini-Micro Systems. Nov. 1982.

2 - W.E. Iversen. "Vision systems gain smarts". Electronics. April 1982.

3 - T.O. Binford. "Survey of Model-Based Image Analysis Systems". The
    International Journal of Robotics Research.

4 - K. Arm bruster. "A very fast vision system for recognizing parts
    and their localization and orientation". 9 International Symposium
    on Industrial Robots, 1979.

5 - P. Malineu, A. Niemi. "Reduction of visual data by a program con-
    trolled interface for computerized manipulation". 9 International
    Symposium on Industrial Robots, 1979.

6 - R.C. González, R. Safabakhsh. "Computer Vision Thecniques for Indus
    trial Applications and Robot Control". Computer. December 82.

7 - J.F. Jarvis. "Research directions in industrial machine Vision: A
    workshop summary". Computer. December 82.

8 - W.K. Taylor, G. Ero. "Real time teaching and recognition system for
    robot vision". The industrial robot. June 1980.

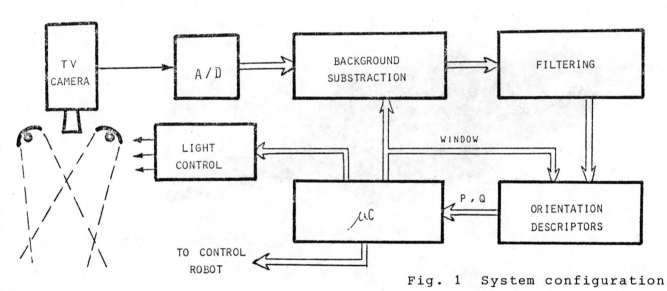

Fig. 1   System configuration

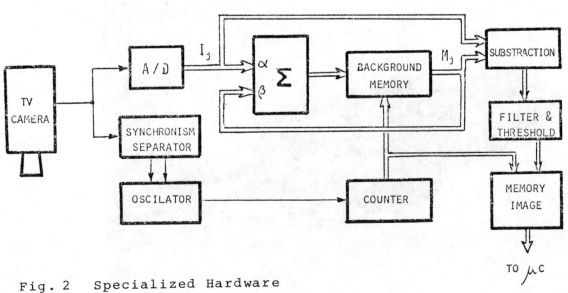

Fig. 2   Specialized Hardware

Fig. 3 Digitized
     image

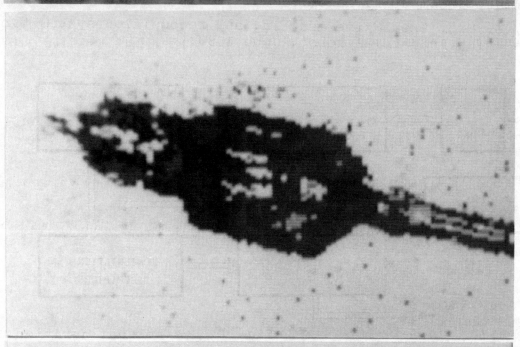

Fig. 4 Result afte
     substractic

Fig. 5 Image afte
     the filter-
     ing process

228

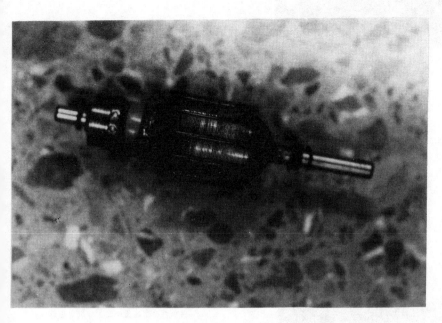

Fig 6.Vew of the part visua
lized in figure 3 on
a non homogeneous
background.

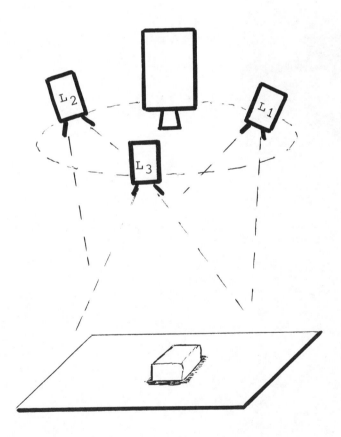

Fig. 7   Distribution of the
lights in order to
produce shadows around
the part to facilitate
its extraction.

Fig.8 Scene containing
a white part on a
background with
strips without
special lighting

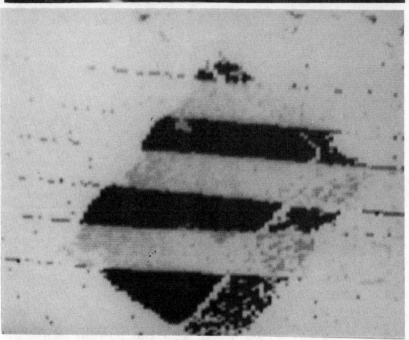

Fig.9 Bad results obtained
for this case after
background substrac-
tion.

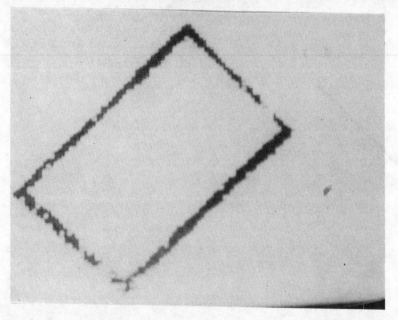

Fig.10 Boundary obtained
after the projectic
of the three source
of light and the
filtering of shadov

Sensor technology with Robot-Mounted Sensors

C. Meier

Siemens AG, W. Germany

Introduction

The extension of the field of application of industrial robots is very
dependent upon the provision within the robot control of freely adaptable sensor signal processing techniques. /1, 2/ describes how these
sensor functions can be realised to be able to provide very specific
functions (speed regulation, on-line path correction, adaptable signal
processing) which are at the same time simple and uncomplicated for
the user to implement. It is important to bear in mind the on-line
sensor functions are part of a closed position control loop or other
general process control loop. The loop is formed by the sensors, the
control, the sensors with the drives and the robot /3, 4/.
The use of sensors mounted to the hands of robots adds another dimension to the subject; such a sensor will be expected to work anywhere
within the working space of the hand and the resulting geometrical
factors must be taken into consideration. This paper shows using three
examples how hand-mounted sensors can be used to tackle the problems:
- recognition and location of parts
- three dimension-search
- path control

1.   Hand mounted optical sensor

Whether an optical sensor, such as a camera, is mounted on the hand of
the robot or mounted to some non-movable fixture influences very
greatly the complexity of the exercises that can be tackled with such
a visual sensor. A camera mounted on the hand can view an object from
a wide range of angels, and the distance from camera to object can be
selected to provide a favourable picture resolution.
The procedure for commissioning an hand-mounted optical sensor will
now be described. This procedure would be applied to the sensor system
"Videomat" when working in conjunction with the Robot Control RCM.
The commissioning is divided into the following phases
- calibration of the sensor-robot system
- teaching the sensor
- programming the robot
- test and automatic phase

In calibration, the position of the camera with respect to a point in
the hand co-ordinate system of the robot is determined and stored in
the control.
In the teach phase the sensor will be taught what the object, the
sensor will be later required to identify, looks like.
In the programming phase the robot control programm will be produced.
The distances in this programm are relative to optical center of the
part which has been taught.

In the test and automatic phase the part can be placed anywhere within the permitted area and the sensor should be able to determine where exactly the part is. This phase can be explained by means of an example in which the object, whose position and orientation is to be determined, is situated on a sloping plane. The angle of the slope is known (Figure 1.1).

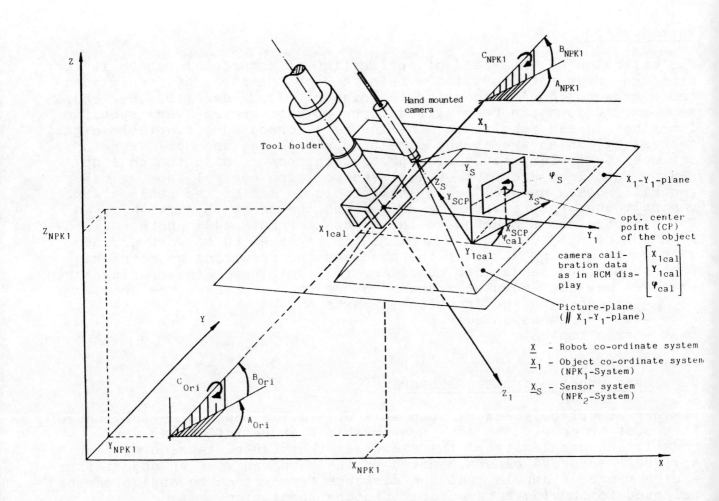

Fig. 1.1: Hand-mounted camera with a free orientated view in the world space

The robot brings the sensor to the viewing position. The viewing position is such that the line of symmetry through the sensor is at right angles to the surface of the object presented to the sensor. The collected sensor data will all be with respect to this viewing position. To enable the Robot to pick up the object, for example, the sensor data must be transformed into the main co-ordinate system of the robot.

This is accomplished with two zero point offsets with corresponding orientation twists. The first transforms the data to the co-ordinate system of the hand and the second from the hand to the main co-ordinate system of the robot.
The sensor provides the particular data for the part, that is offset and twist, for the first transformation. The data or the viewing point itself, also offset and twist, are fixed known quantities. These data ar required in the second transformation.

2. Rapid scan

If a process is to be carried out on a object whose exact position is
not known, it is necessary to scan the area when the part is likely to
be to determine its exact position. Once a robot is equipped with some
kind of sensor, either tactile or non tactile, able to detect the pre-
sance of an object in its vicinity, the robot can be used to make such
a scan. The time for the scan is of great importance and therefore the
speed of the scanning movements should be as high as possible, particularly if the robot has to make several scans.

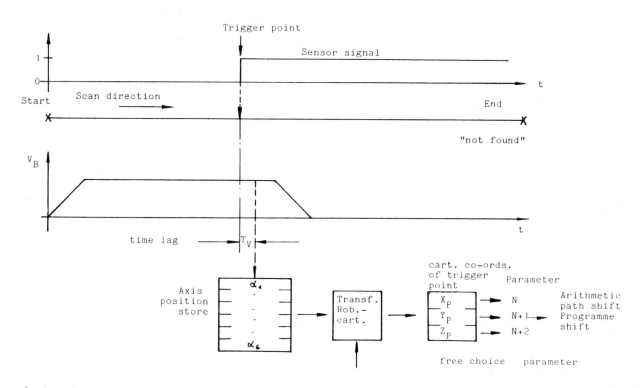

Fig. 2.1: Diagram to show the procedure for determing the co-ordinates
of the sensor trigger point

Figure 2.1 shows how such a scan is made. The change in status of the
sensor signal leads, after a small time lag which is inevitable, to the
storage of the current robot co-ordinates and to a halt in the scan.
The time lag must be kept as small as possible, the error introduced
by the time lag is the product of the lag and the linear speed V_B

$$F = T_V \cdot V_B \qquad\qquad (2.1)$$

The maximum value of V_B is defined once the time lag and maximum acceptable error F are known. Assuming that the position of the sensor
in relation to the robotflange is known, the robot co-ordinates "caught"
when the sensor triggered can be applied to the co-ordinate transformation to determine the X, Y, Z co-ordinates of the point at which the
sensor triggered. These co-ordinates can then be used by the user
program for the workpiece.
(The orientation data are usually not important.)

By combining several scans it is possible to determine the position
edges and to calculate surface areas which in turn makes it possible
to determine the position of points which cannot be directly measured,
for instance, corners see figure 2.2 and table 2.1

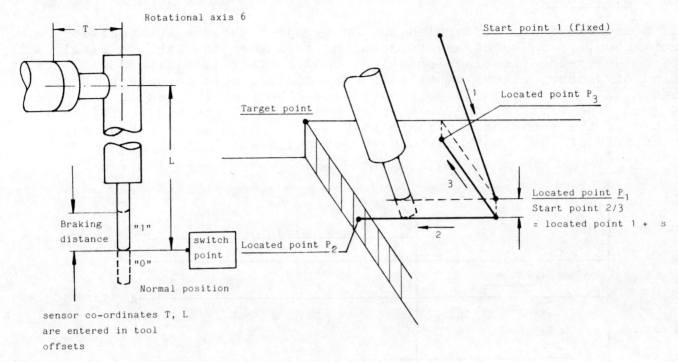

Fig. 2.2: 3 D scan to find a corner point on a flat surface.
In the scan area the edges are straight.

		Parameter information
1.	Scan 1	$\underline{P}_1$
2.	Calculation for start point 2/3	$\underline{St}_{2/3}$
	$\underline{St}_{2/3} = \underline{P}_1 + \underline{S}$	
	$\underline{S}$ can be easily set by the user	
	Return to start point 2/3	
	Scan 2	$\underline{P}_2$
	Return to start point 2/3	
3.	Scan 3	$\underline{P}_3$
4.	Calculation for target point	
	$\underline{P}_{target\ point} = \underline{P}_1 + \underline{P}_2 + \underline{P}_3 - 2\,\underline{St}_{2/3}$	
5.	Set the nero offset to $P_{target\ point}$	

Table 2.1: Sequence of events for 3 D corner search

3. On-line path correction

3.1 Sensors without advance control

In many technological applications the on-line regulation or
modification of a contour is necessary. /2, 4/ shows how to
realize a sensor system which can subsequently be programmed by
the user for path correction. Such a system has two independent
sub-systems, one correspondig to a regulation in a vertical plane
(higher-lower correction), and the other correspondig to a regulation
in a horizontal plane (further to the left, further to the right
correction).

The dynamic behaviour of a closed control loop for a sensor measurement is characterized by a system delay-time T_I, as can be seen by
taking approximations at the working point. The time T_I includes the
control clock T_{IPO} (usually the interpolation clock) and the substitute time T_{RS} of the optimized speed control loops; it can be approximated by

$$T_I \approx T_{IPO} + T_{RS}. \qquad (3.1)$$

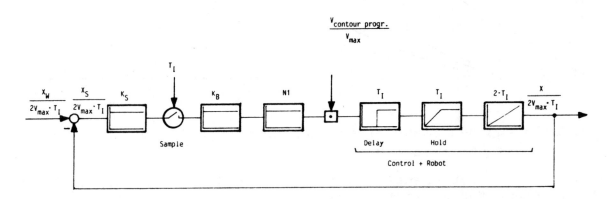

Fig. 3.1: Structural diagram for the on-line contour correction

Fig. 3.1 shows the structural diagram of the position control loop,
where K_B is a free amplification factor and

$$N1 = \tan\alpha_{max} \qquad (3.2)$$

$$K_S = \frac{V_{max} \cdot 2 \cdot T_I}{X_{s\ max}} \qquad (3.3)$$

(X_{smax} = linear measuring range of the sensor

V_{max} = maximal contour velocity)

With the adjustment

$$K_{So} \cdot K_{Bo} \cdot N1 = 1,$$ (3.4)

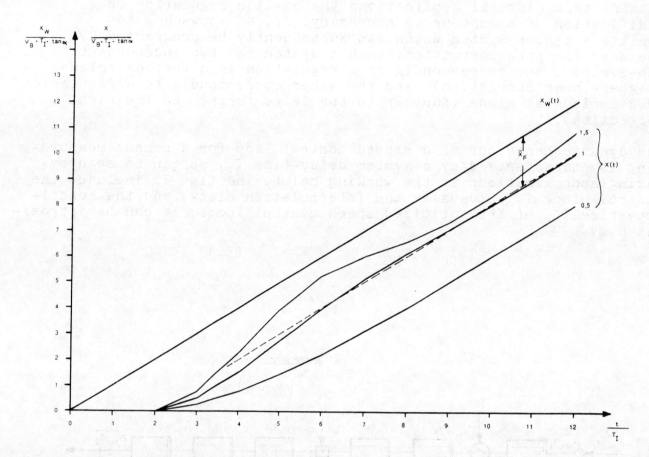

Fig. 3.2: Follow-up behaviour of the sensor correction with
parameter $K_S \cdot K_B \cdot N1$

Curve 1 of Fig. 3.2 represents the follow-up behaviour of the working
point of the robot for a limit curve

$$x_w = V_{max} \cdot t \cdot \tan\alpha_{max}$$

of slope $\tan\alpha_{max}$. Here the quasi-stationary contour error, in terms
of and the actual contour velocity V_B, is

$$\bar{x}_{F_o}\Bigg|_{\frac{dx_w}{dt} = const.} = V_{max} \cdot 2 \cdot T_I \cdot \tan\alpha_{max} \cdot \left[\frac{\tan\alpha}{\tan\alpha_{max}} \cdot \frac{V_B}{V_{max}} \right]$$ (3.5)

Assuming that the drive range of the sensor is unlimited, the change
in the amplification factor K_B is

$$\bar{x}_F = \bar{x}_{Fo} \cdot \frac{K_{Bo}}{K_B} \cdot$$ (3.6)

With decreasing K_B, the effectiveness of the control loop diminishes;
the transient behaviour of the control quantity turns out to be
determined by a strong damping.

3.2 Sensors with advance control

In many applications, sensor measurements cannot be taken at the working point itself but only at an advanced test point. This has significant effects on the regulation, which are to be studied next.

Suppose that the orientation of the tool with respect to the programmed contour does not change because of sensor corrections. Then the actual position of the sensor can be represented by $x(t)$, where $x(t)$ denotes the current deviation of the working point from the programmed contour. Given a constant lead distance S_{Vor} and a contour velocity V_B, which provisionally is assumed to be constant, too, the sensor measures the desired contour by a lead time

$$T_{Vor} = \frac{S_{vor}}{V_B} \qquad\qquad (3.7)$$

Hence the sensor signal becomes

$$x_s(t) = x_w(t + T_{Vor}) - x(t). \qquad\qquad (3.8)$$

First suppose that the processing of the sensor signal is invariant. Then the behaviour can be described by adding an ideal lead component with negative delay-time $-T_{Vor}$ to the command branch: Fig. 3.3, mechanical lead.

For varying T_{Vor}, we obtain a family of curves for $x(t)$, where the error in the quasi-stationary state is

$$\overline{x}_F = V_B \cdot T_I \cdot \tan\alpha \cdot \left[2 - \frac{T_{Vor}}{T_I} \right] . \qquad\qquad (3.9)$$

An ideal behaviour with $x_F = 0$ is obtained if

$$\frac{S_{vor}}{V_B} = T_{Vor} = 2\,T_I; \qquad\qquad (3.10)$$

the advanced control precisely compensates for the delay-time of the system.

The actual contour leads the desired path as soon as $T_{Vor} > 2\,T_I$ so that x_F becomes negative. For $T_{Vor} \gg 2\,T_I$ the error can no longer be tolerated.
Introducing a pseudo-command contour x'_w and a pseudo-error x'_F, we see from equation (3.6) that x'_F is inversely proportional to K_B. If K_B is decreased to

$$K_B = K_{Bo} \cdot \frac{2 \cdot T_I}{T_{Vor}} , \qquad\qquad (3.11)$$

we obtain

$$\overline{x}_F{}' = \overline{x}'_{Fo} \cdot \frac{T_{Vor}}{2T_I}$$

or

$$\overline{x}_F = 0.$$

In the quasi-stationary state; Fig. 3.4.

(For the above equations to be valid, the linear measuring range of the sensor operating in advance control has to be enlarged by

$$X_{S\ Vor\ max} = X_{s\ max} \cdot \frac{T_{Vor}}{2T_I} \ , \qquad (3.12)$$

compared to the measuring range $x_{s\ max}$ of a sensor without lead.)

A large ratio T_{Vor}/T_I has a negative influence on the transient behaviour of $x(t)$, since a change of the desired contour is reacted on too early. This effect can be compensated for by adding a delay-time component T_{Komp} to the control loop (Fig. 3.3).

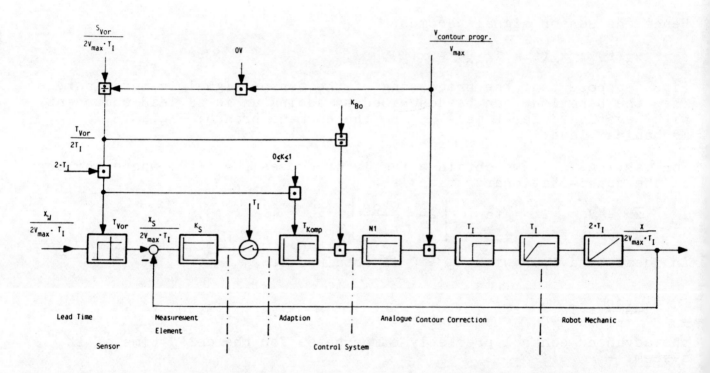

Fig. 3.3: Structural diagram for the sensor correction under advance control, constant sensor orientation and adjustment of K_B an T_{Komp}.

The delay- time T_{Komp} does not influence the stationary behaviour of the control quantity. For

$$0 < T_{Komp} < \frac{1}{2} \cdot \left[T_{Vor} - 2\ T_I \right] \qquad (3.13)$$

the transient behaviour is nearly optimal (Fig. 3.4).

For varying contour velocities the amplification factor K_B and the delay-time T_{Komp} have to be adapted subsequently.

Realization of the Control Functions
Robot Control RCM 2 provides arithmetic operations on variables by means of basic functions. Thus, the amplification K_B can, for instance, be adjusted as a function of the actual override.

238

To obtain delay times of $K \cdot T_{IPO}$, the user may run the path correction through a chain of storage registers. In this case,

$$K = \frac{T_{Komp}}{T_{IPO}} \quad . \tag{3.14}$$

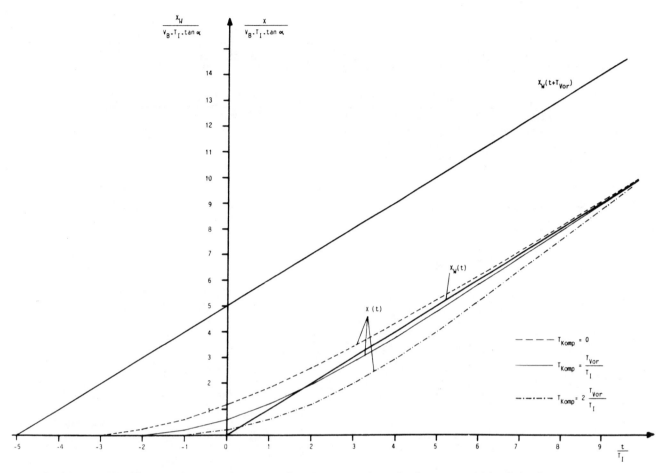

Fig. 3.4: Follow-up behaviour of the control loop with lead
component under various compensation lag times

Summary

Numerous technological tasks require the use of sensors mounted to the
hand of a robot.
With Robot Control RCM 2, parts can be recognized and located by a camera
system connected to the hand. The line of vision is free, and extensive
programming capabilitites for specific applications are available.
Quick measurement functions are provided for searching parts or working
points by simple tactile sensors or, more generally, distance gauges.
On-line path corrections cause significant changes in the behaviour of
the system if direct measurement has to be replaced by sensor measurement under advance control. Of particular interest are the consequences
for the regulation, including optimization of adjustment.

References

1   Bartelt, R.; Massat, H.-J.; Meier, Ch. Verarbeitung von Sensorsig-
     nalen in der Steuerung Robot Control M Siemens Energietechnik 5,
     1983, S. 141 - 144

2   Bartelt, R.; Massat, H.-J.; Meier, Ch. Processing Sensor Signals
     in the Robot Control M System Siemens Power Engineering V,
     1983, S. 183 - 186

3   Hirzinger, G.; Meier, Ch. Robot Control with sensory feedback in
     particular with force or torgue measurements, Proceedings of the
     INTERKAMA, Düsseldorf, Germany, 1983, pp. 354 - 366

4   Meier, Ch. Sensor Technology with the "Robot Control M",
     Proceedings of the 3rd Robot Vision and Sensory Control, Cambridge,
     USA, November, 1983, pp. 481 - 488

An Experimental Very High Resolution Tactile Sensor Array

D. H. Mott

British Robotic Systems Ltd

M. H. Lee

and

H. R. Nicholls

University College of Wales, UK

ABSTRACT

A pressure intensity transduction principle is presented that enables the construction of a tactile sensor that has both small physical size and very high resolution. Two robot gripper mounted sensors linked to a tactile processing module and a tactile work-surface have been designed, built and are being evaluated. They use the pressure intensity transduction principle and are described in this paper; the objectives of the research programme into tactile processing using these sensors are also discussed.

INTRODUCTION

Robot systems need a variety of sensors to obtain a comprehensive picture of the state of a working environment. Each sensor type operates within a particular range, e.g. far vision, near vision, proximity devices and contact sensors, whereas others are mounted on the robot itself, measuring properties such as joint and axial forces, torques etc. Vision processing is often a feature of a robot system, and provides much useful information, but currently has its limitations, for example when lighting conditions are poor or the work has to be performed in confined spaces. The main disadvantage of vision is that it often cannot provide worthwhile information during robot gripper actions because the robot gripper obscures the view of the object being gripped and so really only gives useful data before and after an operation. However, a tactile sensor is capable of detecting the location of contact and possibly the forces of contact with an object and could be used to provide sensory information that vision is unable to produce.

Tactile sensing is defined to be continuous touch sensing over an area within which there is spatial resolution [1] and is distinct therefore from force sensing which measures the totality of forces and torques on an object. If a tactile sensor, or several sensors, were mounted on a robot manipulator, then detailed information about the surfaces in contact with the sensors could be obtained, allowing the monitoring of the gripper/object relationships whilst a manipulative task is in progress.   Tactile images consist of a 2D array of transduced pressure values analogous to visual images, and so tactile sensing can be thought of as 'contact vision'.   Note that some other sensors found in robot systems also operate on contact, for example microswitches triggered by pressure, or force transducers that measure force in any direction, perhaps at the tip of a finger.   These devices can be thought of as contact sensors rather than tactile sensors according to Harmon's definition.

There are various uses for tactile sensors, for example in contour examination, gap detection and other surface inspection tasks, object recognition, grasp verification and grasp error compensation, slip detection, assembly verification and so on.

Harmon has also performed a survey of the requirements for tactile sensing [2], and arrived at a list of desirable characteristics for a tactile sensor.   These include:-

(a)   The sensor surface should be compliant and durable.
(b)   Spatial resolution should be in the order of 1-2mm.
(c)   A range of 50-200 tactile elements would be acceptable.
(d)   The sensor must be stable, monotonic and repeatable with low hysteresis.
(e)   Each element should have a response time of 1-10ms.
(f)   Dynamic range should be 1000:1, threshold sensitivity being 1 gramme.

To this list could be added the need for local tactile image processing, independent of a main computer system, giving a 'smart sensor' capability.

There has been an increasing amount of interest recently in tactile sensing, with a number of prototype sensors under development using various transduction techniques involving carbon fibres, conductive rubber, piezoresistive materials, magnetoelastic materials, electro-optics and so on.   Some of the most promising designs are to be found in [3,4,5,6] who all produce normal-force sensing devices, but an alternative approach to sensor design is presented by Hackwood, Beni and Nelson [7].   Their sensor array uses magnetoresistive techniques to measure local shear and torque and so will be especially useful for manipulative tasks where dynamic gripper/object interaction is important, for example in the detection of slip.   There exists some good designs but no very high resolution tactile sensors have yet been publicised. However, this paper presents a design for a tactile sensor that can have a resolution that matches that of high resolution vision sensors, typically 256 x 256 tactile elements ('tactels'), on a 2cm^2 pad.

THE TRANSDUCTION TECHNIQUE

The transduction method used in the BRSL/UCW sensor uses optical phenomena to form and direct the tactile image onto a camera imaging system, thus allowing the power of existing vision tools and techniques to be exploited.

The tactile sensor utilises the properties of reflection between objects of different refractive index.   The transducing structure is composed of a clear acrylic plate, a light source and a compliant membrane stretched above, but not in close contact with, the acrylic (figure 1).   The acrylic has a higher refractive index than the air surrounding it, and the membrane has a higher refractive index than both air and acrylic.   Light is directed along an edge of the acrylic plate. Most of the light is totally internally reflected inside the plate because it is travelling from a medium of higher refractive index to a medium of lower refractive index i.e. glass to air.   Hence very little of the light entering along the edge of the plate actually leaves it, and so the lower surface of the main plate, which is the imaging area, appears uniformly dark.   However, when the membrane is brought into contact with the plate, as would be the case when an object is placed on the sensor surface, diffuse reflection rather than total internal reflection occurs. This is due to the light travelling from a medium of lower refractive index to a medium of higher refractive index, the reverse of the original situation. Light thus passes out through the lower surface of the acrylic at the points where the membrane has plate contact and this can then be directed onto an imaging device, for example a videcon camera for low cost or a solid state camera for compactness. The image will have a bright patch corresponding in position and size to the part of the object in contact with the sensor surface.

The intensity of the bright patch is approximately proportional to the magnitude of the pressure between object and plate. Considering the area of plate directed onto a single pixel of the imaging camera, the amount of light reaching the pixel depends on the number of elementary contact points between the acrylic plate and membrane microsurface. As relative pressure between acrylic and object increases, more points on the membrane microsurface will come into contact with the plate and thus more light strikes the camera pixel being considered. The light intensity on this single pixel therefore increases with pressure. This principle applies to the surface as a whole. Examples of images obtained in this way are shown in figures 4 and 5.

IMPLEMENTATION OF THE OPTICAL REFLECTION TRANSDUCTION TECHNIQUE

The Tactile Work-surface

A work-surface has been constructed that incorporates a tactile sensitive area employing the optical reflection method of pressure transduction described above. The device consists of a 10mm thick sheet of plate glass with a rubberised material resting on top and a conventional videcon camera underneath to capture the diffusely reflected light. The video representation of the tactile image is passed via a digitiser and framestore into an Autoview Viking vision processing system [8] running on an LSI 11/23. This makes a comprehensive range of vision processing operators available for analysis of the tactile image. The use of a vision package is possible because tactile and visual images have a number of similarities which enable this interchange of analysis techniques. The active surface area of the work-surface is 35cm x 35cm and the video image used by Autoview is 128 x 128, with 8 bits intensity coding per pixel.

It was discovered that a similar device had been independently constructed for an application in medicine. Betts and Duckworth designed and built a device for measuring plantar pressures under the foot [9], and also carried out some useful analysis of the nature of the plastic/glass interface [10].

An assembly workcell could include a tactile work-surface for the detailed examination of an object. For example, a robot manipulator could present different faces of an object to the surface and a tactile image for each face can then be captured and stored. These images can be used to verify the object against a model, possibly finding object surface anomalies that vision systems would not detect.

The tactile work-surface was constructed in the robotics laboratory at UCW to enable the investigation of the properties of various surface materials, but it will also eventually be incorporated into a workcell containing a number of different sensor devices. The work-surface was designed with this in mind, and so is the right height and size for the laboratory robot. Evaluation of the device is currently underway, involving investigations into hysteresis, sensitivity, noise levels and environmental effects for a variety of tactile sensor surface materials.

Robot Gripper Mounted Sensor

A tactile sensor array using the optical reflection transduction method has been designed and built that is small and light enough to be incorporated in the finger of a robot gripper. The sensor system includes a degree of local image processing,

243

relieving the main computer system of the burden of analysing a raw tactile image.

The gripper mounted sensor array consists of a clear acrylic plate mounted in a steel finger with a compliant membrane stretched over, but not in close contact with, the plate [figure 3].   Illumination is provided by a simple bulb and any diffusely reflected light from the lower surface of the plate falls onto a mirror which then reflects the image onto a CCD image sensor chip via a focusing lens.

The CCD image sensor chip is a conventional solid state camera and can store a useful image of 145 x 208 elements with 8 bit resolution on each pixel.   The active surface area is 16mm x 25mm.   The only wires needed in this device are those emanating from the chip itself and these can all be enclosed in a single cable.   This sensor, therefore, does not have the problem of routing large numbers of wires away from the finger to the computer, which is often the case in other tactile sensor arrays.

One of the desirable features of a tactile sensor array is that it can be designed to perform low level processing and analysis on the tactile image before an external computer system receives it.   With this in mind, a special purpose hardware module, called the tactile processing module, was designed and built for use in conjunction with the gripper mounted tactile sensor.   It contains the control and drive logic for the CCD image sensor chip, a CRT controller to facilitate the display of the tactile image on a monitor and a processing system using two Z80 CPUs [figure 2]. The Z80s can be programmed to analyse and if required refine the image before it is passed on to an external computer system.   The tactile processing module is capable of handling the data from two tactile sensor arrays, as it was envisaged that two sensors would be operating in parallel on opposing fingers of a robot gripper.

One Z80, known as the sensor driver, is used for controlling the CCD image sensor chips, clocking images out from them.   These images can either be transferred to an external system via an 8 bit bidirectional parallel port, or to the second Z80 via a undirectional interface.   The second Z80, called the tactile image analyser, is used to examine tactile images and update the video display of the image on the monitor. It also has access to the bidirectional interface and uses it to receive commands from the main computer system and to send images and other data back to this system.

Tactile images are stored in RAM areas called framestores; in the present tactile processing module each framestore is 4K bytes (i.e. 64 x 64) but this can easily be upgraded,to 256 x 256 say.   There are four framestores, two for each sensor enabling comparison between current and previous images and also between images from different sensors.

The functions that are being programmed into the tactile processing module include:-

- Image thresholding
- Feature extraction
- Slip detection
- Measurement of total surface forces
- Monitoring of the maximum tactel value
- Transfer of raw images or processed images to external system
- Control of video displayed images
- Enable/disable interrupts on significant events.

It is intended that the two tactile sensors will be mounted on an IPA MTGS proportion controlled gripper and integrated into the existing multiple-sensor environment already established at UCW.   Inclusion in this system makes available the Autoview vision processing package which provides powerful image processing tools. The tactile processing module and two sensors have been built and programming and sensor evaluation trials are underway; however, no performance data is available at the time of writing.

PROCESSING OF THE TACTILE IMAGE

There are certain similarities between visual and tactile images, in that both consist of relatively large arrays of quantization levels that need enhancing and features within them identified. It is therefore possible to exploit the availability of the BRSL Autoview Viking vision system at UCW to evaluate tactile processing requirements and contrast them with those of vision. Autoview is a vision development system providing flexibility via a wide range of image analysis operators and tools; it is intended to incorporate the most useful operators into the software running in the tactile processing module. The tactile system anticipated for use in industry will have a facility for downloading programs into the tactile processing module, drawing from a library of vision-like operators.

A detailed literature study is underway and it is evident that many researchers are using tactile images in a manner similar to that of visual images, for example in object recognition. Whilst this is important, it is apparent that tactile systems can have complementary as well as parallel roles to vision systems as they can provide surface feature information such as texture, hardness, curvature etc. This is an area that needs more attention and research at UCW will include this aspect, making use of the high resolution data available from its new sensor.

Other uses for tactile sensors centre on the object/manipulator relationship. Hackwood et al cites two types of tactile sensors, those that are used to measure shape, and those that are used to monitor the dynamics within the hand [7]. He suggests that normal-force sensing devices are not very useful for measuring object/ manipulator dynamics and as a consequence has designed a torque-sensitive touch sensor. It is hoped to use the BRSL/UCW sensor for both types of operation. Dynamics can be monitored by fast software inside the tactile processing module studying the relationship between successive images obtained over a short timescale, e.g. for detecting incipient slip. The research programme also includes a study of tactile sensors in adaptive grasping and grip verification. As the gripper-mounted sensors hold an object, verification of its surface features on both grasped sides can be performed, and also compensation necessary due to orientation errors can be carried out.

CONCLUSION

The method of pressure transduction presented in this paper enables low cost tactile sensors having high spatial resolution to be constructed. The basic elements needed to make such a sensor are commonly available and the fabrication techniques simple, so a commercial version should be economically viable. The resolution of the device can be very high, and depends on the quality of the camera imaging system used rather than on manufacturing limitations as is frequently the case with other tactile sensor designs. The smart sensor capability of the tactile processing module is also implementation dependent, the level of sophistication and speed depending on how the hardware inside the module has been designed and the standard of software loaded into it. These features will give this sensor a degree of flexibility not previously available.

The pressure intensity to light intensity transduction principle allows the power of existing vision tools and techniques to be applied to tactile processing, whereby a large body of experiences and results can be drawn from. On the other hand, it is important to examine closely the areas in which tactile sensors provide unique information and investigate new applications; the advent of very high resolution tactile sensors opens up fields not previously available to the robotics researcher. Future work involving this sensor will report on contact vision techniques, looking at surface texture features at a detailed level for contact inspection as well as the exploitation of tactile information during manipulative tasks.

REFERENCES

1. Harmon, L.D. "Touch-Sensing Technology: A Review", Society of Manufacturing Engineers, Technical Report MSR80-03, 1980.

2. Harmon, L.D., "Automated Tactile Sensing", The International Journal of Robotics Research, Vol. 1, No. 2, Summer 1982.

3. Christ, J.P. and Sanderson, A. "A Prototype Tactile Sensor Array", Technical Report CMU-RI-TR-82-14, Dept. of Electrical Engineering and the Robotics Institute, Carnegie-Mellon University, 1982.

4. Hillis, W.D., "Active Touch Sensing", AI Memo 629, MIT A.I. Lab, April 1981.

5. Robertson, B.E. and Walkden, A.J. "A Tactile Sensor System for Robotics", Proc. of 3rd Int. Conf. on Robot Vision and Sensory Controls, Nov. 1983.

6. Raibert, M.H. and Tanner, J.E. "Design and Implementation of a VLSI Tactile Sensing Computer ", The International Journal of Robotics Research, Vol. 1, No. 3 Fall 1982.

7. Hackwood, S., Beni, G. and Nelson, T.J. "Torque-Sensitive Tactile Array for Robotics", Proc. of 3rd Int. Conf. on Robot Vision and Sensory Controls, November 1983.

8. Batchelor, B.G., Mott, D.H., Page, G.J. and Upcott, D.N., "The Autoview Image Processing Facility" in "Digital Signal Processing" ed. Jones N.B., pub. Peter Peregrinus Sept. 1982.

9. Betts, R.P. and Duckworth, T., "A Device for Measuring Plantar Pressures Under the Sole of the Foot", Engineering in Medicine, Vol. 7, 1978.

10. Betts, R.P., Duckworth, T., Austin, I.G., "Critical Light Reflection at a Plastic/Glass Interface and its Application to Foot Pressure Measurements", Journal of Medical Engineering and Technology Vol. 4, 1980.

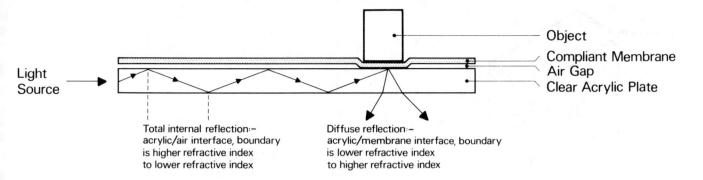

Figure 1: Illustration of the Pressure Intensity to Light
Intensity Transduction Principle

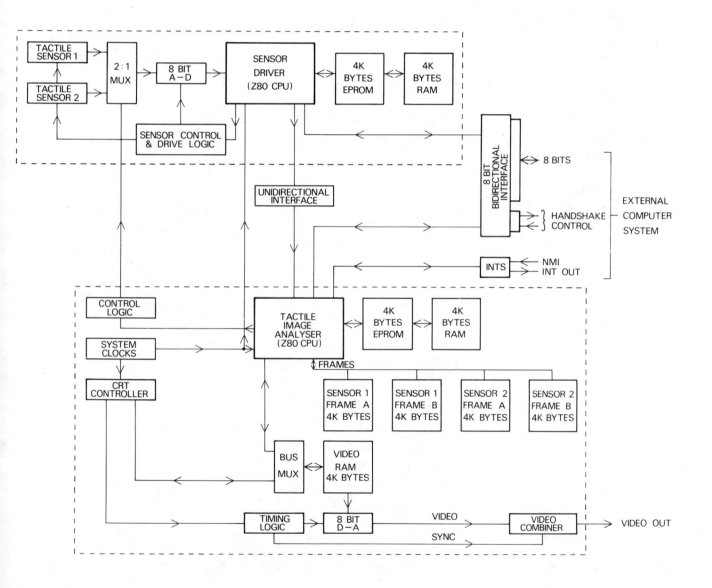

Figure 2: Block Diagram of the Tactile Processing Module

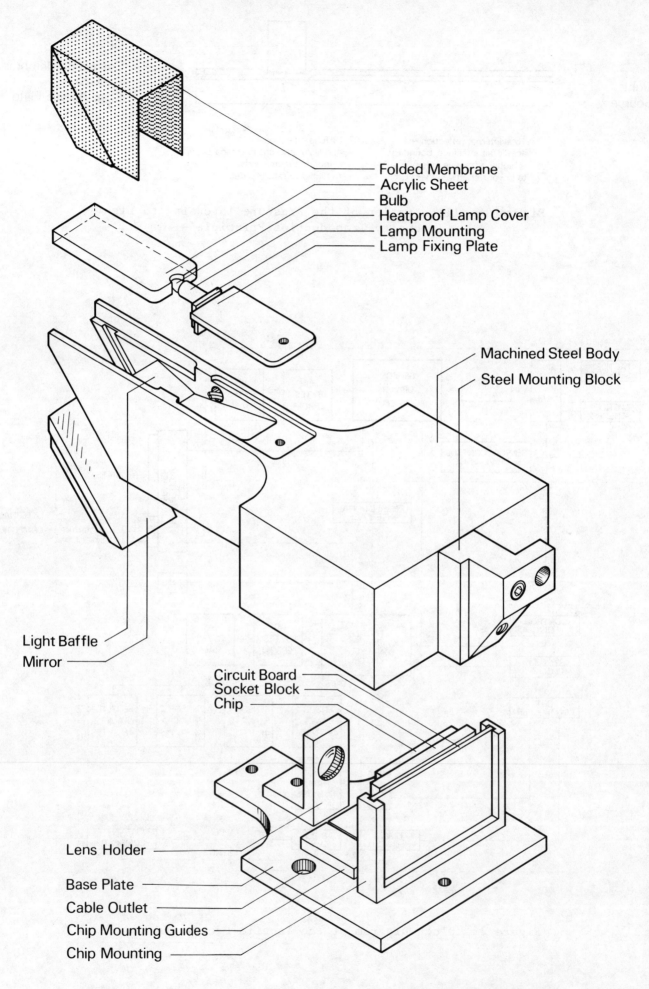

Folded Membrane
Acrylic Sheet
Bulb
Heatproof Lamp Cover
Lamp Mounting
Lamp Fixing Plate

Machined Steel Body

Steel Mounting Block

Light Baffle

Mirror

Circuit Board
Socket Block
Chip

Lens Holder

Base Plate

Cable Outlet

Chip Mounting Guides

Chip Mounting

Figure 3: Exploded view of the gripper mounted tactile sensor

248

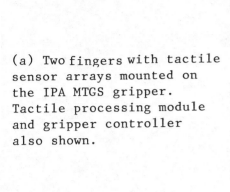

(a) Two fingers with tactile sensor arrays mounted on the IPA MTGS gripper. Tactile processing module and gripper controller also shown.

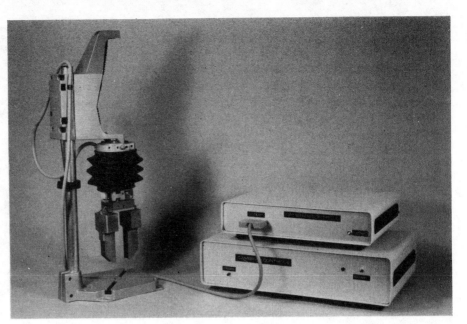

(b) Credit card and rubber stamp. Note the stamp lettering is reversed.

(c) Tactile image of the credit card.

(d) Tactile image of the rubber stamp lettering.

Figure 4

Note: (c) and (d) are photographs of tactile images that were taken directly under the work-surface glass plate

249

(a) Gear Wheel

(b) Top of a Phillips screw

(c) Gear wheel with raised boss

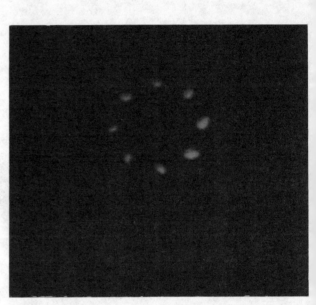

(d) Serrated locking washer

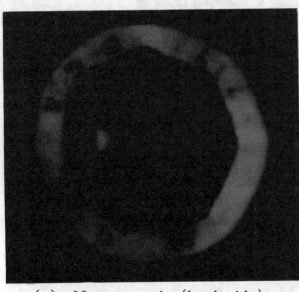

(e) 20 pence coin (head side)

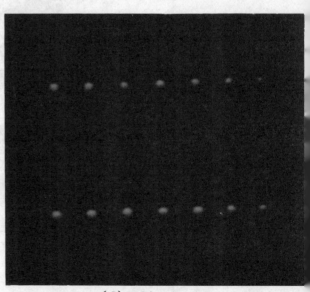

(f) 14 pin IC.

Figure 5: tactile images digitised to 256 x 256

NON-VISION SENSING

A High Resolution Tactile Sensor

K. Tanie, K. Komoriya, M. Kaneko, S. Tachi
and
A. Fujikawa
Mechanical Engineering Laboratory, Ministry of International Trade and Industry, Japan

A high-resolution tactile sensor using pressure-optical conversion technique was developed. The sensor system consists of a transparent acryle plate, an elastic sheet, a light-guide made of plastic fibers and a 32x16 phototransistor array. The elastic sheet, the surface of which is sensing surface, is placed on the acryle plate(56mm x 117mm). The light guided via a light-guide was incident upon one end of the plate. The light conducts in the plate by total internal reflection if no pressure is applied on the elastic sheet. Pressure applied onto the sheet causes an optically active contact between the sheet and the plate whereby the total internal reflection conditions are changed and the light illuminates the sheet. The sheet scatters the light back in the area of the pressure and the patterns of the pressure area can be observed by a phototransistor array. The output of array was transferred to the computer serially for tactile image processing. The evaluation experiment showed the system to work well.

1. INTRODUCTION

High resolution tactile sensors provide a lot of benefits for robot arms to perform tasks dexteriously. Firstly, this type of sensor is useful for a robot to stably grasp an object because it enables a robot to precisely identify the location and orientation of an object which is held in a gripper. Secondarily, it can be effectively used to recognize a profile of an object with complicated surface in the environment where visual sensors can not work well because visibility is poor.

There are several researchers concerned with the development on high resolution tactile sensors[1]-[9]. Simple high resolution tactile sensor is consisted of array of contact like limit switches and produces only contact pattern information via on-off signal pattern [1]. In order to get pressure pattern information , a conductive rubber or a sponge including powdered carbon is most popular material used for constructing the sensors[2]-[7]. The principle of tactile sensing in this type of sensor relates to measuring resistor change caused by the deflection of elastic membrane when it contacts a rigid object. This type of sensor system will give enough information, compared with simple contact array tactile sensors, but the resistive readout sensor requires some cautions to manufacture the array construction with high density.

There are many small size opto-electronic devices. If the pressure information, therefore, is converted to optical information, it may become easy to construct high resolution sensor relevant to detect pressure distribution[8].

This paper proposes a tactile sensor of robot hands using pressure-optical conversion technique with detection ability of complex tactile patterns. The basic principle proposed in this paper has been used for displaying the distribution of pressure in the soles of the foot or measuring the center of gravity of human body during human walking from the displayed pressure distribution in rehabilitation engineering field. This kind of device is widely known as the pedobarograph [10][11] and is one of excellent high resolution imaging touch sensor systems. However, pedobarograph is usually too big to be a tactile sensor installed in a robot system.

The objective of this paper is to develop high resolution tactile sensor system suitable for robot hands by use of the principle adopted in pedobarographs.

2. SENSOR SYSTEM

A sensor system developed includes a transparent acryle plate, an elastic sheet, a light-guide made of plastic fibers, a light source and phototransister array. Fig. 1 shows the general lay-out of the sensor system. An acryle plate is put on a phototransistor array. An elastic sheet is placed between the acryle surface and an object. The elastic sheet has irregular surface at one side which contacts with the acryle surface.

The function of the sensor system is based on the conduction of light by total internal reflexion in a transparent material, acryle plate and the use of an elastic foil as a means of transfer of pressure to the light conducting surface. If no pressure is applied on the elastic sheet, the surface of the sheet is not in optically active contact with the light conducting material and viewing field observed from the free (phototransistor array) side is dark. Pressure applied onto the sheet causes an optically active contact between the sheet and light conducting material whereby the total internal reflexion conditions are changed and the light illuminates the sheet. The sheet scatters the light back in the area of the pressure and the patterns of the pressure area can be detected by the use of optical-electrical signal conversion devices, phototransistor array.

Irregular surface of the sheet will cause an increase of contact area to the acryle plate surface according to an increase of applied pressure. This results in increasing intensity of scattering light observed by an optical sensor. Therefore, the sensor system can measure not only the patterns of pressure area, but also normal force distribution in the sensitive surface via intensity of the light which illuminates the sheet.

3. DESIGN OF SENSOR SYSTEM

3.1 Plate

An optical medium suitable for the conduction of the light by total internal reflection has to be transparent for the light used, optically homogenious, refractive, suitably shaped and of necessary strength. Acryle was chosen as a material which satisfies above conditions. The size was 117mm(length)x56mm(width)x10mm(thickness), which was decided under the consideration of a size of the gripper in which the system will be installed.

Fig. 2 shows the experimental result of optical transparency of acryle plate. To obtain this result, several length plates were prepared. The optical fiber guided light was projected from one side of each length plate and the intensity of light conducted through the plate was measured by a phototransistor placed at the other facing side. Surfaces of each plate except the side where the light was projected and the side where a phototransistor was placed were silver vacuum vapour coated to get the uniform reflexion in the plate. From the observation of the result, it can be found that the acryle plate is favarable in optical characteristics for this sensor system.

The plate used in the developed sensor system is shown in Fig. 3. Each surface was silver vaccum vapour coated except surfaces where an elastic sheet and a transistor array were attached. At the side where the light was projected, the surface was silver vacuum vapour coated except the area with which the light guide fiber contacted directly.

3.2 Light Source and Light Guide

An intensity adjustable halogen lamp was adopted for light source. The light source used was too large to be attached to the sensor directly. Therefore, a plastic optical fiber guided the light to the plate from the light source which was placed away from the sensor. Fig. 4 shows light source and light guide. As shown in the figure, the optical fiber has linear cross section at the end contacted with the plate, which enables the projection of flat beam to the plate. This configuration of optical fiber is effective to obtain good condition for total internal reflection in the plate. The selection of the projection angle of flat beam to the plate is important for favarable total internal reflexion. In this case, the flat beam was projected at the right angle on the surface of the proper side of the plate.

3.3 Elastic Sheet

The elastic sheet has to be capable of reflecting back light which comes out from the plate, and light-coloured materials are therefore suitable. In the developed sensor system, a white silicon rubber sheet was used.

The surface structure of the elastic sheet determines the resolution of the reflecting pattern. The irregular structure needs to be fine enough to present a continuous reflecting pattern to the phtotransistor array. With the problem in mind, a conic irregular surface structure were employed. The cross section is shown in Fig. 5. The pitch of the irregular pattern is approximatly 1 mm and the vertical angle of the cone is 118 degrees. The maximum thickness of the sheet is about 2mm. Fig. 6 shows an example of the reflecting pattern occurred when the sheet was pressed by the fingers.

3.4 Phototransistor Array and Electronic Circuits

The phototransistor array structure has to be carefully designed because it determines the resolution of the sensor. In order to get high-resolution sensor system, miniature phototransistors should be chosen for an element of the array. From the investigation of phototransistors commercially available, TPS 603 phototransistors were selected, each of which has 3mm diameter circular sensitive surface. They were put in holes which were arranged on a base plate made of a plastic material at intervals of about 3.5mm and it allowed constructing a 16 x 32 sensor matrix on a sensitive area of the plate. The Fig. 7(a) shows the phototransistor array manufactured.

In the development of a high resolution sensor system, one of the most important problem to be solved is how signal lines should be drawn. It is unfavarable to get the output signal from each phototransistor independently because it requires complex wiring and networks. To reduce the number of necessary connection, a scan circuit of the array was designed. Fig. 7(b) shows the schematic diagram of the circuit. The array was scanned by applying a voltage to one column at a time. In Fig. 7(b), each column is connected to a fixed voltage source(10 V) through a switching transistor which is included in a TTL inverter with open collector output. If the decorder selects one column, the switching transistor attaching to the column will be open and a fixed voltage is placed on the column, while all other columns are held at ground potential by closing switching transistors attaching to those columns. The output of each element in the selected column was multiplexed and was sent to a computer(PDP 11/44) after AD conversion. The selection of a column and the switching of multiplexer were performed by a computer software.

4. EVALUATION EXPERIMENT

The experiment setup used is shown in Fig. 8. Before experiments, calibration process for each phototransistor was performed. In order to achieve the calibration, the sensitive area of the sensor system was pressed uniformly with several pressures and the relation between the applied pressure and the electrical output for each phototransistor was measured. Each relation was approximated using a linear function described by two parameters and was used for later data processing. An example of calibration function is shown in Fig. 9. The vertical axis is indicating the illuminance instead of the direct output of phototransistor for convenience and is normalized by the illuminance for $1kg/cm^2$ pressure. From the figure, it is found to be valid to describe the relation using a linear function.

Several evaluation experiment were conducted with varying applied pressure and elasticity of silicon rubber. A hemispherical object with a diameter of 35 mm was used as an object, the profile of which was measured. A sample image of the top of the object is shown in Fig. 10. This was obtained by putting a 4.2 kg weight on a sensitive surface of the sensor system via the hemispherical object. The vertical axis is normalized by the maximum pressure value. Cross points of the mesh in the bottom surface correspond to the locations of each phototransistor. Dotted lines indicate the equal contour pressure curves. From the result, it is found that the developed sensor system gives information necessary to recognize a three-dimensional profile of the object.

5. CONCLUSIONS

In this paper a tactile sensor system using a pressure-optical conversion technique was discussed. Some basic evaluation experiments were conducted and showed the sensor system worked well. In the proposed system, a phototransistor array was used for a pressure-optical conversion device, in which the size of each phototransistor determined the resolution of sensor. If a smaller pressure-optical conversion element can be employed, the resolution will be improved. Solid state image sensors like CCD or CID device are attractive for this purpose. In the present stage, however, these devices are too small to be closely attached to the transparent plate which was used in the proposed system. In future, the appearance of a contact type solid state image sensor may contribute to the improvement of the resolution of the sensor system.

ACKNOWLEDGEMENTS

We would like to acknowledge the contribution of Messrs. Takefusa Ohno, Shoichiro Nishizawa and Hideo Iguchi, who are our colleages at the Mechanical Engineering Laboratory(MEL). We would also like to thank Mr. Osamu Saito for his assistance during experiments, who was a Visiting Researcher of the MEL.

REFERENCES

[1] Takabe,M., Iguchi,H., Komoriya,K., Tanie,K. and Fujikawa,A. "Development of a Touch Sensor System with Multiple Contact Sensors for Robots". Journal of Mechanical Engineering Laboratory(Japan), Vol.37, No.6, pp.236-243(November 1983)(Japanese)

[2] Clot,J., Rabischong,P., Peruchon,E. and Falipou,J. "Principles and Applications of the Artificial Sensitive Skin(ASS)". Proceedings of the Fifth International symposium on External Control of Human Extremities, pp.211-220(August 1975)

[3] Bejczy,A.K. "Smart Sensors for Smart Hands". Paper 78-1714, AIAA/NASA Conference on "Smart" Sensors, pp.275-304(November 1978)

[4] Purbrick,J.A. "A Force Transducer Employing Conductive Silicone Rubber". Proceedings of the 1st International Conference on Robot Vision and Sensory Controls, pp.73-80(April 1981)

[5] Larcombe,M.H.E. "Carbon Fiber Tactile Sensors". Proceedings of the 1st International Conference on Robot Vision and Sensory Controls, pp.273-276(April 1981)

[6] Raibert,M.H. "An All Digital VLSI Tactile Array Sensor". International Conference on Robotics, pp.314-319(March 1984)

[7] Hillis,W.D. "A High-Resolution Imaging Touch Sensor". The International Journal of Robotics Research, Vol.1, No.2, pp.33-44 (Summer 1982)

[8] Rebman,J. and Trull,N.W. "A Robust Tactile Sensor for Robot Applications". Lord Library of Technical Articles, LL-2142 (September 1983)

[9] Boie,R.A. "Capacitive Impedance Readout Tactile Image Sensor". International Conference on Robotics, pp.370-378(March 1984)

[10] Chodera,J.D. and Lord,M. "Retraining of Standing Balance Using a Pedobarograph". Proceedings of the Sixth International Symposium on External Control of Human Extremities, pp.333-341 (September 1978)

[11] Chodera,J.D. and Lord,M. "The Technology of the Pedobarograph". BRADU Report, Roehampton, London, Department of Health & Social Security, pp.159-179(1978)

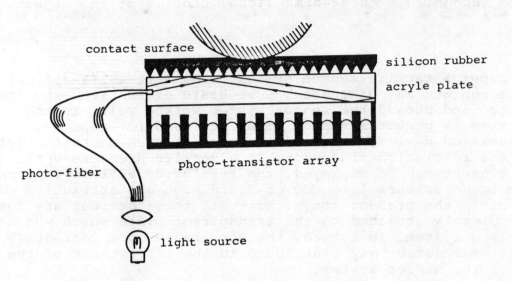

Fig. 1 General Lay-out of the Sensor System

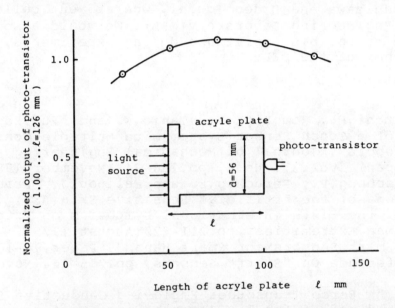

Fig. 2 Experimental Results of Optical Transparency
of the Acryle Plate

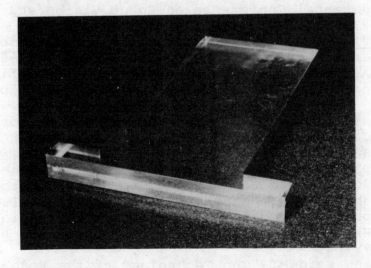

Fig. 3 Plate Used in the Developed Sensor System

256

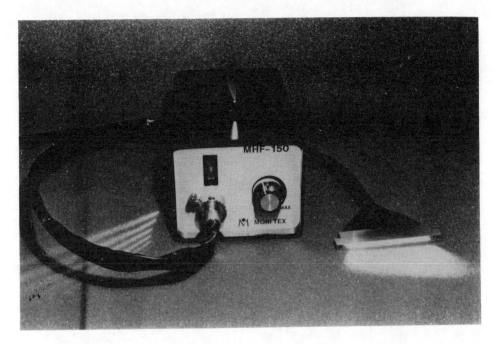

Fig. 4 Light Source and Light Guide

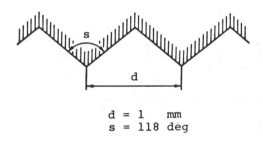

d = 1   mm
s = 118 deg

Fig. 5 Cross Section of the Elastic Sheet

Fig. 6 An Example of Reflecting Pattern

(a) Phototransistor Array Attached with the Acryle Plate

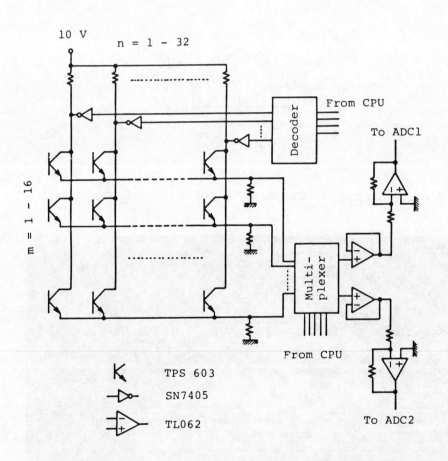

(b) Electronic Circuit for Scanning the Array

Fig.7 Phototransister Array

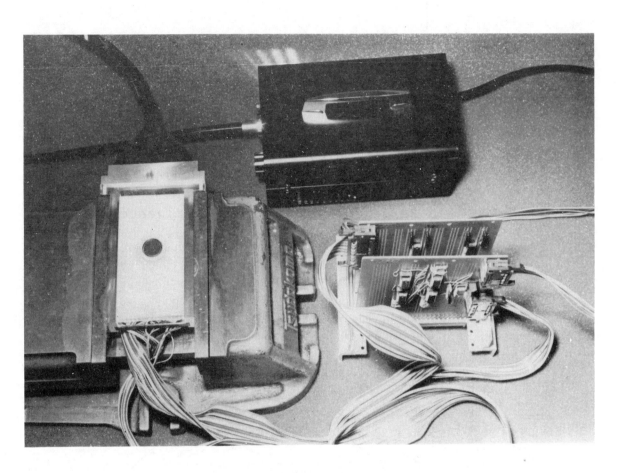

Fig. 8 Experimental Setup

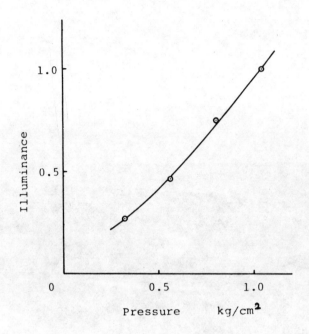

Fig. 9 A Relation between the illuminance
and Applied Pressure

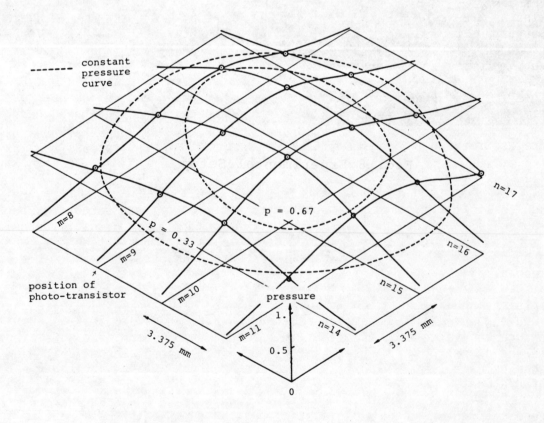

Fig.10 A Sample Image of a Object Observed by the Developed
Sensor System

Acoustic Imaging in Robotics Using a Small Set of Transducers

K. A. Marsh, J. M. Richardson, J. S. Schoenwald
and
J. F. Martin
Rockwell International Science Center, USA

ABSTRACT

A method is described for obtaining 3-dimensional acoustic images of simple objects in air using an array containing only a small number of transducers. Each transducer provides a pulse-echo measurement in an appropriate incident direction. The algorithms involve probabilistic imaging procedures, use entire waveforms (i.e. they are not limited to simple time-of-flight measurement), and are based on the Kirchhoff approximation together with the assumption that the scatterer is a rigid body. The results of preliminary testing of these techniques are presented, using both synthetic and experimental data. The experimental data were obtained using transducers fabricated from a piezoelectric polymer which provided an operating bandwidth of 7-16 KHz, appropriate for imaging objects of centimeter dimensions.

INTRODUCTION

Although considerable success has been obtained with optical techniques for 3-dimensional imaging in robotics, there are some situations in which acoustic techniques might be more appropriate. Examples include smoky environments, undersea, and objects involving highly polished or transparent surfaces. One approach to acoustic imaging is to scan mechanically a single transducer (see for example Jarvis[1] and Schoenwald[2]), although this technique is rather slow, and results in poor lateral resolution. Although in principle the mechanical scanning system could be replaced with a phased array, such an array would require a large number of transducers in order to achieve a sufficiently large aperture to overcome diffraction limitations, and to provide a sufficient diversity of viewing directions. However, successful 3-dimensional acoustic imaging can be accomplished using relatively modest arrays provided one makes efficient use of a priori information.

The present approach is a probabilistic one, based on measurement models using the Kirchhoff approximation for the scattering of acoustic waves. The measured data consist of pulse-echo waveforms taken in a small number of incident directions. The a priori information is in the form of a statistical distribution of possible acoustic impedance values. For imaging of solid objects in air, it is assumed that the acoustic impedance of the object is infinite (i.e. all acoustic energy is reflected), and hence everywhere in space the image can be represented by a 3-dimensional characteristic function with only two possible values at each point, which can be defined as 0 (in air) and 1 (in the object). In many cases, it is not necessary to know this function everywhere, but rather it may suffice to know the elevation function of the visible surface, seen from some viewpoint. In this paper, two inversion techniques are discussed, one involving the characteristic function representation, and the other involving an elevation function which is single valued at each horizontal location.

IMAGING ALGORITHMS

In the case of the characteristic function representation, a measurement model of the following form is used:

$$f(\omega,\underset{\sim}{e}) = \alpha\omega^2 p(\omega) \sum_{\underset{\sim}{r}} [\Gamma(\underset{\sim}{r}) + \Gamma_{sh}(\underset{\sim}{r},\underset{\sim}{e})] e^{i\frac{2\omega}{c}\underset{\sim}{e}\cdot\underset{\sim}{r}} \delta\underset{\sim}{r} + \nu(\omega,\underset{\sim}{e}) \tag{1}$$

where $\Gamma(\underset{\sim}{r})$ represents the characteristic function of the object as a function of the position vector $\underset{\sim}{r}$ in 3-dimensional space, $f(\omega,\underset{\sim}{e})$ represents the measured waveform as a function of frequency ω and direction $\underset{\sim}{e}$, $p(\omega)$ represents the acoustic system response (including the transducer and propagation path), $\delta\underset{\sim}{r}$ is a volume element, $\nu(\omega,\underset{\sim}{e})$ represents the noise, assumed Gaussian and stationary, and α is a constant. The summation is over all space, with respect to some suitably defined grid. Also, $\Gamma_{sh}(\underset{\sim}{r},\underset{\sim}{e})$ represents the characteristic function of the shadow zone, determined (in the geometric optics limit) for the incident wave propagating in the direction $\underset{\sim}{e}$. The geometry of the shadow zone is illustrated in Fig. 1. It is to be emphasized that Γ_{sh} also depends on the geometry of the object. This fact introduces difficulties into the inversion procedure and so far we have only approximate methods for dealing with it. One such method involves the artificial assumption that the shadow zone is the same for all incident directions. On the basis of this assumption, one can treat Eq. (1) as a problem in classical imaging, in the sense that each measurement $(\omega,\underset{\sim}{e})$ provides one spatial Fourier component of the scatterer distribution, where the spatial frequency $\underset{\sim}{q}$ is given by $(2\omega/c)\underset{\sim}{e}$, and the spatial frequency function $\zeta(\underset{\sim}{q})$ is $f(\omega,\underset{\sim}{e})/[\alpha\omega^2 p(\omega)]$. The image $\Gamma(\underset{\sim}{r})$ is related to $\zeta(\underset{\sim}{q})$ by

$$\zeta(\underset{\sim}{q}) = \sum_{\underset{\sim}{r}} [\Gamma(\underset{\sim}{r}) + \Gamma_{sh}(\underset{\sim}{r})] e^{i\underset{\sim}{q}\cdot\underset{\sim}{r}} \delta\underset{\sim}{r} \quad . \tag{2}$$

The image of the object plus shadow zone, $[\Gamma(\underset{\sim}{r}) + \Gamma_{sh}(\underset{\sim}{r})]$, can thus be obtained from $\zeta(\underset{\sim}{q})$ by simple Fourier inversion, provided $\zeta(\underset{\sim}{q})$ is known at a sufficient number of points in q space. To sample the image on a rectangular grid in real space, of dimensions L_x, L_y, L_z at intervals of Δx, Δy, Δz, it is necessary to know $\zeta(\underset{\sim}{q})$ on a corresponding grid in q space, of dimensions $2\pi/\Delta x$, $2\pi/\Delta y$, $2\pi/\Delta z$ and intervals of $2\pi/L_x$, $2\pi/L_y$, $2\pi/L_z$. The image reconstruction technique can be thought of as performing an interpolation of the measured spatial frequencies onto the unmeasured portion of this grid. In the conventional imaging approach, grid cells lying near measured spatial frequencies are set at the mean value within some averaging window, and unsampled cells are set at zero. If the averaging window used is a simple box, the technique is known as box convolution. One noteworthy property of box convolution is that the resulting image $D(\underset{\sim}{r})$ is related to the true image $[\Gamma(\underset{\sim}{r}) + \Gamma_{sh}(\underset{\sim}{r})]$ via a linear operator, representing convolution with the theoretical response to a point scatterer, $B(\underset{\sim}{r})$. $D(\underset{\sim}{r})$ and $B(\underset{\sim}{r})$ are often referred to as the dirty image and synthesized beam respectively.

One technique for deconvolving the synthesized beam from the dirty image is the CLEAN algorithm, used extensively in radio astronomy (Hogbom[3]). CLEAN performs the deconvolution iteratively to yield an approximation to $[\Gamma(\underset{\sim}{r}) + \Gamma_{sh}(\underset{\sim}{r})]$. The 50% contour of this image would provide a reasonable estimate of the boundary of the scatterer.

A preliminary investigation of the above technique ("CLEAN + thresholding"), as applied to acoustic imaging in robotics, has been made by Marsh and Richardson[4]. It has been found that this technique can produce aesthetically pleasing images in the sense that there are no artificial sidelobes present in the image, but that CLEAN is not capable of producing superresolution, i.e. resolution beyond the diffraction limit. An alternate technique for inverting Eq. (1) involves the conjugate vector method proposed by Richardson and Marsh,[5] and considered within a robotics context by Marsh and Richardson.[4] This algorithm achieves a certain degree of super-resolution by making explicit use of the fact that we are dealing with a hard surface. The large amount of computation involved in the 3-dimensional version of this approach, however, has led to consideration of a variant of the algorithm in which the scatterer is represented in terms of a single-valued elevation function. As with the "CLEAN + thresholding" algorithm, we are at present unable to treat the shadow zone exactly. In this case we assume that the shadow zone is nonexistent, i.e. that all parts of the object are visible from all of the incident directions used. Although this situation is physically realizable if the surface to be imaged is sufficiently flat, it is not a situation likely to occur in robotics. However, as with the "CLEAN + thresholding" algorithm, it might still be expected to produce a reasonable image of the illuminated portion of the object. This representation forms the basis of the second algorithm used in this current work, and we now discuss the inversion procedure.

In this case, it is more convenient to work in the time domain, and we consider a stochastic measurement model of the form:

$$f(t,\underset{\sim}{e}) = \frac{\alpha c}{2\underset{\sim}{e}\cdot\underset{\sim}{e}_z} \sum_{\underline{r}} \delta\underline{r}\, p'(t - 2c^{-1}\underset{\sim}{e}\cdot\underline{r} - 2c^{-1}\underset{\sim}{e}\cdot\underset{\sim}{e}_z Z(\underline{r})) + \nu(t,\underset{\sim}{e}) \qquad (3)$$

where $Z(\underline{r})$ represents the elevation function at location $\underline{r}$, δr is the area of a pixel, and f, p, and ν are the time domain analogs of the corresponding quantities in Eq. (1). The vector r takes values on a 2-d grid in the xy-plane (in contrast to r in Eq. (1) which is 3-dimensional) extending over a specified localization area outside of which $Z(r)$ is known to vanish. The function $Z(r)$ is assumed to be random a priori and statistically independent of $\nu(t,\underset{\sim}{e})$. We also assume that $Z(r)$ and $Z(r')$ are statistically independent when $r \neq r'$. The a priori probability density of $Z(r)$ at a single point r will be assumed to be a specified function $P(Z(\underline{r}))$ having a form independent of r for all r in the localization domain.

The problem of determining the most probable elevation function $Z(\underline{r})$ given the measurements $f(t,\underset{\sim}{e})$ involves the maximization of the expression

$$\ln P(\Gamma,f) = -\frac{1}{2\sigma_\nu^2} \sum_{t,\underset{\sim}{e}} [f(t,\underset{\sim}{e}) - \frac{\alpha c}{2\underset{\sim}{e}\cdot\underset{\sim}{e}_z} \sum_{\underline{r}} \delta\underline{r}\, p'(t - 2c^{-1}\underset{\sim}{e}\cdot\underline{r} - 2c^{-1}\underset{\sim}{e}\cdot\underset{\sim}{e}_z Z(\underline{r})]^2$$

$$\qquad (4)$$

$$+ \sum_{\underline{r}} \ln P(Z(\underline{r}))$$

with respect to $Z(\underline{r})$. Using the conjugate vector method we are led to consider the function

$$\psi(\Gamma,w,f) = \sum_{t,\underset{\sim}{e}} \left[\frac{1}{2}\sigma_\nu^2\, w(t,\underset{\sim}{e})^2 - w(t,\underset{\sim}{e})f(t,\underset{\sim}{e})\right]$$

$$\qquad (5)$$

$$+ \sum_{\underline{r}} \left[\phi(w,Z(\underline{r}),\underline{r}) + \ln P(Z(\underline{r}))\right]$$

where

$$\phi(w,Z(\underline{r}),\underline{r}) = \delta\underline{r} \sum_{t,\underset{\sim}{e}} w(t,\underset{\sim}{e}) \frac{\alpha c}{2\underset{\sim}{e}\cdot\underset{\sim}{e}_z} p'(t - 2c^{-1}\underset{\sim}{e}\cdot\underline{r} - 2c^{-1}\underset{\sim}{e}\cdot\underset{\sim}{e}_z Z(\underline{r})) \qquad (6)$$

263

If ψ is minimized on w, then the original function defined by Eq. (4) is recovered. However, if ψ is first maximized on Z and then minimized on w (assuming that the reversal of the order of minimization on w and maximization on Z is valid), then one is led down an entirely different path. Explicitly we consider the minimization of $\chi(w,f)$ on w where χ is defined by

$$\chi(w,f) = \sum_{t,\underset{\sim}{e}} \left[\frac{1}{2} \sigma_v^2 \, w(t,\underset{\sim}{e})^2 - w(t,\underset{\sim}{e})f(t,\underset{\sim}{e}) \right] + \sum_{\underline{r}} g(w,\underline{r}) \tag{7}$$

in which

$$g(w,\underline{r}) = \max_{Z(\underline{r})} \left[\phi(w,Z(\underline{r}),\underline{r}) + \ln P(Z(\underline{r})) \right] \; .$$

It is readily shown that χ is a convex function of w, and hence the minimization can be accomplished by the standard gradient method. The maximization on $Z(\underline{r})$ has to be done by direct search, although this does not represent a significant computational problem since the maximization involves only a 1-dimensional search, and is conducted independently for each pixel.

TESTS WITH SYNTHETIC DATA

Both algorithms were tested with synthetic data for a sphere. The reference waveforms corresponded to a set of transducers whose frequency range covered the ka range 0.6 - 1.8, where a is the radius of the sphere, and k the wavenumber. The response was in the shape of a Hanning window. For a sphere of diameter 1 cm in air, the corresponding frequency range would be 6.6 - 19.8 kHz. Data were generated for a set of 4 incident angles, corresponding to the 3 corners of a cube, plus the body diagonal, the latter defined to be the z axis. In terms of the coordinate system defined in Fig. 1, the directions were:

(1) $\theta = 54.7°$, $\phi = 0°$
(2) $\theta = 54.7°$, $\phi = 120°$
(3) $\theta = 54.7°$, $\phi = 240°$
(4) $\theta = 0°$

In generating these data, we actually used the impulse response for a void in a solid. As with a solid object in air, a large impedance discontinuity is involved, so the results should be very similar, except for the change of sign which was taken into account.

The results for the two algorithms are presented in Fig. 2 in the form of hidden-line representations of the 3-dimensional surface, above an imaginary base plane passing horizontally through the center of the sphere. It can be seen that satisfactory reconstructions were obtained in both cases, except for the pair of spurious peaks in the case of the second algorithm. These peaks are probably due to errors in thge Kirchhoff approximation. The errors in deduced radius (from the vertical extent of each plot) were + 1% and -12.5% for the two algorithms, respectively. At this stage, the results for the second algorithm should be considered preliminary, and considerably more testing with synthetic data will be necessary before it can usefully be applied to experimental data.

EXPERIMENTS

The experiments were designed to obtain pulse-echo scattering data in air from isolated small objects of dimensions ~ 1 cm. A convenient type of transducer for this purpose was fabricated from a piezoelectric polymer, and with the appropriate electronics, provided an operating bandwidth of 7 - 16 kHz at the 10% level. A closely spaced pair of such transducers was used, one for sending and one for receiving, providing a good approximation to a pulse-echo setup. The scatterer was hung from the ceiling with a thin thread. The same set of 4 incident directions was used as for the synthetic data, and which are defined in the previous section. These incident directions were provided by attaching the transducer pair to a PUMA 560 robot arm, and positioning the arm sequentially in the 4 positions with respect to the hanging object. The position repeatability of this robot approximately 0.01 cm, an error that is a satisfactorily small fraction of a typical wavelength ($\sim$ 3 cm). The received waveform was digitized with 8 bits of amplitude quantization with a time sampling rate of 25 MHz, considerably finer than actually required. After digitization, the waveform was passed to a VAX 11/780 for processing.

The objects were a metal sphere of diameter 1.25 cm, and a plastic tetrahedron with sides of length 2 cm. The tetrahedron was oriented such that one of the flat surfaces was in the xy-plane, and the apex was towards the transducer for the vertical incident direction. It was desired to measure the reference waveform (acoustic system response function) $p_n(t)$ for the nth incident direction by means of measurements of a smaller sphere 0.475 cm in diameter, comfortably in the Rayleigh scattering regime, and there-fore with a known theoretical response. The scattering amplitude for the pulse-echo mode is, in fact, given by:

$$A(\omega) = \frac{2}{3}(\frac{\omega}{c})^2 \, a^3 \tag{8}$$

where a is the radius of the scatterer.

In practice it was found that the response of the small sphere was insufficient to enable a satisfactory calibration, hence an indirect procedure was necessary. The shape of $p_n(t)$ was measured accurately by means of the reflection from a flat plate, except for an unknown time shift and vertical scaling factor. The latter quantity was obtained from the measurements of the small sphere (using the one good waveform obtained), and the time shifts were obtained from the waveforms of the large sphere. For this reason, the center of the large sphere defines the spatial origin for the entire experiment. The reference waveform together with a sample waveform from the large sphere is shown in Fig. 3. Image reconstruction was performed using the CLEAN + thresholding algorithm.

RESULTS

Figure 4a shows a hidden-line representation of the image of the sphere. It can be seen that the reconstructed shape is satisfactorily spherical, and the inferred diam-eter, 1.2 cm, is very close to the true value of 1.25 cm. Note that since the sphere waveforms were used to obtain the time delays for obtaining $p_n(t)$, the center of the sphere defines our spatial origin, and the image contains no information on the absolute spatial location of the sphere.

In the case of the tetrahedron (Fig. 4b) the imaging algorithm has produced a sur-face whose base area is similar to that of the original object but whose vertical ex-tent is smaller by a factor of 2. The appearance is significantly non-spherical, and consistent with a severe spatial smoothing of the original object. Since CLEAN does not provide superresolution, this behavior was not unexpected. The vertical displace-ment of the imaged surface above the base plane almost certainly represents the true physical displacement of the tetrahedron from the sphere. The fact that only the upper surface of the object has been imaged is a consequence of the Kirchhoff approximation in which the rest of the object is considered to be in the shadow zone.

CONCLUSIONS

The results of imaging the sphere with experimental data using the "CLEAN + thresh-olding" algorithm gave satisfactory results. In the case of the tetrahedron, however, the spatial resolution was insufficient for a proper characterization of the object. This may be overcome in the future by the super-resolution capability of the elevation-function algorithm.

ACKNOWLEDGEMENTS

This work was supported in part under DARPA contract number N00014-84-C-0085 and in part by Rockwell International Internal Research and Development funding.

REFERENCES

1.   Jarvis, R.A. Computer Vol. 15, p. 8 (1982).

2.   Schoenwald, J.S. and Martin, J.F., "Acoustic Scanning for Robotic Range Sensing and Object Pattern Recognition," Proceedings of 1982 IEEE Ultrasonics Symposium.

3.   Hogbom, J.A. Astron. Astrophys.Suppl. Vol. 15, p. 417 (1974).

4.   Marsh, K.A. and Richardson, J.M., "Application of the Kirchhoff approximation to acoustic imaging in robotics", to appear in Proceedings of 1983 IEEE Ultrasonics Symposium.

5.   Richardson, J.M. and Marsh, K.A., in Proceedings of SPIE Vol. 413 "Inverse Optics", p. 79 (1983).

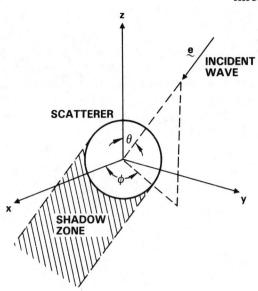

SC83-24730

Fig. 1. Geometry for scattering.

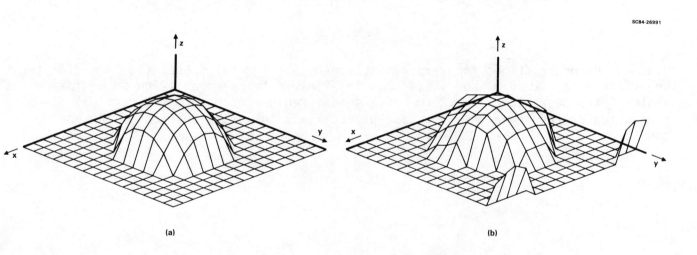

(a)

(b)

SC84-26991

Fig. 2. Reconstruction of the surface of a sphere from synthetic data, using (a) the "CLEAN + thresholding" algorithm, and (b) the elevation-function algorithm. The assumed sphere had a diameter equal to one half of the width of the base plane.

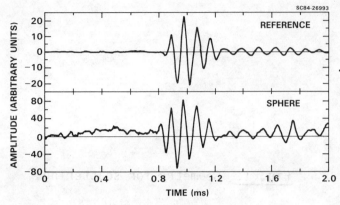

Fig. 3. Experimental waveforms. The upper plot is the second time derivative of the reference waveform, corresponding to the response to a Rayleigh scatterer. The lower plot is the response to the 1.25 cm diameter metal sphere.

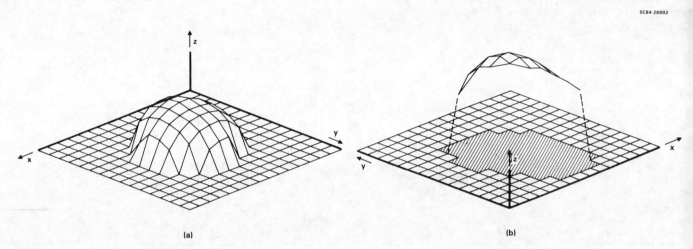

(a)　　　　　　　　　　　　　　　(b)

Fig. 4. Reconstructions from experimental data. (a) Metal sphere (1.25 cm diameter). The width of the base plane is 2.4 cm. (b) Plastic tetrahedron (2 cm on a side). The width of the base plane is 3.2 cm. The shaded region represents the vertical projection of the image onto the base plane, i.e., the cross-section of the shadow zone.

Magnetoresistive Skin for Robots

J. M. Vranish

National Bureau of Standards, USA

A tactile imaging skin for robot grippers based on magnetoresistive technology is proposed. In the design considered here, the skin would consist of a thin film magnetoresistive array with sensor elements 2.5 mm apart (density can be increased an order of magnitude) covered by a sheet of rubber and a row of flat wires etched on a mylar film. Linear pressure and compression relationships are expected over a 20 dB range. By varying rubber stiffness, the pressure range could be set anywhere between 30 N/m^2 to 3000 N/m^2 for applications requiring sensitivity and 2.0 X 10^4 N/m^2 to 2.0 X 10^6 N/m^2 for more rugged industrial uses. By varying rubber thickness the skin could be constructed to detect different compression ranges. For example, a thin skin (2.5 mm) could sense compression from .0025 mm to .25 mm whereas a thick skin (7.5 mm), which is more compliant and conformal, could sense compression from .025 mm to 2.5 mm. This paper describes design, operation and expected performance of the skin.

I. INTRODUCTION:

The design of a magnetoresistive skin for robot grippers which yields high performance (great sensitivity, wide dynamic range, and linear response) from a simple, high density sensor array appears feasible. A linear response is expected over a 20 dB range. A design with sensor elements 2.5 mm apart and 2.5 mm thick is considered herein. The principles of operation of such a skin and its theoretically expected performance are described in this paper.

II. DESIGN GOALS

The overall design goal is to explore the practical performance limits of magnetoresistive skin for robots. Figure 1 shows the proposed construction of the skin. As an interim measure it is useful to create a functional specification for the first prototype, as follows:

Dynamic Range

A dynamic range of 20 dB i.e., maximum allowable pressure divided by minimum detectable pressure = 100.

Sensitivity

a. Force sensitivity

A threshold pressure of 40 N/m^2 is detectable using .2 mm thick open celled sponge rubber between the magnetoresistive elements and the flat wires generating the field (dynamic range of 40 N/m^2 to 4000 N/m^2). By making rubber of stiffer material the threshold pressure can be increased to 2.0×10^4 N/m^2 with a corresponding dynamic range of 2.0×10^4 N/m^2 to 2.0×10^6 N/m^2.

b. Spacial sensitivity

A threshold detection of .025 mm is desired. The skin should be able to detect an encounter with an object when it has been compressed .025 mm. This assumes a skin thickness of .45 mm from the surface of the magnetoresistive elements to the external surface of the skin. Making the skin thicker will cause a proportionate decrease in spacial sensitivity.

Element Spacing And Skin Thickness

Magnetoresistive sensing element are 2.5 mm apart with 64 elements on a 25 mm x 25 mm area and the skin can be made as thin as 2.5 mm including the rubber, sensing elements and alumina (Al_2O_3) substrate. As explained in the section on spacial sensitivity above, the rubber layers can be made as thick as possible but with the tradeoff of decreased spacial sensitivity.

Power Dissipation

One hundred (100) mA and 5 watts are the maximum allowed current and power in the 25 mm x 25 mm chip.

Cross Talk

Less than 5% magnetic field intensity cross talk between adjacent elements is desired.

Durability

The system must be able to withstand 10,000 cycles of 33% compression of the rubber with 10% or less plastic deformation.

III. CONSTRUCTION

The proposed construction of "Magnetoresistive Robot Skin" is shown in Fig. 1. 50 ohm active elements of the magnetoresistive material "permalloy" 81-19 Ni-Fe are fabricated by etching on a substrate of Al_2O_3 along with gold shorts and thick film edge conductors to form an active element network array. Above this array is a thin film of rubber typically on the order of .22 mm thick. Over the top of this rubber sheet is thin film mylar which in turn has a pattern of flat wire copper conducting wires etched on it. These wires are typically 6 μm wide and provide the magnetic fields which will selectively affect the permalloy elements on the substrate located underneath. The covering sheet of rubber which includes the tread pattern (and if necessary magnetic shielding material) can be thicker, perhaps on the order of 2.5 to 5.0 mm thick. Fig. 1b shows a top view of the raster scan geometry. Fig. 2 shows a photograph of an array fabricated with the elements spaced 2.5 mm from each other for an array of 64 elements on a 25 mm X 25 mm square. Fig. 3 shows a photographic enlargement of this array. The small chevrons are the magnetoresistive sensor elements and the larger rectangles are shorts connecting the elements.

IV. DATA ACQUISITION

Fig. 4a shows a possible strategy for achieving a raster scan and extracting maximum sensitivity from the permalloy elements. In this electronic scanning scheme, the permalloy rows are all excited continuously by a square wave signal. Since the permalloy elements require a warm up to prevent noise spikes, a continuous wave signal is required. The 6 μm wide flat conducting wires can be pulsed, however, and these are on a row by row basis. That is, the first wire may be pulsed 8 times, then the second and on through all 8 of the wires (or columns). Each time a flat wire is pulsed, the first element of each permalloy row is affected. Thus, if the permalloy network is sampled row by row during the time a flat wire is being pulsed, the array can be read on an element by element basis. Each individual permalloy row is amplified before being multiplexed out thus making the system as sensitive and noise free as possible. Fig. 4b shows the electro mechanical operation of the system. Fig. 5 shows details of the electronic circuit.

Each permalloy element encounters an $\vec{H}$ magnetization field from the flat wire above it equal to:

$$H = \frac{I}{2\pi S} \left\{ \tan^{-1} \frac{[b+S/2]}{R} - \tan^{-1} \frac{[b-S/2]}{R} \right\} \quad (1)$$

where: H is in amps/meter
         S = width of conductive strip in meters.
         R = distance of the conductive strip above the permalloy in inches.

271

b = distance of center of the conductive strip from the center of the permalloy element at the permalloy. surface in meters.

I = current in amperes for the selected skin geometry.

We also know that the permalloy element shows a change in resistivity that is linear with respect to the H field impinging on it up to $\pm \frac{1.5 \times 10^3}{4\pi}$ A/m (R/R_O = 1%).[2]   A 10% variance from linearity is expected at fields up to $\pm \frac{1.5 \times 10^3}{4\pi}$ A.m. Since $\frac{.4 \times 10^3}{4\pi}$ A/m is the maximum magnetic field intensity expected at the magnetoresistive elements, linearity variance should be in the 2% range.

Experience with the magnetoresistive material to be used shows that $\pm \frac{1}{4\pi}$ A/m should be easily detectable.

For the skin to yield a linear response with respect to pressure, it is essential that the magnetic field at each permalloy element be inversely proportional to its distance from the flat wire above it or $\dfrac{H(R)}{H(R_O)} = \dfrac{R_O}{R}$ (2)

R_O = distance of the conductive strip above the permalloy element in mm with no pressure applied.

H_O = magnetic field at the permalloy element with no pressure applied.

The skin is constructed with each permalloy element having a flat wire directly above it thus b=o and equation (1) reduces to

$$\frac{H(R)}{H(R_O)} = \frac{\tan^{-1}(\frac{S}{2R})}{\tan^{-1}(\frac{S}{2R_O})} \qquad (3)$$

[1]The concept and initial studies of a crosstie random access memory (Cram) by L.J. Schwee, R.E. Hunter, K.A. Restorff and M.T. Shepard J. Appl Phys 53(3) P. 2T762 March 1982

[2]Private communication from Len J. Schwee.
To determine the limits of permalloy linearity set H = H_k sin ϕ where ϕ is the angle between the easy axis and the direction of magnetization. Easy axis is the direction M (magnetization will take if there is no field applied).
H_k is the anisotropic field of the film. $R = R_O + \frac{\Delta R}{2} \cos 2\theta$; θ = angle between current and magnetization. Set θ = 45° to easy axis. Solve for R and if $\theta = 45 - \phi$. We can now solve the 2 equations simultaneously. Thus $R \approx R_O + \Delta R\, H/H_k$ for H <.25 H_k [since $H = H_k\phi$ for small angle approximation.]

In the design proposed here

S is typically on the order of $6 \mu m$

R_O is approximately .225 mm

R can vary from .150 mm to .225 mm (33% allowable compression of the rubber).

Thus the small angle approximation can be used and

$$\frac{H(R)}{H(R_O)} \approx \frac{\frac{S}{2R}}{S/2R_O} = \frac{R_O}{R} \qquad (4)$$

The maximum error introduced by using this small angle approximation for our design is that of a 1.2° angle or 1.5/100 of 1%.

V. RUBBER FORCE ANALYSIS

The rubber structures (Fig. 6) play a key role in linking the force encountered by the robot gripper and resistivity effects in the permalloy elements. Rubber acts essentially like an incompressible fluid in that its volume must be conserved. However, unlike an incompressible fluid which passes pressure to the walls of its container, rubber experiences internal shear stresses. For the model shown in Fig. 6b, the rubber is free to change its shape (no confining side walls) so that all force applied to the top of the rubber block goes into increasing the internal stresses in the rubber. Accordingly, the following equation may be applied.

$$\frac{G}{\lambda^2 - \frac{1}{\lambda}} = K_{oust} \qquad (5)$$

G = shear stress (psi)
λ = h/ho where
h = rubber thickness under compression.
h_O = rubber thickness before compression.

In the tactile sensor array the rubber can be compressed as much as 1/3 without degredation. The discussion that follows examines the relationship between the deformation of the rubber and the force per unit area causing the deformation. A linear relationship is desired.

Since the rubber of the skin has no confining side walls, G (shear stress) is directly proportional to downward tactile force per unit area.

G = KF/A; where F/A = Force per unit area.

Thus $K_{oust} (\lambda^2 - 1/\lambda) = KF/A$;

And it is clear that the force and hence the pressure F/A F is a function of λ .

The linearity of this relationship will now be explored.

Expanding in a Taylor's series about λ_o we have,

$$\frac{F(\lambda)}{A} = \frac{Koust}{\lambda_o}(\lambda_o^2 - 1) + \frac{Koust}{K}\frac{\partial(\lambda^2 - \frac{1}{\lambda})}{\partial\lambda}\bigg|_{\lambda_o = \lambda}(\lambda - \lambda_o)$$

$$+ \frac{Koust}{K}R_n \quad (6)$$

Where R_n = remainder term.

But $\dfrac{F(\lambda)}{A} = \dfrac{Koust}{K}(\lambda^2 - \dfrac{1}{\lambda})$ $\quad\quad$ (7)

So $(\lambda^2 - \dfrac{1}{\lambda}) = (\lambda^2 - \dfrac{1}{\lambda}) + \dfrac{\partial}{\partial\lambda}(\lambda^2 - \dfrac{1}{\lambda})\bigg|_{\lambda = \lambda_o}(\lambda - \lambda_o) + R_n$ $\quad$ (8)

and the first two terms represent the region of linearity and R_n is the deviation from linear.

We know $\lambda_o = \dfrac{h}{h_o} = 1,$

and $\quad \lambda = \dfrac{h}{h_o} = 1 - \delta$ where $\delta = \dfrac{\Delta h}{h_o}$. $\quad\quad$ (9)

Substituting the conditions of equation 9 into equation 8 and solving for Rn we get $1 + \delta + \delta^2 - \dfrac{1}{1-\delta} = Rn \quad$ (10)

Recalling δ max = 1/3, Rn max or % deviation from linear = 5.6%.
Since the sensor elements measure λ , the signal processing can assume a linear relationship between $\dfrac{F(\lambda)}{A}$ and λ and be at most

5.6% in error through 1/3 deformation of the rubber.

VI. PHOTO LITHOGRAPHY PROCESS

Some of the essential details of the photo lithography process (Fig. 7) can now be discussed. The objective of the photolithography process in this instance is to provide smooth, quiet, low cost junctions between the permalloy elements, the gold shorts and the electronics. This is accomplished by first putting thick film pads on the edges of the Al_2O_3 substrate (to carry the signals on and off the permalloy/gold columns). Ultimately the electronics shown in Fig. 5 will be in the reverse side of the Al_2O_3 substrate). Following this, the permalloy elements will be deposited on the substrate, etched to the proper shape and magnetically oriented with a hard and easy axis. The final step will be to vacuum deposit and etch gold film shorts which form low noise junctions with both the permalloy elements and the thick film edge pads. This permits consistent low noise current flow.

---

[3]Louis J. Zapas NBS, private communication

VII. EXPECTED PERFORMANCE

In this section the theoretical capabilities of the skin are compared with the design goals to give expected performance. As shown in Figs. 1, 2 and 3, the permalloy magnetoresistive sensor elements are 2.5 mm apart and an array of 64 elements is on a 2.5 cm x 2.5 cm Al_2O_3 board. Experience with the 81-19 Ni-Fe "permalloy" has shown that magnetic field changes of $\pm \frac{1}{4\pi}$ A/m are detectable. If 30 mA is pulsed through a flat wire .22 mm above a permalloy sensor element, that element will receive a magnetic field of 20.885 A/m. If the wire is moved 33% closer through tactile compression of the skin, the element will now receive a field of 31.324 A/m for a maximum $\Delta \vec{H}$ of 10.439 A/m. Thus the skin has a theoretical dynamic range of 10 log $10.439/1/4\pi$ = 21.2dB. Tests have indicated that open celled sponge rubber can be ground to a thickness of .22 mm and can be repeatedly compressed 1/3 of its unstressed thickness without physical degredation. This 1/3 compression is a linear force-compression relationship to within a 10% variance. The same tests on the sponge rubber indicate that in its linear region it compresses .25 mm for each pressure increment of 1370 N/m^2. Thus the sensor element can detect pressures ranging from 29.3 N/m^2 to 3843.6 N/m^2 with pressure sensitivity of 29.3 N/m^2.

As was mentioned in Section III above, a stiffer rubber can be used. If rubber of 65 durometer hardness is used and grooves cut in it to allow it to compress without sidewall constraint (Fig. 6), the operating range of the skin is approximately 2.0 X 10^4 to 200 X 10^6 N/m^2 with 2.0 X 10^4 N/m^2 being the threshold value. These values are calculated assuming .22 mm rubber thickness and the grooves as shown in Fig. 6 are cut the length of the 2.5 cm strip. These grooves leave blocks of rubber 2.5 cm x 2 mm x .22 mm.

$$\propto \; = \; \propto 55 \; \frac{E_{55}}{E} \; \frac{h\,\beta}{A} \;)^{2/3} \quad \text{is an} \qquad (10)$$

Equation relating rubber compression to stress and the geometry of the rubber assuming no sidewall constraints.[5]

$\propto$ = percentage of deflection of rubber used in skin

$\propto_{55}$ = percentage of deflection of 1 inch cube of 55 durometer rubber used as a standard.

E_{55} = compression modulus of elasticity in psi of 1 inch cube of 55 durometer rubber used as a standard.

E = compression modulus of elasticity in psi of rubber used in skin.

---

[5]Pages 223,4 design of machine elements, MF Spotts Fifth Edition, 1978. Prentice-Hall, Inc.

h = skin thickness (inches).

β = ratio of length to width of rubber blocks used in skin.

A = area of rubber block normal to force (inches).

For the skin, α_2 = 33.3% maximum, h = .22 mm (.009 inch), β = 1/.08 and A = .5 cm^2 (.08 in^2).

$$\alpha_{33.3} = \alpha_{55} \frac{(105)}{145} \left[\frac{(.009)1/108}{\sqrt{.08}}\right]^{2/3}$$

α_{55} = 55.8% corresponding to a load of 2.0 X 10^6 N/m^2 (300 psi) (shown in a graph in Spotts page 224). Thus the maximum load the skin can oppose and still be in the linear region (33.3% compression) is 2.0 X 10^4 N/m^2 (300 psi). With its 20 dB dynamic range, the minimum load will be 2.0 X 10^4 N/m^2 (3 psi).

Spacial sensitivity is a function of the minimum magnetic field change that the mangetoresistive elements can detect. As such it is independent of rubber stiffness; but dependent on rubber thickness, the location of the flexible wires with respect to the magnetoresistive elements and the amount of current flowing through the flexible wires. The minimum detectable change in magnetic field is $\pm \frac{1}{4\pi}$ A/m. The magnetic field at a sensor element when the skin is uncompressed H = 262.45/4π A/m, and thus the field threshold of spacial sensitivity is $H = \frac{263.45}{4\pi}$ A/m

Using the equation (1) and recalling b = o for a sensor element directly below flexible wire we have

$$H = \frac{I}{\pi S} \tan^{-1}\left(\frac{S}{2R}\right) \qquad (11)$$

$$I = 30 \text{ mA}$$
$$S = 6 \mu\text{m}$$
$$H = \frac{263.45}{4\pi} \text{ A/m}$$

R calculates to .224250 mm. Thus the maximum spacial sensitivity possible is .00075 mm assuming that the entire skin rubber is only .22500 mm thick. As rubber tread is added on top of the mylar the rubber compresses as shown in Fig. 6b. Using the example of Fig. 6d.

$$\frac{d_2}{d_1} = \frac{h_2}{h_1} ; \qquad \frac{h_2}{h_1} = \frac{22425}{.2250}$$

And assuming we wish to begin detecting at a spacial deformation of .025 mm we have: h_1-h_2 = .025 mm. h_1 calculates to be 7.5 mm. Thus the skin can be approximately 7.5 mm thick, with the flat wires embedded in the rubber .225 mm from the sensor elements, and still detect a deflection as small as .025 mm at its surface. It can compress as much as 7.5 mm (33.3%) and give readings in which tactile force is linearly related to skin compression to a 10% or less variance. Using the tread pattern shown in Fig. 6a, pressure at the skin is transmitted directly down toward the sensor element below and does not dilute in a conical manner. This serves to preserve skin spacial sensitivity despite increasing rubber thickness.

VIII. CROSS TALK

Magnetic field intensity cross talk between adjacent sensor elements is calculated below. The worst case cross talk occurs when the flat wire generating the magnetic field is .225 mm above the sensor element directly below it. The nearest sensor element in a neighboring column is 2.5 mm to one side and so the relationship between the field it encounters and that which the sensor directly below the flat wire encounters is:

$$H_1 = \frac{I}{2\pi S} \left\{ \tan^{-1} \frac{(b_1+S/2)}{R_1} - \tan^{-1} \frac{(b_1-S/2)}{R_2} \right\}$$

$$H_2 = \frac{I}{2\pi S} \left\{ \tan^{-1} \frac{(b_2+S/2)}{R_2} - \tan^{-1} \frac{(b_2-S/2)}{R_2} \right\}$$

where H_1 = Magnetic field at sensor element directly below wire

H_2 = Magnetic field at nearest sensor element in adjacent column

b_1 = 0

b_2 = 2.5 mm

S = 6 μm

R_1 = R_2 = .225 mm

$$\frac{H_1}{H_2} = 124.4$$

Thus the nearest element in an adjacent column encounters cross talk of .8%. Cross talk in the two elements adjacent to the element being measured is even less. This is because the magnetic field encounters these adjacent elements at an angle not parallel to their surface specifically $\theta = \tan^{-1}\frac{(.009)}{.1}$ = .09 radians. Thus the cross talk is reduced to 2(1.6%) sin θ = .14%.

IX. CURRENT AND POWER

In this section the maximum current and power in the skin are calculated. In a 2.5 cm x 2.5 cm section, there are eight (8) columns of eight (8) permalloy elements, each of which is a 50 ohm resistor. Each column of 50 ohm resistors is balanced by a 400 ohm resistor to complete the bridge circuit into a differential operational amplifier (Fig. 5). Since 5 volts are used to drive the electronics, there are $5V/800\,\Omega$ = 6.25 mA in each column and 50 mA in the total 8 columns. Since the 400 ohm balancing resistors are located off the skin, the total power dissipation in the permalloy is 1.25 watts, with 2mW dissipated in each sensor element. The flat wires on the mylar film are $6\mu m$ wide, 2.5 cm long and $2\mu m$ thick. Made of copper they are 48 ohms each and consume $(30 \times 10^{-3})^2\ 48$ = 43.2 mW. Each line has a duty cycle of 1/8 so the effective heating is much less than the 43.2 mW might imply. The lines will end in a common resistor which is variable, set at approximately 120 Ω (since 30 mA current and 5V are used). This resistor is off the skin, however, so the total power dissipation in the skin is <u>1.25 watts</u> in the permalloy and <u>43.2 mW</u> in the flexible wires. The total current is <u>50 mA</u> in the permalloy and <u>30 mA</u> in the flat wires.

X. SHIELDING

Magnetic shielding does not appear to be necessary. It seems simplest to use a signal processing technique in which the resistance of a column is measured just before the flat wire is pulsed, then during the pulse. The difference between the two readings is a measure of the force on the skin above the element regardless of stray magnetic fields. This is provided that the stray filds plus pulsed signal stay within the $\pm \dfrac{1.5 \times 10^3}{4\pi}$ A/m

linear region of the permalloy the anticipated stray fields should not greatly exceed $\dfrac{10^3}{2\pi}$ A/m, earths field, and the pulsed signal $\dfrac{10^2}{\pi}$

A/m. Thus the total will be well within the permalloy linear region. The speed of pulsing, 1 kHz is much faster than robot movement so stray fields can be considered constant during a pulse.

XI. SUMMARY

It has been shown how magnetoresistive technology might be used to develop a skin for robots. This skin is theoretically able to perform tactile imaging with a linear dynamic range of 20dB and a threshold of .025 mm compression. The pressure threshold relating to the .025 mm compression can be made as low as 30 N/m^2 with a dynamic range of 30 N/m^2 to 300 N/m^2. Using a stiffer rubber, this threshold and dynamic range could be increased to $2.0 \times 10^4\ N/m^2$ to $2.0 \times 10^6 N/m^2$. The proposed thin film magnetoresistive array has 2.5 mm spacing between sensor elements; but easily can be made an order of magnitude more dense. Construction appears simple and economic and a skin which could survive repeated use (10,000 cylces to 33% compression) with 10% or less permanent deformation should be achievable. Further development and prototype construction are being undertaken.

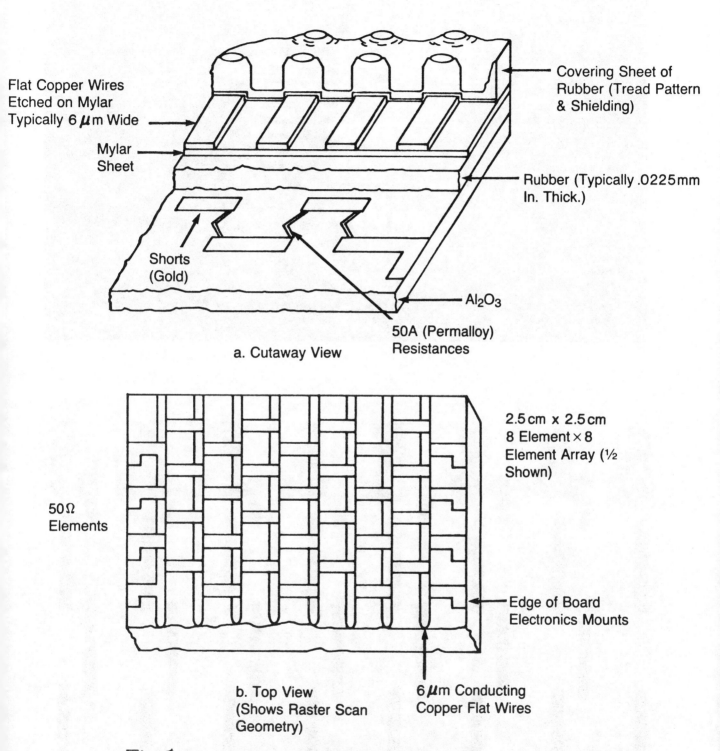

Flat Copper Wires
Etched on Mylar
Typically 6 μm Wide →

Mylar
Sheet →

Shorts
(Gold)

Covering Sheet of
Rubber (Tread Pattern
& Shielding)

Rubber (Typically .0225mm
In. Thick.)

Al₂O₃

50A (Permalloy)
Resistances

a. Cutaway View

50 Ω
Elements

2.5 cm x 2.5 cm
8 Element × 8
Element Array (½
Shown)

Edge of Board
Electronics Mounts

b. Top View
(Shows Raster Scan
Geometry)

6 μm Conducting
Copper Flat Wires

Fᴵᴳ 1 Construction of Magnetoresistive Skin

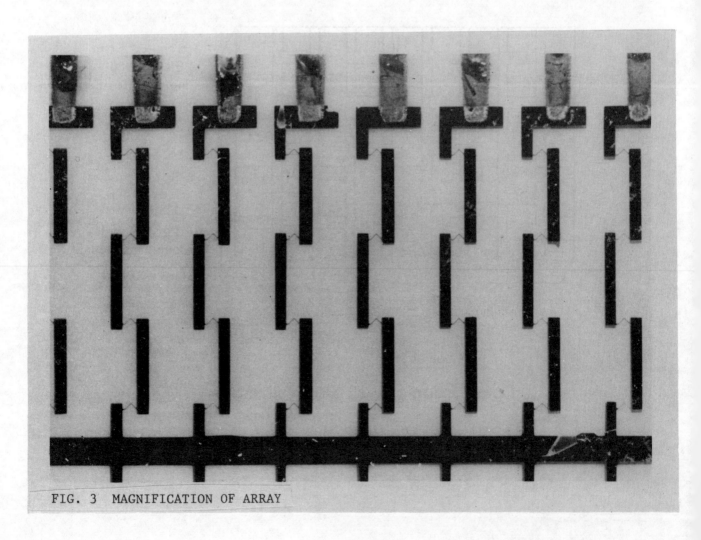

FIG. 3 MAGNIFICATION OF ARRAY

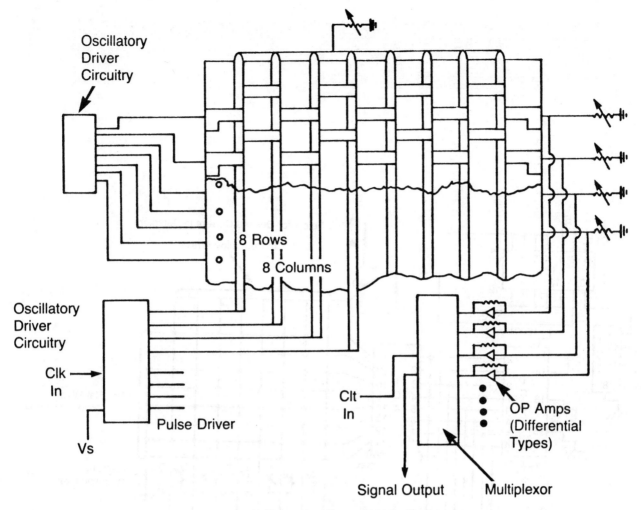

a) Raster Scanning Scheme

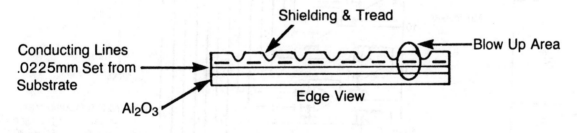

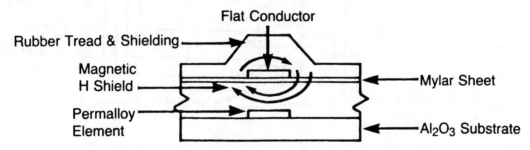

FiG 4 How the Magnetoresistive Skin Works

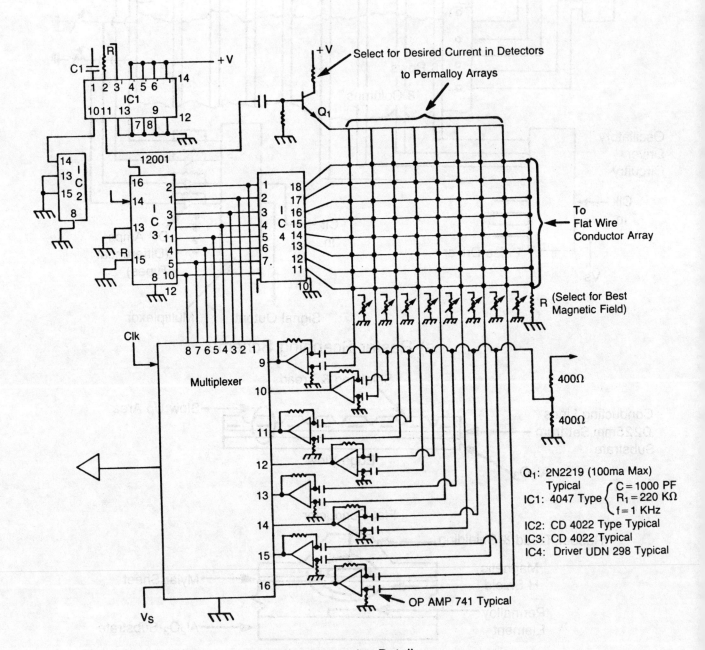

Fig 5 Electronics Details

282

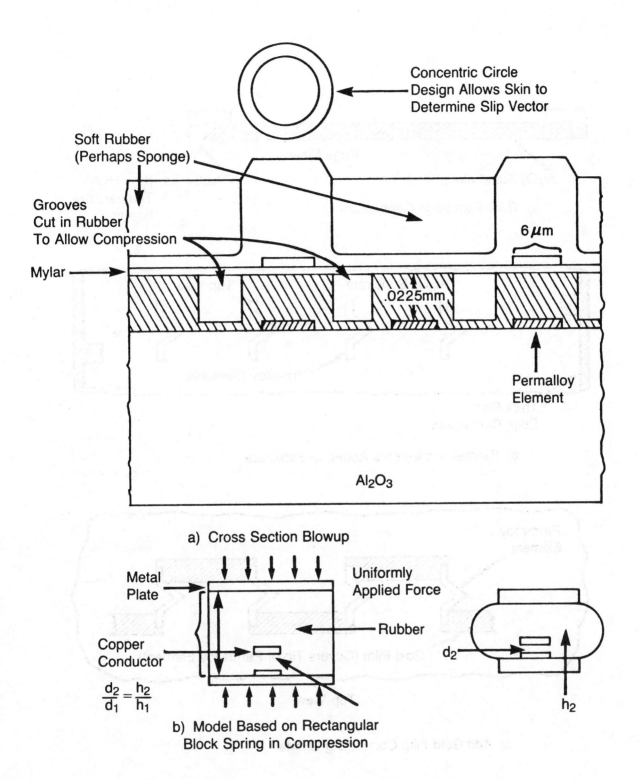

Concentric Circle Design Allows Skin to Determine Slip Vector

Soft Rubber (Perhaps Sponge)

Grooves Cut in Rubber To Allow Compression

Mylar

6 μm

.0225mm

Permalloy Element

Al₂O₃

a) Cross Section Blowup

Metal Plate

Uniformly Applied Force

Rubber

Copper Conductor

$$\frac{d_2}{d_1} = \frac{h_2}{h_1}$$

d_2

h_2

b) Model Based on Rectangular Block Spring in Compression

Fᵢɢ 6 Tread and Rubber Details

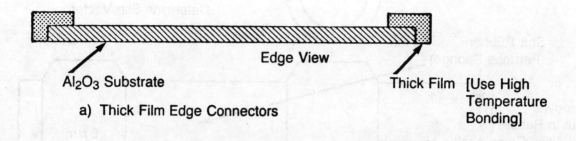

Edge View

Al$_2$O$_3$ Substrate

Thick Film [Use High Temperature Bonding]

a) Thick Film Edge Connectors

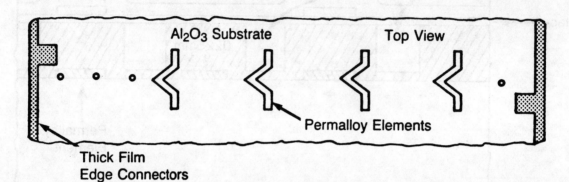

Al$_2$O$_3$ Substrate

Top View

Permalloy Elements

Thick Film Edge Connectors

b) Permalloy Elements Added to Substrate

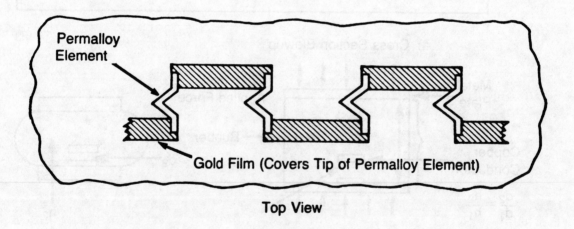

Permalloy Element

Gold Film (Covers Tip of Permalloy Element)

Top View

c) Add Gold Film Connecting Shorts

Fig 7 Photo Lithography Process and Current Flow

Application of Sensory Modules for Adaptive Robots

A. N. Trounov

Nikolaev Institute of Shipbuilding, USSR

Three types of sensory modules such as contact pressure, slip displacement and distance and inclination angle modules have been developed for adaptive grippers and control system of industrial robots. These modules may be applied as an individual sensors at various levels of feedback and as combined into total control system. Main characteristics of sensory modules are described. New material for tactile sensors with wide dynamic range of tactile and force feedback which does not require amplifiers and as flexible material may be tailored to any shape of manipulator body is presented. Comparison with carbon fibre and silicon rubber tactile sensors is given. Design approach for single-sensor and modules integrated into complex control system of adaptive robots is discussed.

New generation of industrial robots, which can rearrange and adapt themselves to the changes of environment, attract the growing interest of engineers dealing with CAD/CAM systems, flexible production systems and especially completely computerized systems for assembly of units and various purpose machines. The adaptability problem in today's robots of this generation is solved with the help of sensors which can be devided into three types: vision sensors detecting the general state of the environment; tactile sensors detecting the relation to the object and object fixation sensors detecting the orientation of the object when it is at close range [1 - 3].

The structures of sensors existing nowadays utilize various physical phenomena and are realized in a great variety of structural design. In most cases different kinds of sensors in actual production systems have been used separately. Unfortunately the standard has not been established yet which would determine the parameters for information exchange between sensors [1, 3].

For several years the author has been carrying on both experimental and theoretical research of three sensor types. The experience of empirical research conducted in the Department of Physics laboratory at the

NSI and at the Edinburgh University, as well as practical application of sensors in making adaptive robot elements helped the author to develop and select the most sophisticated ones as modules of a total sensory system. It is worth mentioning that those sensory modules, in author's opinion, can be used in the system of high intelligence and adaptive or simply sensory feedback.

1. SENSORY MODULES AND THEIR PARAMETERS
1.1. CONTACT PRESSURE MODULE

This module comprises an element changing its resistance under load conditions, and the operating circuit. The primary transducer of contact pressure into varying resistance is essentially an element having soft working surface that closely fits any shapes of robot's elements, and provides direct contact with a workpiece (Fig. 1). Such parameters allow it to be mounted on working surfaces having direct contact with workpieces and other environmental elements and adjacent facilities. It can be put into operation by the direct current source, the e.m.f. being 1-5v. The output signal level ranges from 1 to 20 mA (Fig. 1b). The highest value of mean square error in this range of pressure change, with e.m.f. being equal to 5v, approximates 10% and the current in the pressure range $(0.1-2.5) \cdot 10^5$ Pa actually increases linearly with the pressure rise. It's worth mentioning that this module is able to function from the pulse source of one pole voltage. In this case the power consumed by modules, sharply decreases as well as the quantity of heat generated in the sensor, which results in stabilizing its parameters.

1.2. SLIP DISPLACEMENT MODULE

Primary transducer of this module is made of thin conductive films with a system of 12 contactors connected with the operating circuit. In the displacement detection mode this transducer has a direct current power supply connected with the output contactors 1 and 2 and each of 10 operating circuit inputs is linked with the outputs 3-12, having a joint input, connected with the output 1 (Fig. 2a). It is under such connection that both slip availability and its rate through the rate of potential variation, and contact point coordinate through the current value, are determined. The least force, required for reliable operation makes 0.5 H. This primary transducer can be also used for measuring a coordinate of force application point only. In such mode the oscillator of square pulses as an e.m.f. source is used with maximum 0.1v which is equal to the e.m.f. of the source in the mode of slip detection. In this case the operating circuit comprises a one-channel input, linked with the output contactors 1 and 2 (Fig. 2b).

It should be mentioned that this transducer which makes it possible to determine coordinates of 10 points, can be arranged into a matrix for determination of force application coordinates and contact area since their length is 10 times their width.

1.3. DISTANCE AND INCLINATION ANGLE MODULES

These modules are designed for determining the proximity of an object or obstacles, particularly for determining the distance and inclination angle. Sensors of such type are especially useful in practice. There can be inductive and eddy currents sensors as well as pneumatic and optical sensors. The most universal are optical sensors, which are certainly less sophisticated than vision sensors, but gather more information than tactile sensors. In this aspect they are the most important for application in industrial robots[3].

Most of optical sensors are based on measurement of reflected light signal intensity. It suggests the difficulty in their application to real conditions, since the reflectivity of workpieces is not the factor which is taken into account. The application of the interference principle well-known in wave optics can overcome this drawback of simple optical sensors.

Operating principle of the module which detects the distance to an object is based on measuring the phase difference of two light signal sources arranged asymmetrically in relation to the receiver (fig. 3). Asymmetrical arrangement of light emitting diode allows us to obtain the phase shift as a value depending on the distance to the point of reflection.

$$\psi = \frac{\omega}{2c}\left[\sqrt{4x^2 + b^2} - \sqrt{4x^2 + a^2}\right] + \frac{\pi}{4}. \quad (1)$$

Together with distance determination this module can be used for detecting the inclination angle of a normal to the workpiece surface, if the distance is given, or it can perform the functions of determining distance and inclination angle simultaneously. If, for example, we consider such universal module as the system of four emitters, arranged on two mutually perpendicular straight lines with the common photo-transistor, placed in the point of their intersection (Fig. 4), then the distance is determined with the first pair through the phase difference (1), and the inclination angle of the normal to the surface or of the surface itself, counted in relation to the module plane, can be determined with the second pair through the phase difference.

$$\psi = \frac{\omega}{4}\left[\sqrt{4x^2 + b_1^2} + \sqrt{4(x - b_2)^2 + b_1^2} - \sqrt{4x^2 + a_1^2} - \sqrt{4(x + a_2)^2 + a_1^2}\right] + \frac{\pi}{4};$$

$$a_1 = a\cos\tilde{o}; \quad a_2 = a\sin\tilde{o}; \quad b_1 = b\cos\tilde{o}; \quad b_2 = b\sin\tilde{o}. \quad (2)$$

It's necessary to note that if the second pair of light emitting diodes is placed symmetrically, i.e. $a = b$, then the relationship becomes linear for small angle values.

$$\psi = \frac{\omega}{4c}\left[\sqrt{4x^2 + b_2^2 + b_1^2} - \sqrt{4(x + a_2)^2 + a_1^2}\right] + \frac{\pi}{4}. \quad (3)$$

This sensory module can detect the proximity range from 5 to 70 mm with accuracy $\pm$ 1mm, and inclination angle range $\pm$ 40deg. with accuracy $\pm$ 3 deg. The total time of processing the results equals 5.3 ms.

2. EXAMPLES OF SENSORY MODULES APPLICATION
2.1. MODULE OF GRASP RELIABILITY CONTROL
The module comprises contact pressure sensors as well as sensors of slip displacement and force application area, placed on the surface of the robot hand. It is also provided with the sensors of distance and inclination angle.

The latter three sensors determine the workpiece orientation relatively the robot hand, and the former three sensors provide the grasping force control and fixing the workpiece in the robot hand. This module can perform the following tasks:
  a) hard grasp of an object;
  b) grasp of an object with the given force;
  c) grasp of an object with zero force;
  d) grasp of a workpiece with certain orientation relatively the robot hand.

2.2. MODULE OF GRASPING FORCE CONTROL
The module comprises three or more contact pressure modules determining forces along three axes. Problems of grasping force control during workpiece assembly and outside force effects are solved.

2.3. MODULE OF OUTSIDE EFFECTS

It consists of the distance and inclination angle modules. This module is designed for detecting outside objects in the operation area and is used to solve the following problems:

a) stopping the robot;
b) change of motion trajectory;
c) elimination of obstacles.

3. SELECTION OF SENSOR MODULE AND DETERMINATION OF CONTROL LEVEL

Selection of a sensor module mainly depends on economic factors. Effectiveness of its application depends on its flexibility and the level of control. At the sensory feedback level, for example, sensory data are directly used as the feedback signal which controls the effectors automatically without computer processing. An adequate software of control algorithms in compliance with the used sensor modules is necessary at the adaptive level even for simple decision-making.

Use of sensor module for teaching is commonly executed in the dialogue mode of commands by means of application of such sensors and human intellect. Besides it can be used as adaptive sensor control as well as simple sensory feedback. For example in order to determine wrist position, the inclination angle sensor to fix the object proximity can be used, and the selection of trajectory travelled by the hand can be taught depending on the wrist director.

To obtain higher flexibility, the modules of contact pressure and slip are mounted on the robot hand. Distance, inclination angle and other vision sensors are, on the contrary, taken away from the hand, because its flexibility is determined, as a rule, through the abilities of information processing of information data. Modules fixing the proximity of the elements are mounted both on the moving and fixed elements of the robot, according to its designation.

4. PROBLEMS

Though the process of making sensory system puts forward many problems, the main of them is the problem of information processing for more flexible use of feedback signals. The experience of developed sensory modules application has put forward the problem of developing more complex hierarchical control systems and their algorithmic procedures that are available at present and developing programming languages for adaptive robots.

Unfortunately deterministic models as well as consequent processing of results are of little advantage for such tasks. In this connection it is necessary to develop the techniques of parallel processes, the theory of parallelisms and their application to signal processing directly inside sensors, perhaps, even in the analog form. Direct peripheral processing allows to reduce the load on the central processor, speed-up the information processing which will help to carry out adaptive robot control in the actual time scale (Fig. 5).

All these cause the necessity to develop and improve tactile sensors. Despite the existing data on parameters of tactile sensor materials, experimental research of carbon fibres sensors, silicon rubber and tensors during their direct contact with the workpiece demonstrated that devising material for sensors is a serious problem. In the majority of materials under investigation the dispersion of parameters during touches and shocks is so high, that sensors are able to detect only a binary signal (Fig. 6). The material developed at the department of physics of NSI represents an elastic, composite conductive material readily fitting any shapes. The output signal level approximates 20 ma, which allows to exclude intermediate amplifiers in the circuit of signal processing (similarly for the sensors made of carbon fibres). Its parameters are more stable to temperature and shock effects than all the materials under investigation. However mean square error of sensors made of such material attains 10 % which will require further improvement of search in this direction.

The problem of wear in protection materials surface arose during the maintenance of the developed modules, since all these materials must have high friction characteristics, be elastic and strong together with high wear resistance.

CONCLUSION

The application of the above approach made it possible to develop the system of robot control having six degrees of freedom. As the main regulator the 8-bit microprocessor and high-speed digit-analog transducers with servomechanisms, supplied with algorithms of motion trajectory control, were used. The effectiveness of the system was proved during the solution of workpiece grip problem, determining the main features of the workpiece, and planning the trajectory.

REFERENCES

1. Kozirev J.G. Industrial Robots. Mashinostroyenie Publishing House, Moscow, 1983.

2. Kostyuk N.Y. et als. Industrial Robots in Assembly. Tekhnika Publishing House, Kiev, 1983.

3. Makarov Y.M. Sensory Control of Robots System. Nauka Publishing House, Moscow, 1983.

Table 1

Parameters of contact pressure modules

type	range of pressure, $P \times 10^5$, Pa	Sensitivity, Am^2/H	Value of mean square error, %
P-A	0.05 - 2.5	900	11
P-B	0.1 - 25	470	15
P-C	1 - 500	85	7

Table 2

Parameters of slip displacement modules

type	number of coordinats	level of input signal, A	dimensions, $L \times W \times H$, mm $\times$ mm $\times$ mm
SD-A	10	0.1	$15 \times 5 \times 2$
SD-B	10	0.5	$15 \times 5 \times 2$
SD-C	100	0.2	$20 \times 20 \times 3$

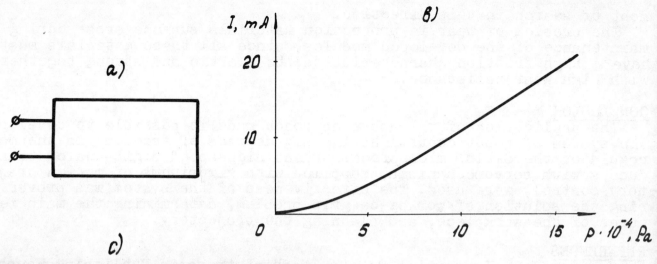

Fig.1. Contact pressure
module:
a) general view;
b) load characteristics;
c) example of mounting the
module on the gripper.

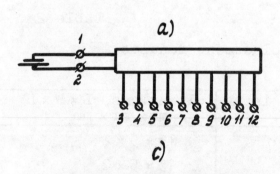

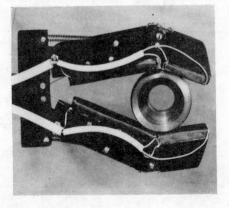

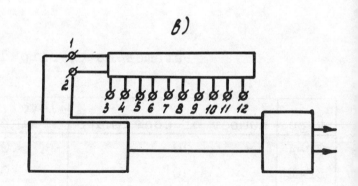

Fig.2. Module of slip displacement:
a) circuit of slip displacement module connection;
b) circuit of determining the point
of force application;
c) application example in the grip.

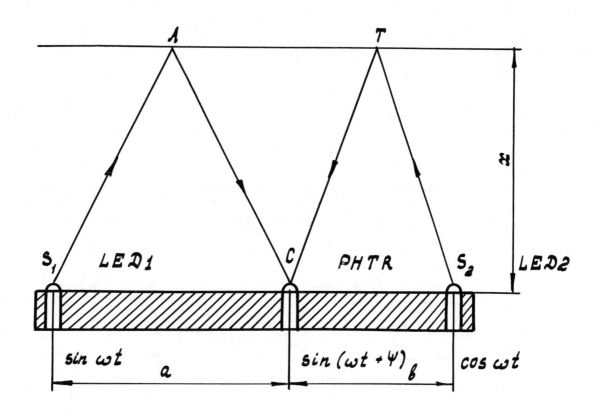

Fig.3. Principal scheme of distance module.

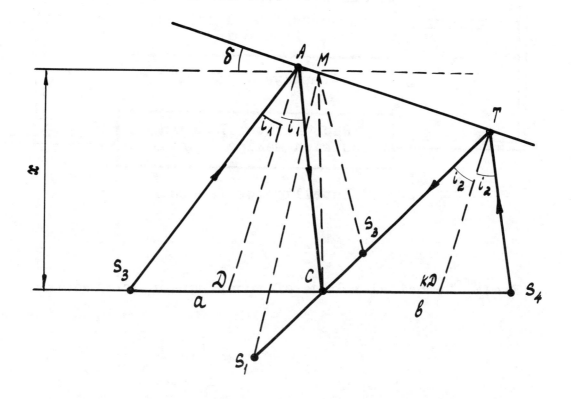

Fig.4. Principal scheme of inclination angle module.

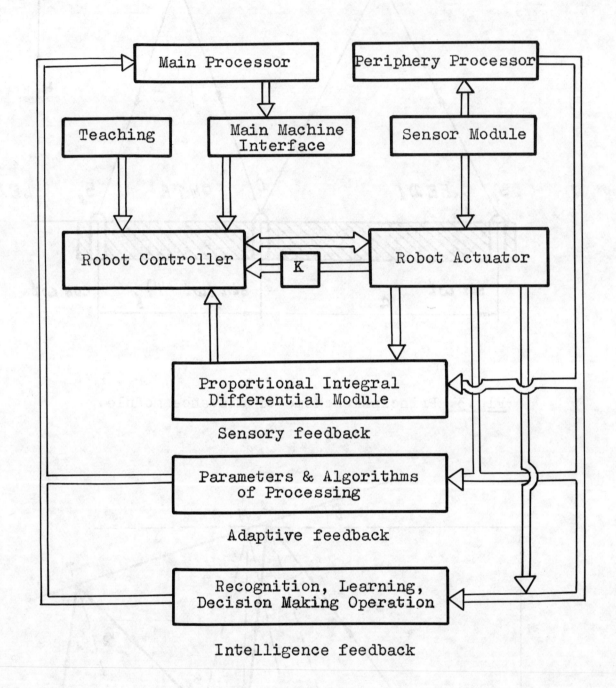

Fig.5. Principal scheme of robot sensory control system.

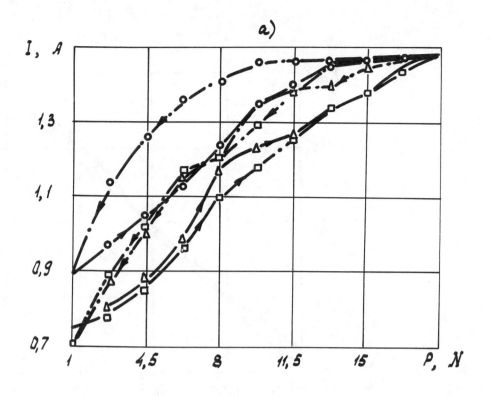

a)

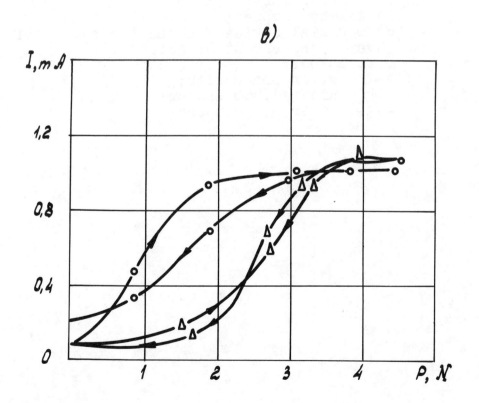

в)

293

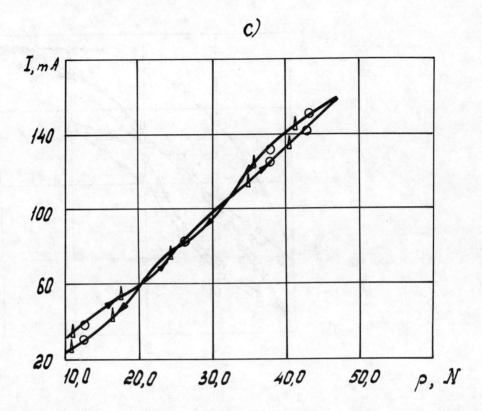

Fig.6. Characteristics of sensor materials:
a) carbon fibre;
b) silicon rubber;
c) material devised in the laboratory of
physics department at NSI.
Definitions:
—o— after production;
—□— after 1,000 shocks;
—△— after 2,000 shocks.

Carbon Fibre Sensors

J. B. C. Davies
Heriot Watt University, UK

A tactile sensor has been constructed utilising the variation in contact resistance between carbon fibres and a metal contact. Repeatability and accuracy are not high, but with suitable encapsulation and circuitry a very robust independent sensor has been produced. Feed back can take the form either of a continuous variation in resistance or discrete changes in LED illumination.

INTRODUCTION

The electrical properties of the element carbon have been known and exploited for many years, notably in the early carbon microphones and in present day motor brushes, commutator slip rings etc. With the development of carbon fibres by Phillips at the Royal Aircraft Establishment, Farnborough, in the 1950's a new arena of application was opened up before the carbon molecule. The main applications of carbon fibres as strength additives in various aeronautical applications, along with its well publicised inclusion in golf clubs, tennis racquets, etc. are well known. In the mid 1950's a Scientific Adviser in the Navy, recognising the limitations of the carbon brushes as used in electric motors, viz. a current transmission area of only approximately 30% of the actual brush, suggested that a compressed group of carbon fibres, with potentially individual filamental contact, may have superior current transmission characteristics. The brushes (a more accurate description than hitherto) developed, did not, for various reasons, exhibit the gains expected of them, but amongst the tests carried out, it was shown that the electrical resistance between the brush and the commutator was reduced as the axial load on the brush increased. The relationship is typified in Figure 1.

It is this relationship that forms the basis of the carbon fibre sensor, and some simple tests have been carried out, both at Heriot-Watt University and by Dr M. Larcombe of Warwick University. In those tests it was observed that the simplest of sensors, viz. two lengths of carbon fibre laid at right angles, exhibited a change in resistance of significant proportions when the load at the junction was varied.

Despite the initial usefulness of the sensors of the type shown in Figure 2, for engineering applications the sensor exhibited certain major drawbacks.

1. Carbon fibres whilst possessing high tensile strength, are, by definition, extremely fibrous and prone to abrasive decay.

2. Repeatability of the sensor was poor. Variations of the order of 10% of Full Scale Deflection (F.S.D.) being observed.

3. Drift of reading during constant applied load was also significant, and of the order of 5% of F.S.D.

4. Hysteresis losses were in region of 6% F.S.D.

To reduce these disadvantages to acceptable proportions the following steps were proposed.

1. The sensor should be encapsulated in a manner that could satisfy the following criteria.

    a) Adequately protect and support the carbon fibres

    b) Provide insulation

    c) Provide suitable resilience within sensor. Consequently the spring rate of the sensor could be varied by varying the encapsulating material.

    d) Stable and non toxic

    e) Compatible with carbon fibres!

    f) Easily handled

    g) Cheap and readily available.

After a great deal of searching and hunting, Bath Sealant Silicon Rubber was discovered to be an almost ideal material. In addition to satisfying the above requirements, it was also relatively quick setting, transparent, and arrived in a dispenser ideally suited to filling simple moulds.

In parallel with the above task, several experiments were conducted using modified carbon fibre resistance junctions.

1. The standard fibre/fibre resistance junction was replaced by a fibre/metal contact junction. This exhibited properties similar to the original sensor and was considerably easier to manufacture. The electrical characteristics were not appreciably altered.

2. Early carbon microphones constructed around the turn of the century suffered from similar drift and repeatability problems to those encountered with the carbon fibre sensors. The microphone utilised carbon granules and it was found to be beneficial to contain the carbon granules above a series of small cones, rather than a flat plate. These cones encouraged the granules to pack together more predictably and so improved the overall performance of the microphone. Whilst the carbon fibres used in the sensors were not granules, the addition of a locating groove in the metal contact considerably enhanced the performance of the sensor. (See figure 3)

Sensors could now be constructed reasonably easily to produce the characteristics typified by Graph 1, the average overall resolution of the sensor was still in the region of 15 - 20%, but it was extremely robust in construction and could tolerate a considerable amount of misuse and rough handling. It was considered appropriate, therefore, to attempt to develop the sensor along these lines.

Photograph (1) depicts the sensor at the stage of development.

SENSOR DEVELOPMENT

To utilise the sensor output, a small board was made up to allow the output from the sensor to drive a 10 unit LED chip. Thus, the number of lights illuminated upon the chip would be proportional to the load applied to the sensor. This arrangement was chosen because it was felt that the hostile environment of subsea conditions may be a suitable application for the sensor. The circuit was manufactured and driven by the sensor as indicated in Photograph (2). A resolution or step change of 10 units was obtained and Figure (4) depicts the typical characteristics of this sensor. However, whilst the sensor itself was of a very rugged nature, the actual electronics involved appeared quite vulnerable and detracted from the sensors usefulness as an inherently robust tool. Consequently, considerable effort was directed at miniaturising the whole circuit and incorporating within the sensor encapsulation. The final arrangement is shown in Photograph (3) where the only connection to the sensor would be from the power supply. The illumination received from the LED's is such that they could either be viewed by a diver, or possibly monitored by a remote, free swimming TV camera.

CONCLUSION

The sensor is a robust, useful tool, the output from which does not necessarily have to be utilised to drive a 10 bar L.E.D. The analogue output could easily be conditioned to provide decision making information for a robot or manipulator control system.

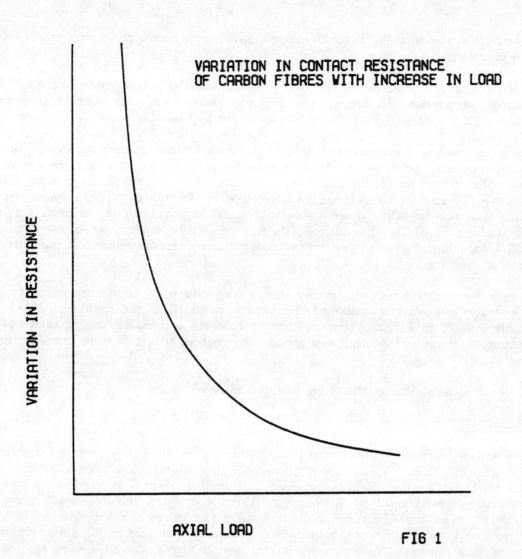

VARIATION IN CONTACT RESISTANCE
OF CARBON FIBRES WITH INCREASE IN LOAD

VARIATION IN RESISTANCE

AXIAL LOAD

FIG 1

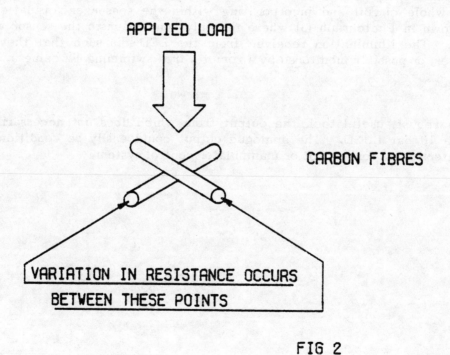

APPLIED LOAD

CARBON FIBRES

VARIATION IN RESISTANCE OCCURS
BETWEEN THESE POINTS

FIG 2

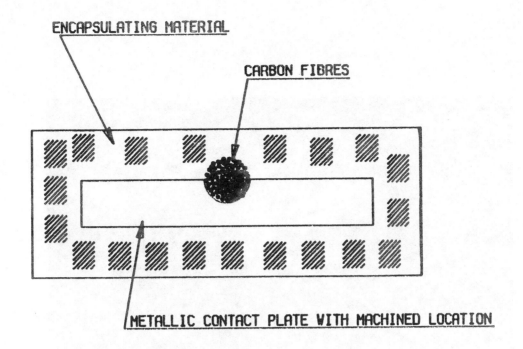

ENCAPSULATING MATERIAL

CARBON FIBRES

METALLIC CONTACT PLATE WITH MACHINED LOCATION

FIG 3

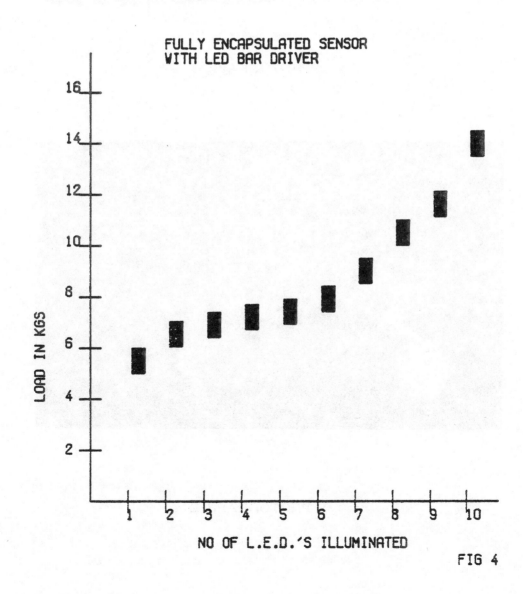

FULLY ENCAPSULATED SENSOR
WITH LED BAR DRIVER

LOAD IN KGS

NO OF L.E.D.'S ILLUMINATED

FIG 4

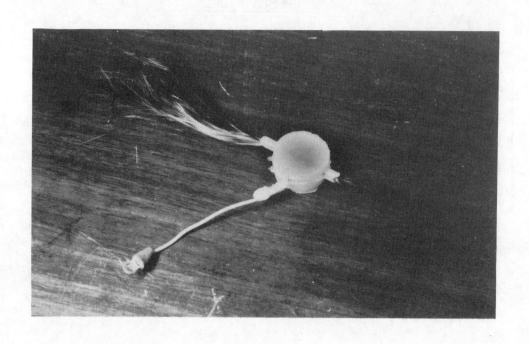

PHOTOGRAPH 1

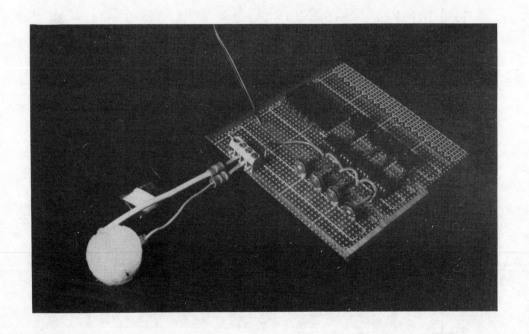

PHOTOGRAPH 2

PHOTOGRAPH 3

KNOWLEDGE-BASED SENSORY SYSTEMS

Development of an Expert Vision System for Automatic Industrial Inspection

J. Foster, P. M. Hage

Joyce Loebl (Vickers) Limited, UK

and

J.Hewit

University of Newcastle-upon-Tyne, UK

ABSTRACT

The Joyce-Loebl Magiscan system has been used for the inspection of complex industrial assemblies for many years. The knowledge base and experience so obtained is now being used to develop an expert system for the automatic inspection of surface texture and finish of industrial components.

Two examples are given of the application of these techniques. These are, the inspection of explosive detonator material and the classification of wooden staves.

The system makes automatic measurements of a number of key parameters and by intelligent assessment decides upon the quality or grade of the component.

Self learning is used to allow the system to accommodate different components and ranges of materials from which the components can be manufactured.

1. INTRODUCTION

The surface finish and texture of many materials may usefully be classified according to the presence or absence of cracks and other defects and by variations in colour. Features such as these can be readily detected and measured using standard image analysis methods.

Unfortunately, however, the vast range of possible techniques make it infeasible to combine them into one general purpose package since the time available for the inspection process is limited by the needs of the production environment.

The processes by which a human undertakes inspection of such features, however, does not rely upon a strictly quantitative analysis. Instead the human inspector develops an ability to recognise and assess an overall 'grade' (or 'appearance') factor.

The development of a system introducing the concept of a 'grade' factor to machine visual inspection has been undertaken. Although still in an early form, this expert system has been used to classify surface finish and texture of industrial components. In particular it has been applied in the inspection of detonators for the mining industry and of beech staves.

2. DETONATOR MATERIAL INSPECTION

Detonators used in mining are manufactured automatically on dedicated manufacturing lines. The explosive material of which they are formed can be damaged if subjected to scratching or knocking during the final assembly process. Any defects must be detected visually.

Fig.1. is a machine image of two views of a detonator head. The righthand view is taken directly, the lefthand view is the reflection in a mirror placed behind the detonator. Thus the single image may be used to analyse the whole head surface.

The texture and colour of the surface may vary from one detonator to the next, but defects usually show up as regions of slightly lighter appearance (Fig.1. lefthand view). When perfect the surface has a glossy finish which can cause saturation regions in the image. This must not be allowed to interfere with the classification process.

Apart from the defect detection and classification, checks must also be performed to ensure that the overall amount of detonator material is within the specified tolerances.

Because of the wide range of colours and textures which may be encountered and because of the severely limited processing time available it was found that conventional segmentation routines were too time-consuming and, in any case, unreliable.

Thus the 'grade' factor approach was adopted.

3. CLASSIFICATION OF WOODEN STAVES

Beech staves are manufactured automatically and must subsequently be classified into five classes according to the degree of surface discolouration and the density of cracks and knots.

Fig.2. shows a view of three staves. The centre view is of a top grade stave. The lefthand view is of a stave of lowest grade. The righthand view is of a stave of intermediate grade.

The Magiscan system, using three cameras, analyses the surface of three staves at a time.

To perform analysis of all of the knots and cracks and then to quantify surface discolouration would require a total processing time far in excess of the manufacturing cycle time. Reduction of this processing time with the maintenance of a reliable inspection was made possible by the adoption of the 'grade' factor approach.

4. CLASSIFICATION PROGRAM

4.1 Introduction

A fast versatile, general purpose classication system based upon the Magiscan system has been developed which is capable of gaining knowledge from the human inspector.

A range of parts with quality varying over the entire possible spectrum is presented to the system. During this teaching phase the system builds up a database of knowledge which it can subsequently call upon during the actual classification process.

4.2 'Grade' factor specification

In order to emulate,by machine, the process of human inspection most efficiently it is necessary to give to the machine an ability to take an 'overall' view of the inspected product. This can best be done by analysing the frequency of occurrence of grey levels over the image presented to the machine.

Windows may be defined within the image observed by the visual system. Over these windows the frequency distribution of grey levels may be computed.

In the Magiscan system, which uses a 512 x 512 pixel image it is possible to define n x n pixel windows, each with 64 defined grey levels.

Microcoded system library routines callable from PASCAL can, for example, compute frequency distributions over a window in which $n = 200$, in less than .5 sec.

The mean and standard deviation of each frequency distribution provide an adequate measure of quality for distinguishing between good and bad components in the majority of cases.

For the classification of mining detonators two 'grade' factors have been computed. One of these ignores the effects of image saturation, the other takes them into account.

For the classification of the beech staves a single 'grade' factor is sufficient. This is based simply upon the basic mean and standard deviation values.

In each case the required number of windows has proved to be fairly small.

In the case of the detonators, two windows defined in appropriate regions of the image have proved to be sufficient while for the beech staves two windows per stave or 6 per image are normally used.

In either case if a borderline decision results it is a simple matter to increase the number of windows, of smaller size, and hence decide whether there exists a field whose 'grade' factor is sufficiently low to indicate a defect.

4.3 Teaching the System

Fig.3. shows the menu of options. The first of these is Teach, the mode in which the operator imparts his expert knowledge to the system. In this mode the operator may either specify the 'grade' factor and the range covered by it or may teach the system to recognise good and bad components.

The operator is able to introduce as many images as required to the system and may define, within each image, as many windows as required to highlight or exclude certain features.

For each window the 'grade' factor is computed and stored on disc.

Fig.4 shows a window deliberately selected to outline a detonator head. The number of frames within a given window and the total number of images required to analyse a given component can be demanded by the operator.

4.4 First Stage Automatic Classification

When the system has been set up correctly the automatic classification routine can be selected.

Fig.5 shows the results of an analysis of two surfaces of beech staves. Here, 'Debug' drawing, a facility provided so that the operator can see how well the system is performing, has been selected. Image windows and calculated distributions are displayed.

In the distributions, the abscissa is a measure of grey level (0 - 63) while the ordinate is a measure of frequency of occurrence over the window area of that grey level.

In the example shown (Fig.5) the system has previously been taught to select all grade one wood from the samples presented to it.

The top window and distribution show an example of this top grade (which has a 'grade' factor here of 19.5). As can be seen, the distribution is very narrow. The 'grade' factor training range for this example was 18.6 to 21.1.

Using the same technique on a lower grade of wood with small knots and slight discolouration as shown in the bottom window results in a distribution more widely spread and with a 'grade' factor of 10.7. This sample was therefore classified as being below the acceptable range.

If the small dark knot at the top of the window had been absent the sample might then have been classified as being on the borderline of acceptability. In such a case the system would automatically indicate a second stage of analysis in which the window would be subdivided to determine whether or not there exists an unacceptable region.

4.5 Second stage classification

A human inspector who is undecided upon the quality of a component may 'take a second look'. So does the Magiscan system. The window is subdivided as many times as necessary and 'grade' factor calculation proceeds over each subwindow.

Fig. 6 shows the result of subdividing the original window into 15 subwindows, two of which are outlined. Along the bottom all 15 distributions are displayed. The window containing the dark knot has the distribution shown outlined in black. It may be seen to be wider than the rest and it has, in fact, a 'grade' factor of only 7.1 . The range of 'grade' factors over the whole 15 windows is from 7.1 to 14.4 the higher values being at positions where there is only slight discolouration of the wood.

By detection of the worst windows it is possible to classify border-line cases correctly.

5. EXTENSIONS

The expert vision inspection system is now being adapted for use on the interactive surface polishing of turbine blades for power generation plant.

Many blades are hand polished and the surface finish is assessed by human vision (and to a certain degree by tactile sensing). This polishing/inspection process is very laborious and time consuming. It also requires a high level of skill on the part of the operator.

It is hoped that the use of machine vision of the type described will enable the human expertise to be passed over to the expert system.

Extensions are also underway to enable the system to be used in the process of complex industrial assembly. At present design changes in components often requires extensive reprogramming.

It is hoped to use Magiscan's extensive measurement capabilities to allow a human to specify to the system which measurements can be used to classify an assembly as acceptable or not.

The knowledge so gained will then enable Magiscan to suggest to the system operator which measurement to take to enable expert component classification to take place.

6. CONCLUSION

An expert visual system has been described for the analysis of surface texture and finish of components.

The ability of the system to correctly and reliably classify components is based upon knowledge imparted to the system by expert human operators.

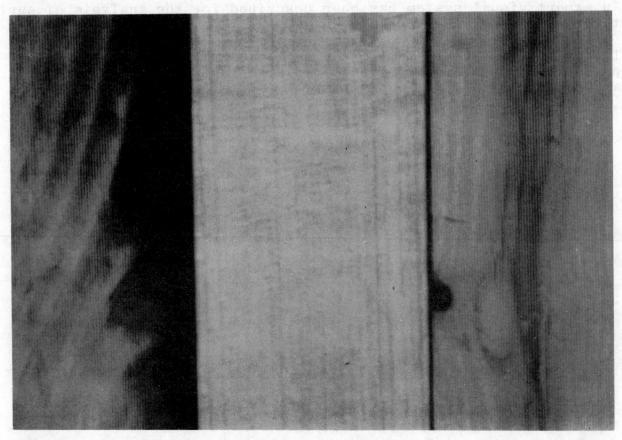

FIG. 2

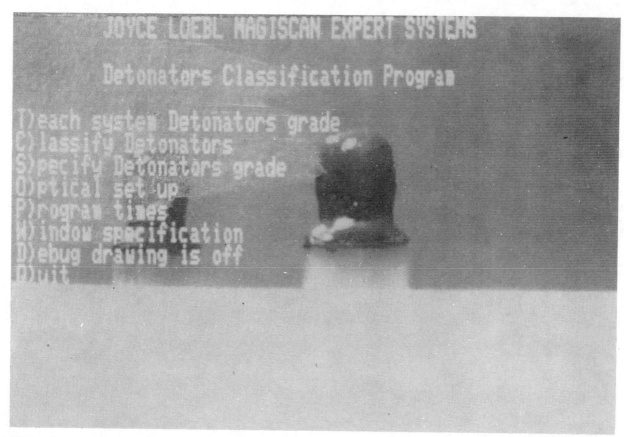

FIG. 3

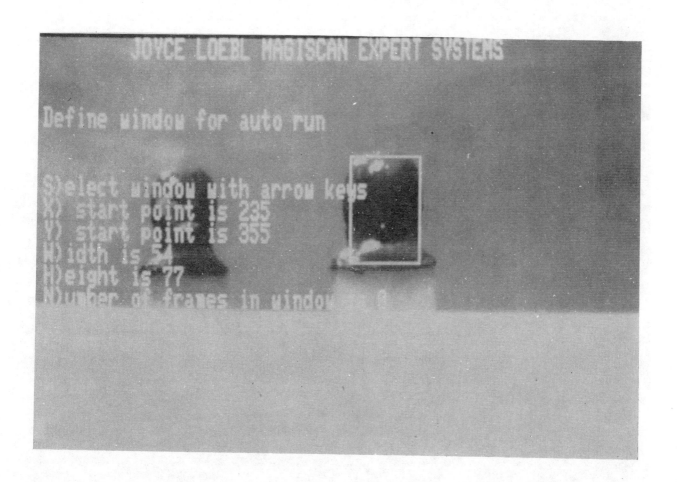

309

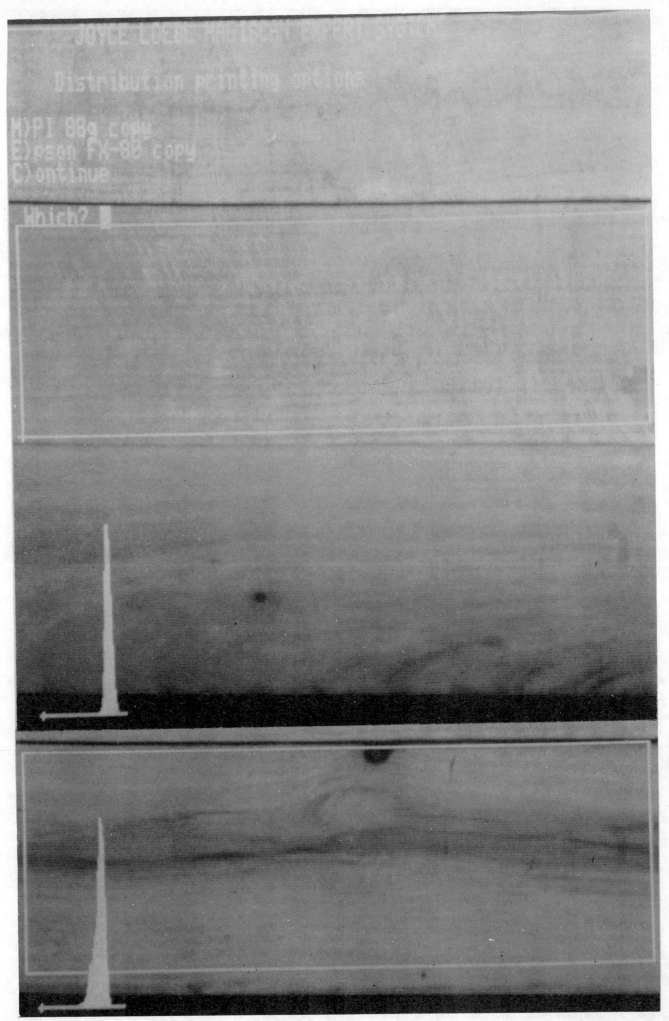

FIG. 5

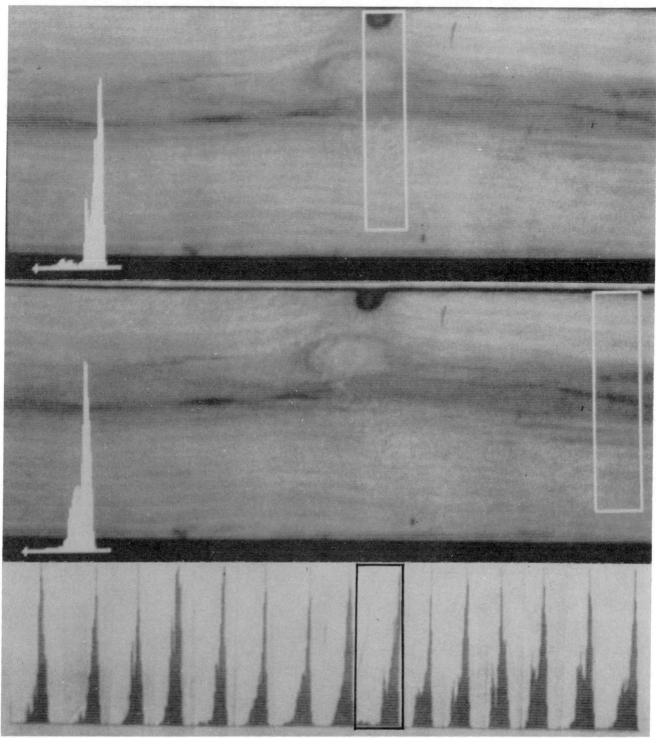

FIG. 6

Pragmatic Approach to the Bin Picking Problem

O. Ledoux

and

M. Bogaert

Centre for Scientific and Technical Research in the Metalworking Industry, Belgium

A pragmatic visual method is proposed to grasp parts from a bin. In this approach the localisation and the recognition of an object are done in two steps. In the first step, we try only to grasp a part by looking for a grasping site. The image processing is based on an edge operator, computed on a 64 grey levels image, to find some parts of the object's outline. Next the vision system models two opposite parts of the object's outline and tries to find a grasping site where the fingers of the gripper fit this modelled outline. In the second step, the part is on a flat surface below another camera where lighting conditions are easily controlled and where the part is isolated. This simplified scene is then analysed by conventional means.

INTRODUCTION

The great difficulty in the bin picking problem is that parts touch and overlap each other and therefore cannot be seen by a conventional vision system as separate objects.

The second problem is the bad contrast between the parts and the background in a bin, so it is very difficult to follow the outline of an object.

The method described cuts the problem into two easier subparts : isolating an object by grasping it out of the bin, and then recognising and locating it in another environment with well-controlled lighting conditions.

In the first step, we don't use the concept of "object", we only try to find a place where we can put the fingers of the gripper in order to grasp something. We don't know what we grasp and it's not our problem in the first step of our approach. It is what we call the search for a "grasping site".
Then we put this object under another camera in order to recognize and locate the part. In the second step of our approach, the concept of "object" appears because the part is isolated and we control the lighting conditions so there is no more ambiguity in the definition of this concept. We assume that the gripper used is one with two parallel fingers. We have no "a priori" knowledge of the parts, but there is one restriction : they must be flat (two dimensions are greater then the third). This restriction is necessary because we use a simple camera and then we only have a two dimensional view of the three dimensional scene. The program was first written for a two-dimensional problem : unloading a con veyor belt.
The definition of the two-dimensional problem is the following : the objects are flat and are put on a flat surface. They may touch each other but cannot overlap. The definition of the three-dimensional problem : the restrictions on the objects are the same but they can overlap each other (the objects are in a box). In a futher paragraph we shall describe how to solve the additional problems met in the three-dimensional case

ORGANISATION OF THE PROGRAMS

Near the robot we have installed a conveyor belt on which the parts arrive, a flat surface where the objects are put in order to recognize and locate them, and a box where they have to be placed in an predifined position (figure 1).
We work with three cameras : the first is placed above the conveyor belt in order to see the loose parts, the two others are placed above the plane to recognize and locate the parts.
These two cameras give us two so-called "recognition stations". It enables the robot to pick an object from the conveyor belt and to put it o the free "recognition station" during the computing of the position and the identity of another part on the other "recognition station".
The two image processing tasks (the "grasping site" task and the "recognition and locating" task) are working in parallel. These two tasks an the robot's movements are controlled by the "supervisor" task as you can see in figure 1.

OVERVIEW OF THE METHOD

The problem of recognition and locating of an isolated object on a plane under good lighting conditions is solved by a lot of commercialized systems, so we turn our attention to the first step of the method : the search for a "grasping site".
The search for the "grasping site" is based on the modelisation of the gripper and on the modelisation of the outline of the object into what w call "elementary primitives", and we try to find a place where the gripper's model and the outline's model fit well.
The "elementary primitives" (E.P.) are defined from the straight lines o a polygonal approximation of the shape. There are three types of E.P. : the straight primitive, the concave primitive and the convex primitive (see figure 2). We will see later how these E.P. are defined. We can see in figure 3 some examples where the gripper fits the outline well.
The input image of the system is a 256×256 image with 64 grey levels per pixel. These 64 grey levels allow us to detect edges in an image with

poor contrast. First we scan the image to find a seed point (to follow the outline of the object) which is a point where the gradient magnitude is large. Starting at this point we follow the outline of the object by means of an edge operator. At the same time, we segment this outline into straight lines to approximate it. This parallel computation enables us to know where to stop the contour following. We have then polygonized a part of the outline of one or more objects. We then look for "elementary primitives" along this outline.
In front of each E.P. we try to find another E.P. and test if this couple of E.P. fits the gripper's model well.

THE EDGE OPERATOR

The most critical step in the image processing is the contour following, so particular attention must be drawn to the edge operator. You can see the type of edge operator used in our program in figure 4.

DETERMINATION OF THE "ELEMENTARY PRIMITIVES"

We define three types of elementary primitives : - the straight primitive,
- the concave primitive,
- the convex primitive.

Each of them is defined by : - one or two vectors (one for the straight primitive and two for the two others),
- the norm of the E.P.,
- the allowed gripper-orientation range, so that the finger fits the E.P. well.

To determine the E.P. we start from the result of the polygonal approximation which is a set of vectors (see figure 5).

Production rules of the E.P.

Let l be the width of a finger,
    e the distance between the two fingers,
    V_1, V_2 two consecutive vectors,
    $||V_1||$ the norm of the vector V_1,
    $||V_2||$ the norm of the vector V_2,
    (V_1, V_2) the angle between V_1 and V_2.

A straight primitive : . the norm of the vector must be greater then the finger's width, $||V_1|| > l$.

A concave primitive : . $-\pi < (V_1, V_2) < 0$
    . $||V_1+V_2|| > l$.

A convex primitive : . $0 < (V_1, V_2) < \pi$
    . $||V_1+V_2|| > l$.

DETERMINATION OF A COUPLE OF E.P.

To find a couple of primitives that fit the gripper's model well we'll

express the mathematical expression of the gripper's model in another representation. Let D,H be a reference frame whose axis D is parallel to
the fingers of the gripper (see figure 6).
In this reference frame we can see that : - the middle of the two fingers
have the same coordinate D,
                                           - the difference between the
coordinate H of the two fingers equals e (e = distance between the two
fingers),
                                           - the angle of one finger
equals the angle of the other finger plus π.

Determination of the set of possible positions of a finger along an E.P.

In fact we must determine two ranges :
   - a range of lateral position (the set of authorized values of D).
   - a range of angle which depends on the position (the set of authori-
     zed values of theta).

For a straight primitive : - the lateral position range : all the posi-
                             tions between the two extreme positions
                             (see figure 7),

                           - the angle range : the angle of the E.P. +/-
                             five degrees (tolerance).

For a concave primitive : - the lateral position range : all the posi-
                             tions between the two extreme positions,

                           - the angle range : for each lateral position
                             there is only one authorized angle.

For a convex primitive   : - the lateral position range : all the posi-
                             tions between the two extreme positions,

                           - the angle range : the angles between the an-
                             gle of the two extreme positions (indepen-
                             dent of the lateral position).

Thus if we work in the reference frame (D,H), the three conditions to have a couple of primitives that fit with the gripper's fingers well are :
   - for both the E.P. of the couple we must find a position of the fin-
     gers that has the same D,
   - the difference between the H coordinate of the two fingers must be
     smaller then e
   - the angle of one finger must be equal to the angle of the other plus
     π (parallel fingers).

Value of the parameters D and H as a function of theta

For the notations see figure 8.

For a straight primitive

$$d_1 = x_1 * \sin(theta) + y_1 * \cos(theta), \tag{1}$$
$$d_2 = x_2 * \sin(theta) + y_2 * \cos(theta), \tag{2}$$
$$h = x_1 * \cos(theta) - y_1 * \sin(theta). \tag{3}$$

For a concave primitive

$$l_1 :: \sin(\text{alpha}) = l_3 :: \sin(\text{phi} - \text{alpha}), \tag{4}$$

$$l_1 = l_3 :: (\sin(\text{phi} - \text{alpha})/\sin(\text{alpha})), \tag{5}$$

$$l_2 :: \sin(\text{alpha}) = l_3 :: \sin(\text{phi}), \tag{6}$$

$$l_2 = l_3 :: (\sin(\text{phi})/\sin(\text{alpha})), \tag{7}$$

$$p_1 = V_1 + (vp_1 - v_1) :: (l_1/||vp_1 - v_1||), \tag{8}$$

$$p_2 = V_1 + (vs_1 - v_1) :: (l_2/||vs_1 - v_1||), \tag{9}$$

$$M = (p_1 + p_2)/2, \tag{10}$$

$$d = Mx :: \sin(\text{theta}) + My :: \cos(\text{theta}), \tag{11}$$

$$h = Mx :: \cos(\text{theta}) - My :: \sin(\text{theta}). \tag{12}$$

For a convex primitive

$$d_1 = x_2 :: \sin(\text{theta}) + y_2 :: \cos(\text{theta}) - 1/2, \tag{13}$$

$$d_2 = x_2 :: \sin(\text{theta}) + y_2 :: \cos(\text{theta}) + 1/2, \tag{14}$$

$$h = x_2 :: \cos(\text{theta}) - y_2 :: \sin(\text{theta}). \tag{15}$$

ADDITIONAL PROBLEMS IN THE THREE DIMENSIONAL CASE

Some additional problems appear when we work with a container of parts.
Firstly, the contrast between the parts and the background is not so good
as in the 2D problem. It lowers the percentage of success but there is
no miracle solution. Secondly, because of the camera used we only know
the position of an object in an horizontal plane (the X,Y coordinates)
but we don't know at which depth the object is (Z coordinate). To solve
this problem the last axis of the robot is flexible and a switch placed
on this axis enables us to know when the robot's hand touches the neigh-
bouring parts. So we know when the robot has to be stopped and when we
must close its gripper.
Another thing is that the X,Y coordinates depend on the Z coordinate (see
figure 9). In reality the set of the possible positions of the object's
center of gravity is a straight line. Thus when the robot goes down into
the container it must follow this straight line.

CONCLUSION

The first results obtained are very cheering. Thanks to our first expe-
rimentations we know that we must improve some things like the contour
following, the "elementary primitives" concept, the method used to find
the second set of E.P. ... A great quality of this method is its independence of the used gripper and of the object to be grasped.
In a near future, it is in our project to apply this method to rangefinder's data. With such vision systems the results should be clearly
better in the three dimensional case.

BIBLIOGRAPHY

1. Robert B. Kelley, Henrique A.S. Martins, John R. Birk and Jean-Daniel
   Dessimoz. "Three vision algorithms for acquiring workpieces from bins".
   Proceedings of the IEEE, Vol. 71, pp. 803-820 (July 1983).

2. Robert B. Kelley, Henrique A.S. Martins, John R. Birk and Richard
   Tella, Departement of Electrical Engineering, University of Rhode Island, USA. "A robot system which acquires cylindrical workpieces from
   bins". Robot Vision, pp. 225-245. Spring-Verlag (1983).

3. Jean-Daniel Boissonnat. "Stable matching between a hand structure and
   an object silhouette". IEEE transactions on pattern analysis and machine intelligence, Vol. pami-4, No.6, pp. 603-612 (November 1982).

4. J.D. Boissonnat and F. Germain de l'I.N.R.I.A. "A new approach to the
   problem of acquiring randomly oriented workpieces out of a bin" IJCAI
   Vol. II, pp. 796-802 (1981).

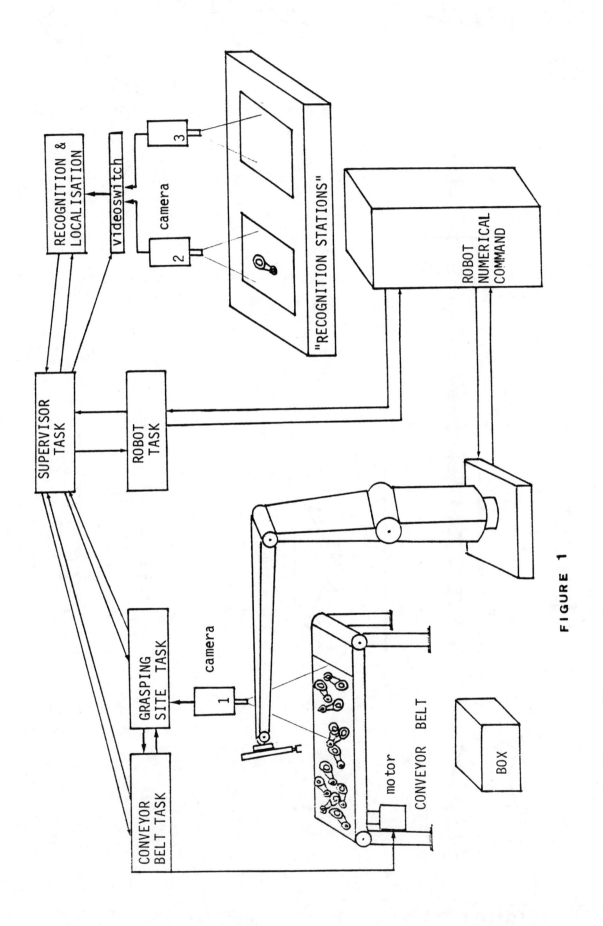

FIGURE 1

319

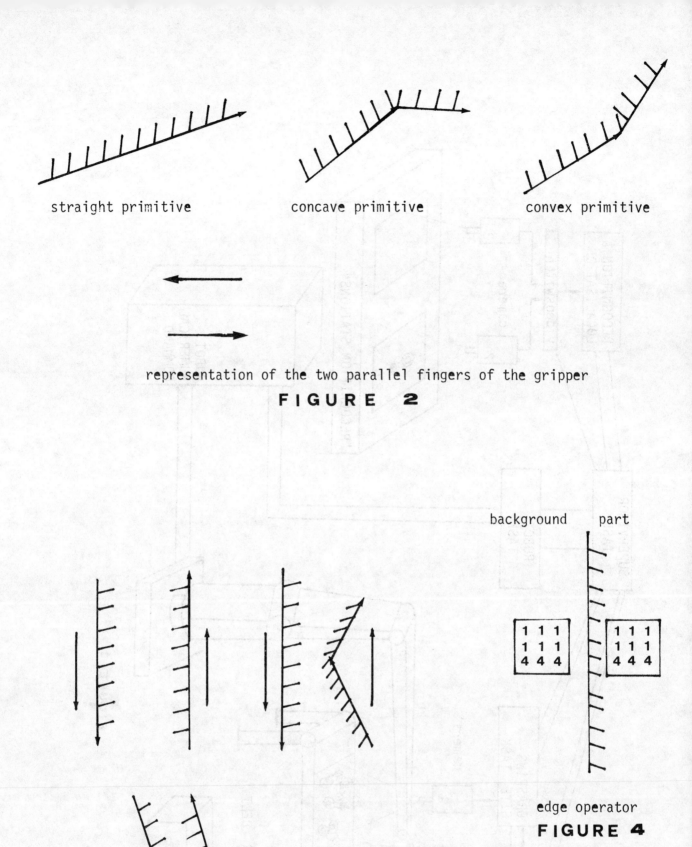

straight primitive   concave primitive   convex primitive

representation of the two parallel fingers of the gripper

FIGURE 2

background   part

1 1 1   1 1 1
1 1 1   1 1 1
4 4 4   4 4 4

edge operator

FIGURE 4

FIGURE 3: Some exemples of grasping site

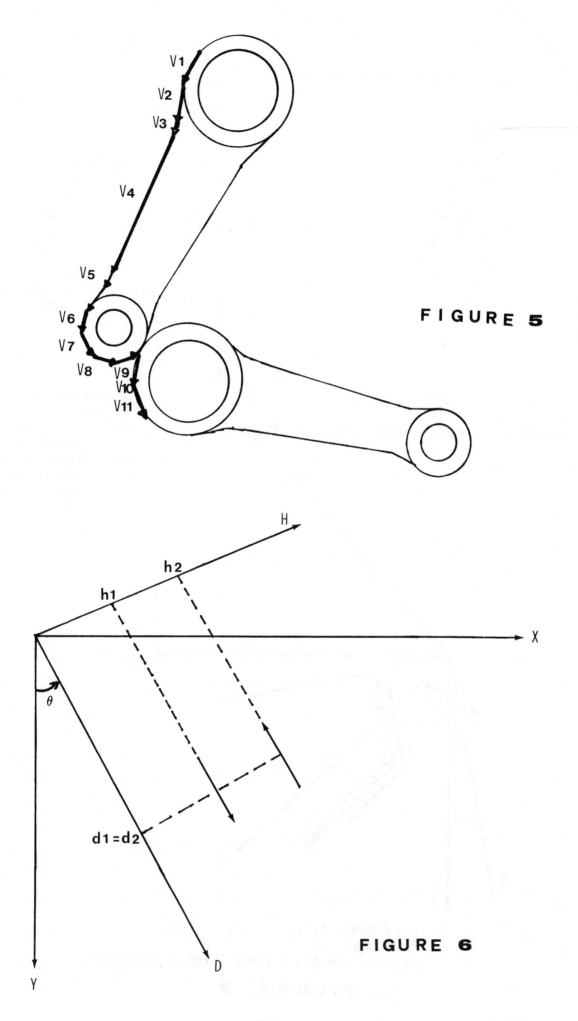

FIGURE 5

FIGURE 6

321

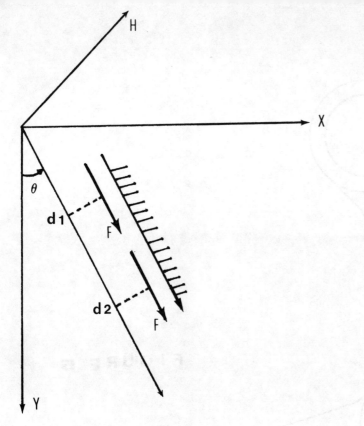

Straight E.P.
F : finger
d1, d2 : the two extreme lateral positions

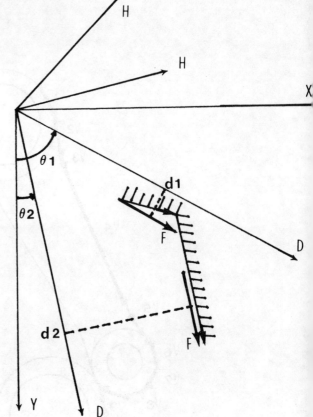

Concave E.P.
F : finger
d1, d2 : the two extreme
           lateral positions
θ1, θ2 : the angle correspond
           respectively to the
           lateral position d1 :

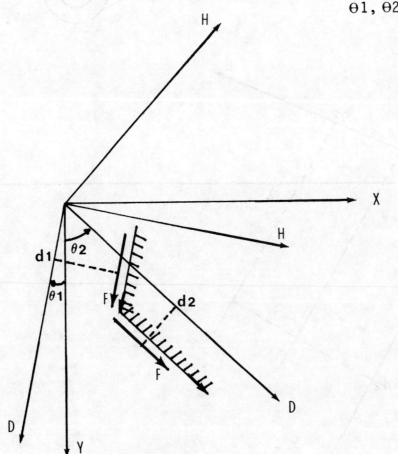

Convex E.P.
F : finger
d1, d2 : the two extreme lateral positions

FIGURE 7

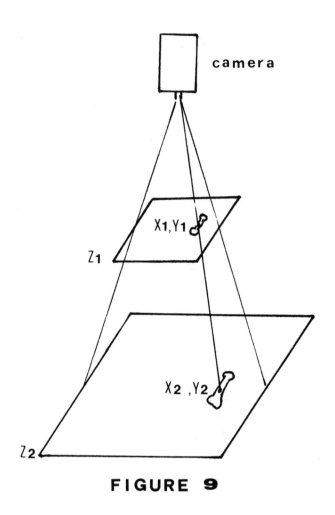

FIGURE 9

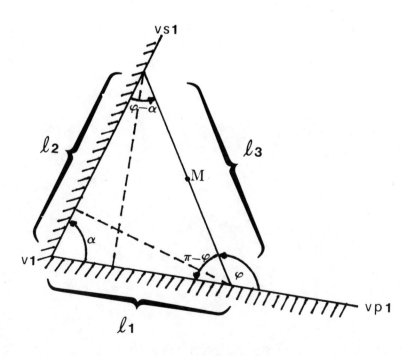

FIGURE 8

Knowledge Based Models for Computer Vision

P. J. Gregory
Visual Machines Limited, UK
and
C. J. Taylor
Wolfson Image Analysis Unit, University of Manchester, UK

Vision systems capable of analysis of grey scale images are now becoming available and offer the potential to fully address the complex inspection requirements that exist in today's manufacturing industry. Their widespread use is however likely to be limited by the cost of developing applications specific software that is sufficiently robust to achieve acceptable inspection performance. By using a "Knowledge Based Model" to direct the analysis, performance can be improved while allowing programming to be performed by non-expert users in a graphical manner. This approach is illustrated by an example from the automobile industry.

KNOWLEDGE BASED MODELS FOR COMPUTER VISION

Introduction

The growth in interest in Computer Vision for quality assurance and machine control has been the result of genuine need coupled with technological advance. The need felt by all sections of manufacturing industry is to increase product quality while reducing unit costs. The availability of low cost, high performance microprocessor elements for Computer Vision should provide the technological tools. How can they be used to satisfy this need?

Although Computer Vision systems are now available from a number of commercial sources, application of the current generation of machines is limited. This limitation has been primarily due to their capabilities and high overall cost. Successful applications have in the main involved major simplification of the problem. The major simplifications have included:

1. Back illumination of parts allowing the use of binary vision techniques for scene segmentation and analysis.

2. Analysis of simple, isolated piece parts.

In general the needs of manufacturing industry are far more complex. Manufacturers are often seeking to apply Computer Vision systems to existing processes where it is not possible to design the component or process line to simplify the job of the vision system. The inspection of complex assemblies, with high added value, is a growing requirement.*

Although the complexity and difficulty of the vision processing tasks is increasing, system costs must be reduced if the use of vision systems is to become widespread. The cost of the computer vision hardware is becoming a smaller element in the overall system cost. Time and effort involved in configuring a system for a particular application, together with the cost of software development and support will become the major component of cost.

In order to expand the range of applications of Computer Vision systems, and to meet the growing needs of industry, progress will thus be required on two fronts. Improvements must be made in the abilities of such systems to interpret and act upon complex images and the cost of doing so must be reduced. Knowledge based software techniques offer a solution to both problems.

Knowledge Based Models

Although manufactured components and assemblies can be visually complex, they are still highly constrained. Even the most flexible manufacturing cell handles a limited range of individual piece parts and assemblies at any one time. Detailed a priori information in the form of manufacturing and assembly drawings, together with material specifications are available and can be incorporated in the model of the world that the vision system must interpret.

Knowledge Based Model software for the inspection of industrial piece parts and assemblies has been developed at the Wolfson Image Analysis Unit at the University of Manchester. This software comprises a set of tools for generating new models, training them with representative examples of parts to be inspected and implementing them as part of a visual inspection or control process. It operates at two levels, a high level containing geometrical information about the construction of the part and a low level containing information describing image features.

The high level geometrical model is used to define the shapes and spatial relationships between components in an assembly. This model is constructed from a series of vectors drawn relative to an origin vector. Origin vectors may be defined as absolute to the frame of reference of a camera image, or relative to some other vector. In this way the system can model the geometry of assemblies that contain components which may move relative to each other.

Geometrical models can be organised as a hierarchy. A simple model may be used to find the location of an assembly within the field of view of a camera. This may then call more complex models defining positions of parts within the assembly.

Given this geometrical model, the system now knows where to look in the image for specific features such as the edges of components, the centres of holes or the textures of surfaces. These features are found by examining the grey scale information within the image either along vectors or within regions. For example the presence or absence of a spring may be found by performing fourier analysis of the grey scale profile taken along the axis of the spring. (Fig.1.)

Although the geometrical information is fixed for a given application, the visual appearance of different examples of the same object may vary. This variation is handled by the second, lower levels of image model. These models must be able to cope with the variations in grey scale information for the same feature from object to object. This is achieved by training each feature model with a test set of typical objects.

Consider finding the exact location of the edge of a stamped metal component within an assembly. The geometrical model is used to place a test vector across the expected position of the edge. A grey scale profile taken along the vector would, in an ideal case, show a set function from which the position of the edge could be found. Changes in the surface finish of the material, stamping shear marks, varying backgrounds and illumination highlights all make the profile less than ideal. By training on a set of profiles a matching template can be generated specific for that edge.

System Hardware and Software

The use of Knowledge Based Model software imposes different compromises on hardware design to those normally accepted in Computer Vision systems. The need for high speed execution of whole image operations such as filtering or edge finding for scene segmentation is reduced. Instead emphasis is placed upon the manipulation of data structures. These may occur at both the high level within the model data base, and at the low level in terms of grey scale profiles and region grey distributions.

The VM-1 Computer Vision system has been designed to efficiently implement Knowledge Based Model software. It consists of a number of loosely coupled processor units sharing a common memory space. High level control and model interpretation is handled by one or more Motorola 68000 32 bit microprocessors, interconnected with each other, memory and peripheral devices via a VMEBus**. The use of a standard high speed data bus simplifies the task of interfacing the vision system to other components within a flexible manufacturing cell.

Low level image point functions are performed in an Intelligent Frame Store (IFS) coupled to up to 8 high resolution 625 line videcon type cameras. The IFS comprises a 1024 x 1024 image memory (8, 12 or 16 Bit resolution), tightly coupled with a microprogramable, pipelined CPU and an independent memory address processor (MAP). The MAP, controlled by a separate field in each microinstruction, can perform two dimensional offset addressing of the iamge memory relative to a current working point. During a single 200 nsecs microinstruction, the unit can perform an arithmetic operation, compute a new image memory address, fetch a grey value from the image memory and conditionally branch to the next microinstruction.

High level software is implemented in PASCAL[I], running under the UNIX[II] operating system. All software is highly structured and makes extensive use of the data typing abilities present in PASCAL to define data structures appropriate to the processing task. A subset of the Graphics Kernel Standard (GKS)[III] is also implemented on the system to provide a unified user interface to the model generation and runtime software.

Automobile Brake Inspection

Volkswagen AG as part of their investment in the new GOLF 2, have installed a new automated assembly line for rear brake drums at their plant in Braunschweig, BRD. The line, based on the modular transfer system manufactured by Robert Bosch GmbH, uses manual, semi-automatic and automatic workstations to perform the complete assembly operation on the drum. (Fig.2.)

Brake assemblies are obviously a critical safety component on a car, and rigorous quality assurance based on 100% inspection is enforced. To further improve these standards and to cope with the increased production made possible by the new line, Volkswagen have installed a computer vision system incorporating Knowledge Based Model software to check finished brake assemblies.

The inspection station, constructed as a module on the transfer line, uses three high resolution cameras to view the part. In addition, the station provides a 180 degree rotation of the part to allow imaging of all four side views together with a plan view. Operation of the station is controlled from the vision computer, making the image analysis and control independent of the remainder of the line.

Correct component illumination is critical to the operation of any vision system in providing good contrast between the features of interest and the background. In this system, three colour structured lighting, matched to filters on the cameras is used to highlight different parts of the assembly in each view. Red diffuse lighting is used for the plan view, while green tangential lighting is used to highlight surface texture on the brake shoes. (Fig.3.)

In operation a brake assembly, mounted on a transfer line pallet, is moved into the station. The plan view is then analysed to find the exact location and orientation of the shoes, which may move relative to the backing plate. (Fig.4.) Specific checks are then made on each item in the assembly to check for the presence of correct components, correct dimensional tolerances between components and for possible damage during assembly. Image processing techniques used range from edge and hole finding to locate positions, fourier analysis for components such as springs to texture analysis for showing surface roughness due to grinding faults.

Summary

Advances in technology are opening up the prospect of applying increasingly sophisticated computer vision techniques in automated industrial inspection. In practice it is essential that this additional sophistication is available to the end user without the costly and inefficient intermediary of a computer expert. This means that the production engineer must be able to communicate prior knowledge about the nature of a product to be inspected via a direct and natural interaction with the vision system. The vision system must be capable of operating according to rules contained in data structures built up by this interaction. Software of this type has been developed and its application to a real production problem has been described. The VM-1 is a low cost computer vision system incorporating this Knowledge Based software.

References

*   Proceedings of the 3rd International Conference on Robot Vision and Sensory Controls – IFS (Publications) Ltd 1983.

**   VMEBus Technical Specification – United Technologies, Mostek August 1983.

Ŧ   PASCAL User Manual and Report – Jensen K and Wirth N 1974 (Spinger-Verlag).

ŦŦ   UNIX is a trademark of A.T.T.

ŦŦŦ   Inroduction to Graphics Kernel Standard – Hopgood, Duce, Gallop, Sutcliffe (Academic Press 1983).

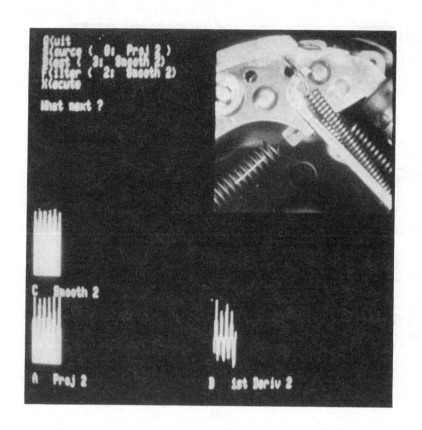

Fig.1. Grey scale profile along the axis of a spring

Fig.2. Automobile Drum Brake assembly

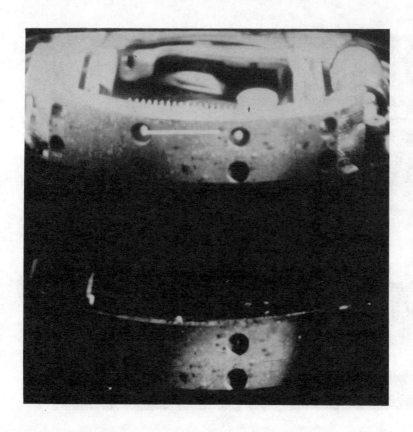

Fig.3. View of shoe lining showing lighting for surface texture measurements

Fig.4. Geometrical model construction for finding thickness of brake shoe lining

A Conceptual Approach to Artificial Vision

G. Adorni, M. Di Manzo, F. Giunchiglia

and

L. Massone

Istituto di Elettrotecnica, University of Genoa, Italy

ABSTRACT

Object description based on geometrical features are not very satisfactory because of wide variations in structure and shape often due only to stylistic purposes. A more functional description is needed, which in turn requires some knowledge about the everyday experienced physical processes in order to be linked to the observed shapes. Some preliminary geometrical operations on visual data are also discussed, and a representation for integrating structural and functional information is also proposed.

1. INTRODUCTION

Recent results from the psychology of human vision and the application of artificial intelligence techniques to machine vision are pointing out that the recognition of complex, 3-D scenes can be approached only by making bottom-up, data driven processes to cooperate with top-down, expectation driven ones. This leads to a hierarchical but feedbacked model, which can be organized in three main modules:

i) Early processing, devoted to extract a set of features, using only very basic assumptions about physics and optics, in a typically data driven way;

ii) Geometric reasoning, devoted to perform geometrical operations as the identification and aggregation of areas, and the recognition of objects based on some kind of structural knowledge;

iii) Conceptual processing, in which the residual uncertainty resulting from ambiguous object shapes, occlusions, special light effects and so on, can be faced only using a more general knowledge and reasoning capabilities about the expected environment. This kind of knowledge typically includes (but is not limited to):

1. environment organizations and spatial relations among objects;
2. naive physics theories, to be able to reason, for instance, about object equilibrium, supporting and so on;
3. functionality of objects, and the derived constraints on shape.

In this paper a more functional approach to object description is suggested as a tool to overcome the difficulty of making clear classifications on the basis of purely geometrical features. In section two the reasons for this approach are discussed and the required cognitive tools are briefly outlined. Section three is devoted to preliminary geometrical operations. In section four some considerations are made about the integration of functional and geometric descriptions and the use of a-priori knowledge to guide searching and matching.

2.   THE SEMANTICS OF OBJECTS

When we address the problem of object representation, the question immediately arises if a description based on purely geometrical features can be powerful enough to capture the inner semantics of the observed objects. Despite the large popularity of descriptions based on primitive volumes, as in ACRONYM (4), or surfaces, as in IMAGINE (6) when we try to use this kind of representations to model an even simple object, as a chair, we get quickly into troubles. Look, for instance, at the set of objects depicted in fig.1. If we describe each of these objects by means of a list of primitive components plus a set of interconnecting relations, all the resulting descriptions are clearly different from each other. However there are some similarities, so that, given some proper restrictions on the overall dimensions, nobody should fail in identifying a and b as two instances of a chair, c and d as stools, e as a shelving and f as an unknown object which resembles a chair but which could not be easily used this way. This kind of clustering can be performed also on the basis of the previously suggested descriptions, relaxing the structural constraints on the matching conditions and putting some threshold on the total allowed mismatching.

However, as shown in (13), it is difficult to tune the matching procedure so that the obtained clusters correspond to our common sense classifications. In fact, depending on the assumed weights and thresholds, we can obtain, for example, the clusters:

$$\{a, c, d, g\}, \{b\}, \{a, e, f\} \quad or \quad \{a, b, c, g\}, \{d\}, \{e\}, \{f\}$$

The reasons for this difficulty to have a truly satisfactory clustering is that common sense classifications are often based on some kind of judgement on the apparent functionality of the observed objects, and, unfortunately, there is no simple way to relate functionality to structure. Very often a lot of even complicate structural details are due only to stylistic reasons, or perform specific tasks, as increase robustness (see object d versus object c) or permit rotation (see object b), which have only secondary influence on the object main functional goal. So, what knowledge is necessary to check the "chairness" of an observed object or, in other words, how can we give a satisfactory intensional representation of it (15)?

A first, naive functional description of a chair could be "an object which can support a human body in sitting position". The geometric meaning of "a human body in sitting position" may be roughly described by means of generalized cylinders, joints and simple constraints on rotations. The functional aspects of the previous definition of chairness are related to the action of giving support; hence, some knowledge about the action of the force of gravity and the rules of statics must be included in the cognitive database of the system.

To define a proper subset of rules, we must take into account the high level of uncertainty which is typical of shapes inferred from visual data; so, a qualitative theory seems better suited than a hard quantitative one, since it allows to reach some conclusions even for ill-defined problems (7).

To briefly outline our qualitative approach to statics, let's consider a simple object as the cylinder C resembling the human trunk. This cylinder has a planar convex base Bc containing the orthogonal projection Pc of the object barycentre. If C must be supported by another object S, having a plane horizontal top surface Ts, the contact area a between Bc and Ts must contain the point Pc and must be "large enough" to ensure the stability of C. "Large enough" is a common sense evaluation which can be formalized defining a Minimum Equilibrium Area M_E_AREA around the point Pc (3). The evaluation of the capability of S to give support to C is basically a placement strategy to find a stable position for C on S; in such a simple case, the placement strategy consists only in checking that the contact area a can contain the M_E_AREA. Some difficulties can arise when there is a pattern of separate supports, as shown in fig.2. Here both the support pattern and the supported object exhibit an axial symmetry which simplifies the following discussions. However symmetry is not really useful, since an easy generalization can be based on the concept of skeleton of the empty space between the supports (8), and moreover it can be cumbersome to look for symmetries in the visual data. In the symmetrical situation of fig.2 the simplest placement strategy is to align the object axis to the support one. In this case the interior contours of each support and the object contour can be divided into a number of intervals, so that for each interval [Xi, Xi+1] and [Yj,Yj+1] we have the situation described in figs. 3a and 3b respectively.

Of course, some situations are mutually exclusive (type a with type A or type b with type B intervals).

Equilibrium positions may be found superimposing object intervals to support ones by means of rules which are specific for each combination of types. For example, one type A and one type b intervals can be used to search for an equilibrium position by means of a rule that can be roughly expressed as:
"put type A on type c and type C on type b so that distance t (see fig.4) is maximized". The supporting area obtained this way is shown in fig.4 (dashed area).

This kind of rules can be extended in order to reason about situations as that depicted in fig.5a, where the top surface of the supporting object is a grid whose holes may be ignored, due to the dimensions of Bc, (this is not true for the grid in fig.5b), so that this apparently complex pattern of interconnected supports can be replaced by a simple convex surface. As a conclusion, the grid in fig. 5a could be the top surface of a chair while this hypotheses is likely to be discarded in the case of fig.5b.

To perform this kind of analysis, a high level of organization of the visual data is necessary, in order to organize (from a functional point of view) the analysis of problems, deriving from the seemingly infinite variety of object structures, such as surface and volume identification, occlusion analysis, surface orientation and coplanarity deduction in perspective projection.

The next section is devoted to a set of preliminary operations on visual data.

3.   PROCESSING GEOMETRIC INFORMATION

This section is devoted to explain how some of most common geometric problems are faced in this work. Problems considered here are: i) occlusions analysis, ii) surface orientation and coplanarity deduction in perspective projections, iii) volume identification. At this stage of our work these problems have been considered within a space environment in which objects have planar surfaces; what is needed to achieve domain generality is an efficient and suitable solid modeler (11,8,10), which is still an open problem and goes far, beyond the scope of this paper.

3.1. Occlusion analysis

One of the crucial points in artificial vision is the presence in images of perceptual ambiguities rising from partial occlusions of objects, depending on the specific observer position; to allow understanding visual data, the information about shape of objects has to be restored. At this purpose a project has been developed (1), called EXPOSE, which infers the shape of occluded surfaces on the basis of some perceptual principles. The reconstruction of hidden parts is performed looking for situations being perceptually instable like, for instance, concavities and discontinuities in boundaries. This strategy refers to Rubin's principle (12) stating that convex regions are usually preferred to concave areas as figures over a ground. This fact, already taken into account in previous works (14), is introduced in EXPOSE, according to Hochberg's "principle of minimum": the perceptual system processes visual data in such a way to keep changes as low as possible. A measure of complexity is then defined on the basis of concavities giving directions for local processing. The occluded shape is recovered by means of a set of rules implementing principles of good continuation, symmetry, minimal distance which can be fairly applied in EXPOSE because boundaries of objects are described by means of the Coding Theory developed at Psychology Labs of Nijmegen University (5), which is able to express the regularity of shapes with a formal language. In this way some local hypotheses on the real shape of regions are triggered and their consistency is subsequently checked within the global structure of the image. A backtracking mechanism allows removing inconsistent solutions. Fig.6a shows the image of a table top covered with various objects and in fig.6b the input data are presented, which are derived by means of a search algorithm (2). Fig.6c shows how surface shape has been transformed by EXPOSE.

3.2. Deducing surface orientation and coplanarity

Projections of objects into two dimensions always introduce distortion and ambiguity. The inverse problem, i.e. recovering spatial information from images, is then ill-defined. Many mathematical methods exist, known as "shape-from" methods, exploiting low-level features such as contours, texture, motion, shading, high light, etc. (9).

Our purpose is to define a set of naive qualitative rules avoiding complex mathematical computations to deduct some rough information about 3-D surfaces which can help cognitive processing as spatial orientation and coplanarity relationships. It is worth noting that this set of rules are far from claiming general validity because of domain restriction; in fact, our goal is not to find a solution to the inverse problem, but just to suggest plausible hypotheses to drive the following steps, where they can be confirmed or rejected. An example is given by the analysis of the image in fig.7. Assuming that the surfaces s1, s2, s3 are approximately rectangles (such an assumption is based on the preference given as long as possible to the interpretation of the percepted as regular figures), a rule as R1 can be used to raise the hypothesis of horizontal lying for s1, s2 and s3. Similarly rule R2 generates the hypothesis that surface s5 lies on a tilted plane and R3 generates the hypothesis of vertical lying for surface s4.

R1) IF   two pairs of not adjacent sides belonging to
         a surface S converge on the horizon line (HL)
    THEN S lays on a horizontal plane

334

R2) IF   a pair of not adjacent sides belonging to
          a surface S converges on the HL
    AND   the remaining sides converge on a set of
          points$\{$pi,i=1 to n$\}$not belonging to HL
    THEN S lays on a tilted plane

R3) IF   two sides belonging to a surface S are
          vertical (i.e. orthogonal to HL)
    THEN S lays on a vertical plane

R4) IF   a pair of not adjacent sides of s1 lays on the
          prolongation of a pair of not adjacent sides of s2
    THEN s1, s2 are coplanar

The rule R4 is used to make the hypothesis of coplanarity (coplanarity is a basic prerequisite when we look for a seat in our block world) of the surfaces s1, s2 and s3. This hypothesis, as the previous ones, will be submitted to a first check by rules working on (apparent) relations between inferred volumes. In the case of fig.7 the three parallelepipeda P1, P2, P3, inferred in the next step, seem to have the same thickness and seem to be connected to the same couple of parallel supports P4 and P5: the situation enforces the hypothesis of coplanarity for s1, s2 and s3. Further checks are based on higher level representations and will be briefly discussed in the next section.

3.3. Volume identification

We refer here to the problem of grouping surfaces into a suitable volume. In the following are very briefly outlined some criteria which could help to perform this task:

i)   presence of occlusion relationships among adjacent surfaces;

ii) presence of totally common sides between adjacent surfaces;

iii) presence of fork points (which can be derived by means of classic labeling algorithms (14)).

Consider, for example, fig.8. According to i) grouping of B1-A1 and B3-A2 is not considered because of the occlusion relationship (see dashed line); following ii) A1-A2, A1-A3, A2-A3 are grouped together because they have totally common lines; grouping of C1-A3, C2-A2 is disregarded because of fork point p1.

4.   REPRESENTATION AND SEARCH

Let's now turn back to the representation of chairness. First of all, we write the simple functional definition suggested in section two into a slightly more formal frame-like structure, as follows:

```
object:       chair
f_goal:       horizontal_support  sitting_human_body
f_cond:       horizontal_support by floor
```

The restriction of the supporting section is obviously introduced to distinguish a chair from a hook which can give support by hanging; the introduction of functional conditions f_cond asserts that a chair usually is not floating in air, so its structure must be able to ensure a stable position on a planar horizontal surface when it is properly supporting a human body.

This definition is too vague, because it can be clearly matched by a number of objects which are not chairs. Further restrictions on structure and shapes can be introduced to distinguish a chair from, for instance, a cube (that is to distinguish the sittingness from the chairness) as follows:

```
object:          chair
f_goal:          horizontal_support  sitting_human_body
f_cond:          horizontal_support by floor


   component: seat
   f_goal:    horizontal_support  sitting_human_body
   dimens_restrictions: ...


   component: leg
   f_goal:    rigid_support seat
   f_cond:    orizontal_support by floor


   component: back
   f_goal:    back_rotation_support

 rigid_connection: seat, back
 rigid_connection: legs, seat
```

So structural and geometric features still play a basic role in building descriptions, but only those which are really distinguishing must be included, and unspecified details do not cause mismatching when an observed object is compared with a prototypical description. Restrictions can be functional as well. In fig.1 chair b can be put into a separate class of chairs if a prototypical frame exists where the "component: leg" has the "f_restriction: vertical_axis_rotation"; this restriction must be interpreted by a suitable theory of rotations which infers the relational properties of any given geometric structure.

Given a set of more or less sophisticated functional descriptions of objects, we need a strategy to match the observed patterns of connected volumes to these prototypes. The first problem is where to look for a specific object in the image of the scene. Any searching procedure, to be cost effective, must take advantage of some a-priori knowledge of the current environment, which can raise expectations about the most likely positions of objects. So, a search for a chair in a room should look firstly for objects on the floor, with proper constraints on the overall dimensions, and secondly for objects on tables or other pieces of fornitures where there could be reasons to put chairs. Knowledge about the most likely relations between objects is embedded in the object frames and is used to build indirect searching procedures. For example, if the goal is to find a lamp, the first step of the search consists of looking for a table or for some other object on which the lamp frame says that a lamp can be. Each object in this chained search defines a local context for the further step. Object ordering depends on the "identifiability", which, in the current implementation, is simply related to their overall dimensions or to distinguishing geometric features as large horizontal or vertical surfaces. The criterion of looking firstly for expected "big" objects works also when the goal is not to locate a specific object but to build a general description of the current scene. In this case the selection of the prototypes to match with each observed object is guided by the content of the frame structure describing the prototype of the assumed environment; when an object is recognized, this can raise new expectations because of the relations that link it to other objects in the current contexts and so on. Only those objects which are placed far from their typical positions, as a chair hanging on the ceiling, or which have no typical positions, as a fly, will escape this phase of analysis and require a further extensive matching.

The second relevant problem is the matching criterion. The preliminary operations described in section three often cannot give sufficient evidence of the fact that a set of apparently connected volumes really is a unique object. In such a case a prototypical description can be matched by a subset of the observed volumes. This situation results in a hypothesis of instantiation which can be rejected later if it is not possible to find a consistent interpretation for the unmatched volumes. Inconsistency can be caused not only by failure in finding a proper prototype, but also by the impossibility of justifying the actual object positions through the knowledge the system has of the equilibrium rules or other physical rules.

On the other hand, some relevant details may be completely missed because of occlusions or perspective. To handle this situation the matching requirements must be suitably relaxed. Let's consider, for instance, the image of fig.9b. Since we see only two legs, the functional description of the leg component of the chair frame is not matched. However the object does stand on the floor, and the system knowledge of the equilibrium rules could not justify this fact without assuming the existence of a more complete set of legs: so, the existence of two occluded back legs is assumed, for symmetry reasons, and this hypothesis can be accepted since it is consistent with the current point of view. For the same reasons this assumption were rejected in the case of fig.9a, and the system would formulate a different hypothesis, as that of a rigid connection with the floor, to justify the object upright position.

5. CONCLUSIONS

Even if most problems addressed in this paper are only scratched, we claim that common sense reasoning about facts which are in our every day experience is a necessary tool to improve object description, hypothesis generation and consistency checking. It were tempting, at this point, to list some major areas which need deeper investigation, as environment setting, shape constraints from functionality, relations between top-down and bottom-up expectance generation and so on. However, the whole area of cognitive modeling is still largely open, and a great effort is necessary in this field, since vision is basically a cognitive process sharing its knowledge domain with other basic human activities as language interpretation or action planning. We feel that a significant improvement of the performances of artificial vision systems will be achieved only by means of a thorough investigation of this cognitive background.

ACKNOWLEDGEMENTS

This work has been supported by the Italian Departement of Education under grant No. 27430.81

REFERENCES

1. Adorni, G., Massone, L., and Trucco, E., "Using perceptual principles in image segmentation", Proc. IASTED Symposium on Robotics, Amsterdam, Holland, (June 1984).

2. Adorni, G., DiManzo, M., and Massone, L., "Coding Theory and Computational Vision", Proc. IASTED Symposium on Applied Informatics, Innsbruck, (February 1984).

3. Adorni, G., DiManzo, M., and Giunchiglia, F., "From Descriptions to Images: What Reasoning in Between?", Proc. 6th European Conference on Artificial Intelligence, Pisa , (September 1984).

4. Brooks, R.A., "Symbolic Reasoning among 3-D Models and 2-D Images", Artificial Intelligence Vol. 17 pp. 285-348 (1981).

5. Buffart, H. and Leeuwenberg, E., "Structural Information Theory", in Modern Issues in Perception, ed. Geissler, H.G., Buffart, H., Leeuwenberg, E. and Sarries, V.,North Holland, Amsterdam (1983).

6. Fisher, R.B., "Using Surfaces and Object Models to Recognize Partially Observed Objects", Proc. 8th IJCAI, Karlsruhe, (August 1983).

7. Forbus,K., "Qualitative Process Theory", A.I. Memo N.664A, M.I.T., Boston, MA (1983).

8. Gaglio, S., Massone, L., and Morasso, P., "Representing Shape for Visuo-Motor Processes in Robots", Proc. Intenational Symposium on Artificial Intelligence, Leningrad, (October 1983).

9. Marr, D., Vision, Freeman, W.H., San Francisco, CA (1982).

10. Massone, L. and Morasso, P., "SCULPTOR-2: Representing, Generating and Editing Natural Planar Shapes.", Proc. AIMSA-84, Varna, (September 1984).

11. Requicha, A.A.G., "Representations for Rigid Solids: Theory, Methods, and Systems", Computing Surveys Vol. 12 N. 4, (December 1980).

12. Rubin, E., Visuell Wahrgendumene figuren, Glyndenhals, Copenaghen (1921).

13. Shapiro, L.G., Moriarty, J.D., Haralick, R.M., and Mulgaonkar,P.S., "Matching three-dimensional models", Proc. IEEE Conf. on Pattern Recognition and Image Processing, Dallas, (August 1981).

14. Waltz, D., "Generating Semantic Descriptions for Drawings of Scenes with Shadows", in The Psychology of Computer Vision, ed. Winston, P.H.,MIT Press, Boston, MA (1975).

15. Woods, W.A., "Foundations for Semantic Networks", in Representation and understanding, ed. Bobrow, D.G. and Collins, A.,Academic Press, N.Y. (1975).

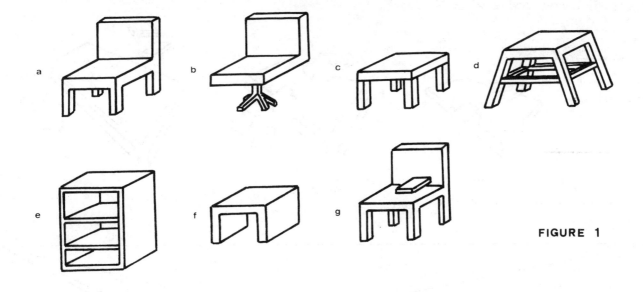

FIGURE 1

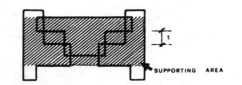

FIGURE 2

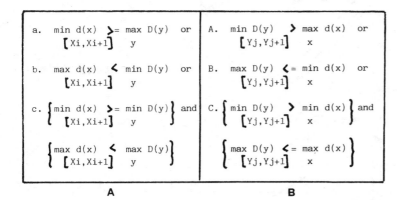

A	B
a. $\min_{[Xi,Xi+1]} d(x) \geq \max_{y} D(y)$ or	A. $\min_{[Yj,Yj+1]} D(y) > \max_{x} d(x)$ or
b. $\max_{[Xi,Xi+1]} d(x) < \min_{y} D(y)$ or	B. $\max_{[Yj,Yj+1]} D(y) \leq \min_{x} d(x)$ or
c. $\left\{ \min_{[Xi,Xi+1]} d(x) \geq \min_{y} D(y) \right\}$ and $\left\{ \max_{[Xi,Xi+1]} d(x) < \max_{y} D(y) \right\}$	C. $\left\{ \min_{[Yj,Yj+1]} D(y) > \min_{x} d(x) \right\}$ and $\left\{ \max_{[Yj,Yj+1]} D(y) \leq \max_{x} d(x) \right\}$

FIGURE 3

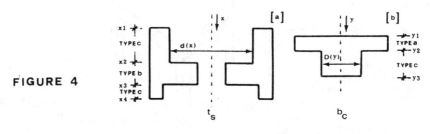

FIGURE 4

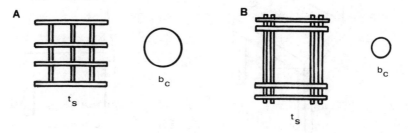

FIGURE 5

339

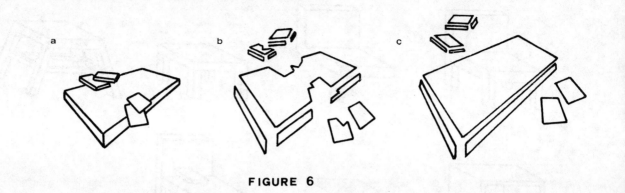

FIGURE 6

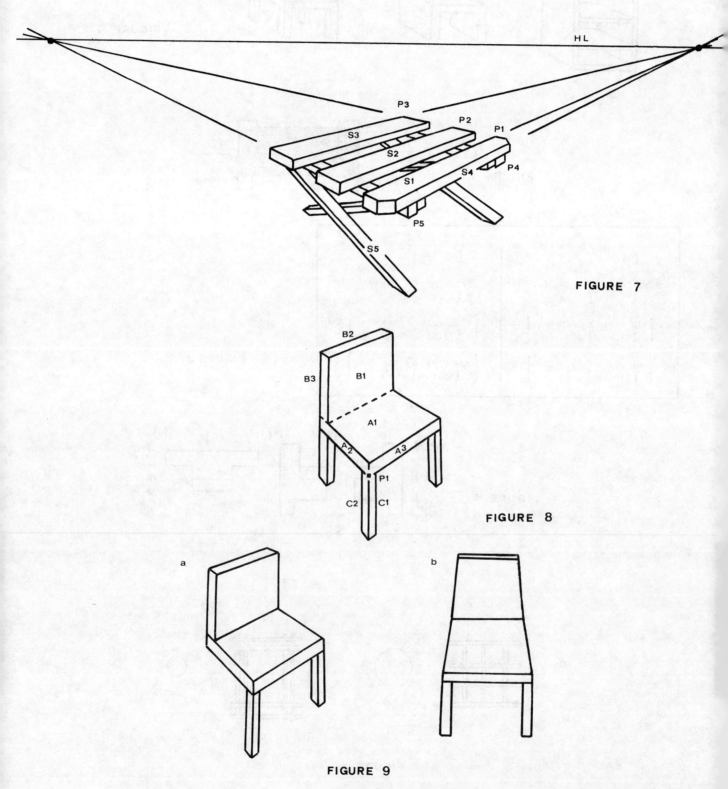

FIGURE 7

FIGURE 8

FIGURE 9

340

ADVANCED VISION TECHNIQUES

Heuristic Vision Hole Finder

R. B. Kelley
and
P. Gouin
Robotic Research Center, University of Rhode Island, USA

ABSTRACT -- In the usual methods, the holes must be of known shape, for example circular, and corrections are needed to account for perspective distortion or hole surface orientation. Heuristic vision algorithms are based on the notion of locating good holdsites on some part of the objects seen by the camera. The power of the heuristic approach lies in the fact that these algorithms can be uninformed about the nature of the object itself. The heuristic vision hole finder seeks edge patterns which correspond to holes -- any promising hole. By using neighborhood operators applied in both horizontal and vertical stripes across the image, the hole centers are located directly. The algorithm requires modest computational resources (it is micro-computer implementable) and is designed to minimze execution times (which depend on scene complexity).

INTRODUCTION

Automation is being considered for most industrial activities as a way to increase productivity by removing the human element from monotonous, hard, or dangerous work. The trend has been to replace manual labor in most steps of a manufacturing or assembly process. Usually, the job of loading or unloading a machine or of transporting or assembling the final product is left to a human operator. These tasks are now being targeted for automation. Reprogrammable robots could be used to handle different parts by changing some software and/or making minimal changes in the hardware, like changing a hand, or a lens for a camera. There are many jobs best done by robotic machines. Picking up raw castings is one of them. A serious problem with raw castings is their appearance is not completely predictable when they are fresh from the mold. Many castings have holes and, although the shape of the holes vary, these holes would permit the castings to be grasped and brought to a desired location. If a robot could grasp castings by locating such holes, then it could be used to segregate raw castings according to type.

Several contributions have been made towards a solution to this problem. In 1978, Rosen et al. [1] developed a system that acquired castings by dragging an electro-magnet through a bin of castings. In 1980, Ferloni et al. [2] used the same technique for billets. In 1977, to stimulate research on this problem area, Baird [3] made available a data base of images of castings in hoppers. When dealing with parts in heap, a non-planar problem, knowledge of the pose (position and orientation) of a part may not be sufficient to guarantee it can be acquired. Even when the pose of a part is known, it may be found that all possible holdsites are hidden or difficult to access. To check on these conditions, adjacent parts may have to be identified and their poses computed, thus complicating the acquisition process. Partially in response to this challenge, what appears to be the first experimental vision-based robot system to acquire a class of parts from a bin was reported by Birk et al. [4] reported in 1979 . In 1981, Kelley et al. [5] reported the first robot system using vision and simple hand sensors to acquire connecting rod castings fron a bin. This paper presents a continuation of the work at the University of Rhode Island aimed at creating more general sensor-based robot systems [6-13]. For this reason, a vision algorithm was developed which can find a hole in an object regardless of the shape and orientation of the hole.

HEURISTIC VISION APPROACH TO ACQUISITION

A vision-based robot part acquisition system needs two types of sensory information -- remote and local. Remote information is provided by a television camera, a laser range finder, or other vision type sensor. This information can be used to direct a robot hand to the vicinity of a part. Local information is provided by both contact and non-contact sensors. This information can be used to modify the control of the robot when the robot hand reaches the vicinity of a part, when the hand comes in contact with a part, or when grasping occurs. It can also be used to decide whether or not to abort an acquisition attempt.

Either type of sensing could provide sufficient information to acquire a part from a heap. A perfect vision system with the ability to determine the exact position of a part could direct a robot hand to grasp a part without any need for local feedback. On the other hand, a robot with only local and no remote sensing, a "blind" robot, could acquire parts by searching the work volume. By scanning until contact with a part was sensed, for example, a robot could use tactile-type sensors to grasp the part. Ideally, both types of sensing should be used. The heuristic vision approach relies on simple local sensors to actually accomplish the acquisition of a part. Remote vision is used to guide the robot hand to the part. The approach presented here tries to overcome the usual problems of using remote sensors to guide the robot hand [11]. Except for a knowledge of the type of hand mounted on the robot (surface attachment- or clamping-type), little part specific information is used.

Heuristic vision algorithms are based on the notion of locating good holdsites on some part of the objects seen by the camera [12]. The power of the heuristic approach lies in the fact that these algorithms can be uninformed about the nature of the object itself. In an image containing randomly oriented objects, this is done by finding image features which correspond to the robot hand geometric requirements. This information is used to guide the hand to a place on the object where it can be grasped. The heuristic vision hole finder is an extension of these ideas to the problem of finding holes in an object which might permit it to be grasped. It does this by seeking edge patterns which correspond to holes -- any promising hole. By using neighborhood operators applied in both horizontal and vertical stripes across the image, the hole centers are located directly.

HOLE FINDER ALGORITHM

The algorithm is a hybrid algorithm which has the best characteristics of its parent algorithms for parallel-finger hands, the "Collision Fronts" algorithm and the "Matched Filter" algorithm [12]. The Collision Fronts algorithm finds parallel edges on a part by creating an edge-direction image using a 4-neighbor edge operator. When the edge-direction pixels move toward each other, they will collide. The collision points are modeled by constant curvature line segments to form collision fronts (see Fig. 1). These segments provide support for the hypothesis that there are parallel edges in the image which might be suitable for a parallel-finger hand to grasp. In

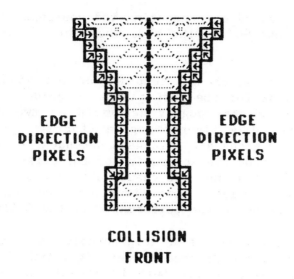

EDGE DIRECTION PIXELS   EDGE DIRECTION PIXELS

COLLISION FRONT

Fig. 1. Collision Fronts algorithm. Sketch of collision front formation through colliding edge direction pixels.

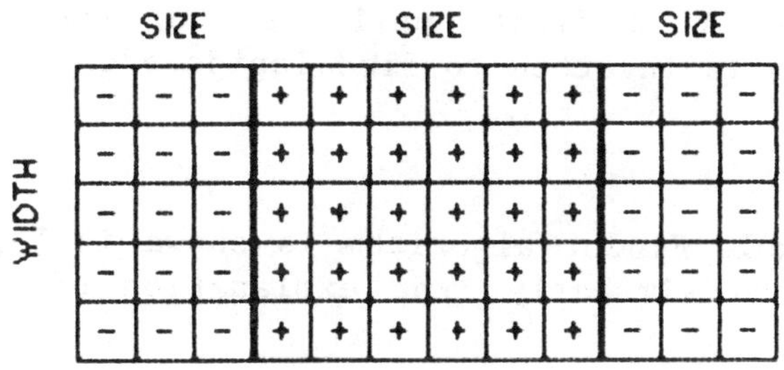

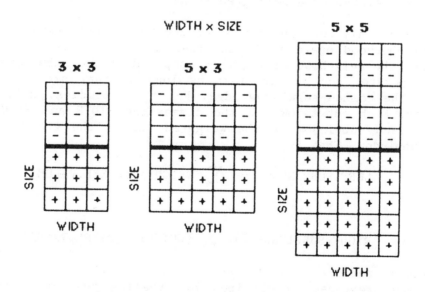

Fig. 2. Matched Filter algorithm. Typical template for parallel-finger hand (5 x (3, 6, 3)).

Fig. 3. Hole Finder algorithm. Typical neighborhood operators, 3 x 3, 5 x 3, 5 x 5.

1980, when this was first tested on connecting rod castings, tests indicated that the hypothesis was correct almost 90% of the time when three attempts per image analysis were allowed.

The Matched Filter algorithm uses a template of the idealized intensity pattern such shown in Fig. 2 to look for the best places to guide a parallel-finger hand to grasp a part. The pattern can be roughly described as two darker regions (for the finger-tips) on both sides of a lighter region (for the part to be grasped). The dimensions of the template regions are both hand dependent (for the size of the darker regions) and part dependent (for the separation of the darker regions). This algorithm has yet to undergo objective testing; but casual observation suggests that, because of the larger aperture of the template, noise is suppressed and the selection of false holdsites is minimized relative to the performance of the Collision Fronts algorithm.

The Hole Finder algorithm utilizes a neighborhood operator which is roughly one half of the template used in the Matched Filter algorithm. The neighborhood operator is symmetric so that the response for both halves of an equivalent matched filter template can be obtained on one pass. Examples of typical neighborhood operators are shown in Fig. 3. Because the responses are not constrained to have a fixed separation distance, it is necessary to match responses which correspond to parallel edge pairs -- as in the Collision Fronts algorithm. Since the operator is applied in only one direction on each pass, the search for pairing is carried out only in that direction. Both horizontal and vertical passes of the neighborhood operator are used -- as in the Collision Fronts algorithm. As a first application of these ideas, the finding of holes, the search direction is inverted to find inside edges.

Algorithm Description

The flowchart for the Hole Finder algorithm is shown in Fig. 4. Since both passes of the operator are treated similarly, the explanation of the different phases is given for the combined case.

1. Acquire Image. A gray scale image of the scene is acquired and stored in memeory as a two-dimensional array (see Fig. 5).

2. Apply Horizontal and Vertical Edge Operators. Two output images are obtained, one for each pass of the neighborhood operator. The horizontal and vertical neighborhood operators are related by a 90-degree rotation.

In the following, only the application of the horizontal operator is described. The output of a neighborhood operator is obtained only when the neighborhood window is within the image. The neighbood operator is composed of two subwindows. The weights of each subwindow operator can be arbitrary as long as they are mirror anti-symmetric with respect to the common subwindow edge. For these studies, uniform weights of +1 and -1 were used. Each subwindow has the dimensions WIDTH and SIZE, as shown. The contribution of each subwindow is the weighted sum of the gray-scale values of the pixels in the subwindow. That is,

$$\text{LEFT.SUBWINDOW} = \sum (\text{WEIGHT})(\text{PIXEL.VALUE})$$

and

$$\text{RIGHT.SUBWINDOW} = \sum (\text{WEIGHT})(\text{PIXEL.VALUE})$$

Holes in general reflect less of the incident illumination than the part itself. Hence, a left edge response is obtained by treating the contribution of the left subwindow as positive and the right subwindow as negative. A right edge response is obtained by simply inverting the sense of the operator output. The edge response operator output is difference of the two subwindow contributions. Then,

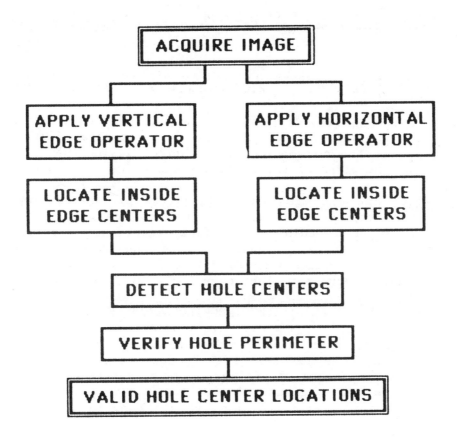

Fig. 4. Hole Finder algorithm. Flowchart.

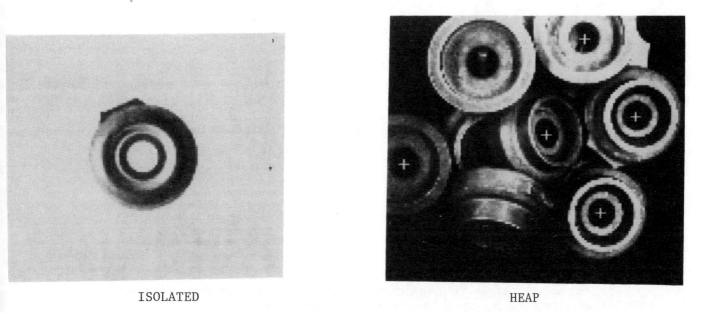

ISOLATED   HEAP

Fig. 5. Digitized gray-scale images of ISOLATED and
        HEAP of gear blank castings. Hole centers
        found by the Hole Finder are marked on the
        HEAP image.

$$\text{LEFT.EDGE.RESPONSE} = \text{LEFT.SUBWINDOW} - \text{RIGHT.SUBWINDOW}$$

and

$$\text{RIGHT.EDGE.RESPONSE} = \text{RIGHT.SUBWINDOW} - \text{LEFT.SUBWINDOW}$$

In this study, the neighborhood operator is applied in non-overlapping, contiguous stripes (see Fig. 6). The width of the stripe is the parameter WIDTH. The location of the operator output is the common subwindow edge. It may be considered to be a vertical line segment the width of the stripe.

3. Locate Inside Edge Centers. Each stripe is processed independently. The neighborhood operator response is examined for local maxima (left edges) and local minima (right edges). When proceding from left-to-right, the left edge locations are put on a stack (initialized at the beginning of the stripe) until a right edge location is found. The edge center location is the average of the left edge location on top of the stack and the current right edge location. That is,

$$\text{EDGE.CENTER.LOCATION} = (1/2)[\text{LEFT.EDGE.LOCATION} + \text{RIGHT.EDGE.LOCATION}]$$

Then the left edge location stack is reinitialized and subsequent right edge locations are ignored until another left edge location is found. Examples of the edge center locations are shown in Fig. 7. (Alternatively, subsequent right edge locations could be paired with the left edge location on top of the stack to obtain additional edge center locations before reinitializing the stack again when the next left edge location is encountered.)

4. Detect Hole Centers. The edge center location line segments are plotted on two image planes, one for the horizontal edges and the other for the vertical edges. When these two images are superimposed, some of the line segments cross. (The common points are found by applying a logical AND operator to the two image planes.) These points are the locations of the proposed hole centers as shown in Fig. 8. Hence,

$$\text{HOLE.CENTER.IMAGE} = \text{HORIZONTAL.EDGE.IMAGE} \text{ .AND. } \text{VERTICAL.EDGE.IMAGE}$$

5. Verify Hole Perimeter. To verify that the proposed hole center corresponds to a valid hole, the perimeter is traced by means of a gray-scale perimeter tracking operator. The operator is called a Max-Diff Direction operator since it seeks to follow the direction of the maximum local intensity differences aroung a hole. A 3x3 operator is defined as follows.

   A. Compute the eight-neighbor differences as defined in Fig. 9. The set of differences is given by

$$\text{DIFFERENCE.SET} = \left\{ A - I, B - I, \dots , H - I \right\}$$

   B. Find the maximum member of the DIFFERENCE.SET.

   C. The operator output is the DIRECTION.NUMBER which corresponds to the maximum difference (see Fig. 8).

   For example, if the maximum DIFFERENCE.SET member is A-I with a DIRECTION.NUMBER of 3, then the operator output is 3.

This operator is used to extract the hole perimeter (see Fig. 10). Valid holes are those whose perimeter is closed and has geometric properties which conform to the needs of the robot hand such as minimum finger opening and finger width. If desired some ordering criterion such as compactness (ratio of the square of perimeter length to area enclosed) could be employed to seek holes with preferred characteristics.

Horizontal   Vertical

Fig. 6. Hole Finder edge response for horizontal
and vertical passes for ISOLATED image. Left and
bottom edge is shown in black. Right and
top edge is shown in gray.

ISOLATED - horizontal   ISOLATED - vertical

HEAP - horizontal   HEAP - vertical

Fig. 7. Hole Finder edge centers. Horizontal and
vertical edge center images shown for
ISOLATED and HEAP images.

347

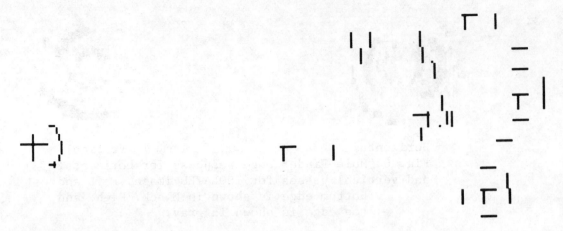

ISOLATED   HEAP

Fig. 8. Hole Finder centers image showing result of
        superimposing the horizontal and vertical
        edge images for ISOLATED and HEAP images.

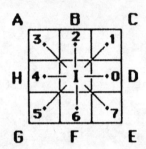

PIXEL.LABELS = A, B, C, D, E, F, G, H, I
DIRECTION.NUMBERS = 0, 1, 2, 3, 4, 5, 6, 7

DIFFERENCES = A - I, B - I, ..., H - I

OUTPUT = DIRECTION.NUMBER of MAXIMUM.DIFFERENCE

Fig. 9. Max-Diff Direction operator definition.

Fig. 10. Perimeter extraction
         using the Max-Diff
         Direction operator
         shown on the ISOLATED
         image.

6. Valid Hole Center Locations. The output of the hole finder algorithm is a list of validated hole center locations. It is now necessary to translate these hole center locations into robot control data. Specifically, the robot hand must be guided along the line-of-sight projection of the image hole center to the hole in the part. This requires a geometric camera model [13,14] and simple sensors [11] on the robot hand to successfully grasp the part.

SUMMARY and CONCLUSION

The heuristic vision Hole Finder algorithm presented tries to overcome some of the problems of handling parts in a heap or a bin by looking at the it from a different perspective. Instead of asking where a part is, the Hole Finder asks where there is a possible hole. The pose estimation problem is left for a later stage of the process, after a part has been successfully acquired. The heuristic approach for the heap of parts problem works best for parts which, from most viewpoints, possess a reasonable number of places which match the selected holdsite patterns -- in this case holes. In the usual methods, the holes must be of known shape, for example circular, and corrections are needed to account for perspective distortion or hole surface orientation. The heuristic vision Hole Finder seeks edge patterns which correspond to holes -- any promising hole. By using neighborhood operators applied in both horizontal and vertical stripes across the image, the hole centers are located directly.

The Hole Finder algorithm requires modest computational resources (it is microcomputer implementable) and is designed to minimze execution times (which depend on scene complexity). For raw castings with odd shaped holes, the Hole Finder algorithm may be the answer to a robot's prayer.

ACKNOWLEDGEMENT

The authors wish to express their appreciation to the members of the Robotics Research Center at the University of Rhode Island for their suggestions and encouragement. This work was supported in part by the National Science Foundation under grant MEA 78-27337.

REFERENCES

[1]   C. Rosen, D. Nitzan, G. Agin, A. Bavarsky and G. Gleason, Research Applied to Industrial Automation, 8th Report, Stanford Research Institute, Menlo Park, CA, Aug. 1978.

[2]   A. Ferloni, I. Franchetti, P. Vincentini and P. Fici, "Ordinatore: a dedicated robot that orientates objects in a predetermined direction," Proc. 10th Intl. Symp. on Ind. Robots, Milan, Italy, pp.655-658, 1980.

[3]   M. Baird, "A computer vision data base for the industrial bin of parts problem," General Motors Research Center, Publication GMR-2505, Warren, MI, Aug. 1977.

[4]   J. Birk, R. Kelley, D. Duncan, H. Martins and R. Tella, "A robot system which feeds workpieces from bins into machines," Proc. 9th Intl. Symp. on Ind. Robots, Washington, D.C., pp. 339-355, Mar. 1979.

[5]   R. Kelley, J. Birk, J.-D. Dessimoz, H. Martins and R. Tella, "Acquiring connecting rod castings using a robot with vision and sensors," Proc. 1st Intl. Conf on Robot Vision and Sensory Control, Stratford-upon-Avon, U.K., pp. 169-178, Apr. 1981.

[6]   J. Birk, R. Kelley and L. Wilson, "Acquiring workpieces: three approaches using vision," Proc. 8th Intl. Symp. on Ind. Robots, Stuttgart, Germany, pp. 724-733, May-June 1978.

[7]   J. Birk, R. Kelley and H. Martins, "An orienting robot for feeding workpieces stored in bins," IEEE Trans. Sys., Man and Cyber., SMC-11, No. 2, pp. 151-160, Feb. 1981.

[8]   J.-D. Dessimoz, J. Birk and R. Kelley, "Integrated robot systems for feeding parts from bins," Proc. 11th Intl. Symp. on Ind. Robots, Tokyo, Japan, Oct. 1981.

[9]   R. Kelley, J. Birk, H. Martins and R. Tella, "A robot system which acquires cylindrical workpieces from bins," IEEE Sys., Man and Cyber., SMC-12, No. 2, pp. 204-213, Mar./Apr. 1982. (Also in Robot Vision, A. Pugh, ed., Springer-Verlag, London, pp. 225-244, 1983.)

[10]   R. Kelley, J. Birk, J.-D. Dessimoz, H. Martins and R. Tella, "Forging: feasible robotic techniques," Proc. 12th Intl. Symp. on Ind. Robots, Paris, France, pp. 59-66, June 1982. (Also in Robot Vision, A. Pugh, ed., Springer-Verlag, London, pp. 285-294, 1983.)

[11]   R. Tella, J. Birk and R. Kelley, "General purpose hands for bin picking robots," IEEE Trans. Sys. Man and Cyber., SMC-12, No. 6, pp. 828-837, Nov./Dec. 1982.

[12]   R. Kelley, H. Martins, J. Birk and J.-D. Dessimoz, "Three vision algorithms for acquiring workpieces from bins," Proc. IEEE, Vol. 71, No. 7, pp.803-820, July 1983.

[13]   H. Martins, J. Birk and R. Kelley, "Camera models based on data from two calibration planes," Comp. Graphics and Image Proc., Vol. 17, pp.173-180, Oct. 1981.

[14]   I. Sobel, "Camera modes and machine perception," Stanford University AI Lab., Memo AIM-121, Palo Alto, CA, May 1970.

Local Ordered Grey Levels as an Aid to Corner Detection

K. Paler
SERC Rutherford Appleton Laboratory, UK
J. Foglein
Institute for Co-ordination of Computer Techniques, Hungary
J. Illingworth
and
J. Kittler
SERC Rutherford Appleton Laboratory, UK

Abstract

This paper presents a novel method for detecting corner points in a scene. It is based on features extracted from the distribution of ordered greylevel values within a local window. The algorithm is effectively orientation independent. Results are presented for an image containing corners of different angles and orientations. A comparison is made between the positions obtained for the corners detected using this method and those obtained by direct measurement. The signal strengths of similar corners at different orientations are also compared. The results of these comparisons show that the proposed algorithm can provide useful corner point information without the need for preprocessing of the image.

1. INTRODUCTION

The corner points of an image are where object boundaries make discontinuous changes in direction. In a practical context corners occur at points with large values of edge gradient (contrast) and rate of change of gradient direction (sharpness). Such points are extremely useful for shape description and image matching. They constitute significant data reduction and an increase in information density. These attributes make corner detection an important aspect of image processing. A recent summary of the current situation is given by Kitchen and Rosenfeld [1].

Generally the approaches to corner detection, except for the filtered projection method of [2], involves an analysis of the boundary features of an object, eg variations in local gradient directions [1,3], variations in boundary shape [4,5,6,7] and chain code values [8,9]. Template matching has been used in cases where the orientation of the corner is limited [10]. However this approach is part of the general area of image

*Now at the Institute for Coordination of Computer Techniques 17, Akademia Str., Budapest, Hungary.

matching and is not specific to corners.

The method proposed in this paper uses the information contained in an ordered array of local greylevel values. In essence it utilises the insensitivity of the median filter to features with dimensions smaller than half the size of the filter window. These features are recovered by generating an output signal equal to the difference of the median greylevel and the input greylevel. The principal advantage of this approach is that it is effectively independent of the orientation of the corner. Complete independence of orientation is obtained only for a circular window. Additional advantages are that no edge detection, edge thinning or boundary encoding are needed and thresholding is only applied to the signals contained in the corner map. It should be noted that the operators proposed by Beaudet [7] are also orientation independent but are based on the values contained in a vector edge map.

In section 2 a simple corner model, upon which this approach is based, is discussed. Section 3 contains a description of the algorithm developed and section 4 presents the results obtained by applying the method. The final section discusses the advantages and shortcomings of this procedure and its domain of applicability.

2. THE CORNER MODEL

The model to be described illustrates the approach being proposed. Although it is ideal and simple it provides a framework within which the method can be discussed. We denote by, G_i, the greylevel at a particular pixel location within the window and, $G(i)$, the greylevel at the ith position in the ordered set of greylevels.

Firstly assume an image with three regions of different greylevel, viz:

I	The object	–	greylevel	=	G_I
II	The edge region	–	greylevel	=	G_{II}
III	The background region	–	greylevel	=	G_{III}

As an example consider a bright object on a dark background. We then have:

$$G_I > G_{II} > G_{III}.$$

Consider a window of size, w x w pixels (w – odd) containing a 90° corner of the object with the centre pixel being that situated at the corner. This situation is shown in figure 1. The median greylevel will be that of region II. The distribution of greylevels, ordered in increasing value, for such a window is shown in figure 2. In this configuration the condition of local monotonicity (i.e. the sequence of pixel greylevels within the median window which corresponds to an object must be monotonic over at least half the total number of pixels in the window) required for the edge preserving properties of the median filter is broken [11]. This means that the difference between the median greylevel, G (median) and the greylevel of the central pixel, G_C can be significant. It is this difference signal which forms the basis of the corner detection method. All the pixels in the hatched area shown in figure 1 will have a difference signal of $G_I - G_{II}$.

An approximate estimate of the magnitude of the difference signal for this particular configuration can be obtained by assuming a linear dependence of greylevel on order number through the edge region.

Thus the number of pixels between the rank of the median and the rank of central pixel is given by:

$$n_{diff} = n_{median} - n_I \tag{1}$$

where n_{median} is given by:

$$n_{median} = (w^2 + 1)/2. \text{ for } w > 3 \text{ and odd} \tag{2}$$

and n_I is the number of pixels in the corner region in the window, ie region I in figure 1, and is given by:

$$n_I = (w + 1)^2/4. \tag{3}$$

The rate of change of grey level in the edge region, r, is given by:

$$r = c/n_e. \tag{4}$$

where c = contrast across the edge region, ie $G_I - G_{III}$

and n_e = number of edge pixels. This corresponds to the number of pixels in region II.

n_e is given by:

$$n_e = (w + 1)e + e^2. \tag{5}$$

where e = width, in pixels, of the edge region. The difference signal is given by

$$S = n_{diff} \cdot r \tag{6}$$

This assumes $n_{diff} < n_e$. If $n_{diff} > n_e$ then $n_{diff} = n_e$.

For a 90° corner oriented along the window axes we have:

$$S_{90}(0) = \left\{ (w^2+1)/2 - (w+1)^2/4 \right\} c/((w+1) \; e+e^2) \tag{7}$$

$$= c(w-1)^2/(4e(w+1+e)) \tag{8}$$

A similar analysis for a 90° corner oriented at 45° with respect to the window axes yields:

$$S_{90}(45) = c(w-1)^2/(4\sqrt{2}ew) \tag{9}$$

A comparison of the signals for corners of 45° and 135° gives:

$$S_{45}(0)/S_{90}(0) = 1 + \left\{ (w^2+1)/(w-1)^2 - 1/2 \right\} \tag{10}$$

$$= 1.5 \text{ for } w \gg 1$$

and $$S_{135}(0)/S_{90}(0) = 1 - \left\{ 1/2 - 2/(w-1)^2 \right\} \tag{11}$$

$$= 0.5 \text{ for } w \gg 1$$

Thus, as could be anticipated, there is a proportional relationship between the signal strength and the angle through which the boundary turns, referred to as 'sharpness' in [12].

The corner signal produced by this approach appears to embody the two main features of 'cornerity' [6,1], contrast and sharpness. The contrast or gradient is contained in the factor, c/e, and the sharpness is related to the difference between half the number of pixels in the window and the number of pixels corresponding to the corner. It is expressed as a function of w.

3. IMPLEMENTATION OF THE ALGORITHM

The aim of the algorithm is to transform the input image into a corner map with the maximal value of signal to noise. The transformation is carried out in two steps:

(i) A Priori Information

a.   A suitable window size is selected.   This depends on how well the corners are defined.

b.   An estimate of the width of the edge region.   This is obtained from the performance of the image system with a reference image.

c.   The effective contrast between background and object.   This is determined as part of the setting up of the vision system.

The parameters obtained from steps a, b and c allow an estimate of the normalisation factor to be made and should be computed for the sharpest corner to be detected to avoid saturation of the corner signal.

(ii) Noise Suppression

In any practical situation the input image will contain noise.   In this present context the noise can originate from a stochastic mechanism or be due to actual, but irrelevant features in the image which have dimensions smaller than half the width of the window. The stochastic noise is suppressed by the algorithm not using greylevels at the extremes of the ordered set of greylevels.   The signal from small features is suppressed proportional to the difference between the number of pixels in the window corresponding to the feature and the number of those of the corner to be detected.   If the image has irrelevant features comparable to the size of the corner in the window then they will not be discriminated against and will produce spurious corner signals. However by selecting a larger window size the predicted position of the corner within the ordered grey levels will move whereas the grey levels corresponding to the small feature will remain fixed.   Thus discrimination can be restored.   The limit to this procedure will be when the window size is equal to the smallest distance separating the corners.   Thus both kinds of noise can be suppressed by using an algorithm which produces a maximum output signal for an input signal which is matched to the greylevel distribution predicted by the corner model and which does not use the extreme grey levels.   The principal features of the predicted greylevel distribution are:

a.   A change in slope at the corner point.

b.   The greylevel of the central pixel should be close to the predicted value at the corner point position, $G(n_I)$.

These features are depicted in figure 3 which shows a schematic representation of the typical distribution of ordered grey levels in the region of a corner.   The notation used in equations 12 to 19 is denoted on fig 3.

Feature (a) is expressed as an enhancement factor, F_1, and is given by:

$$F = 1 + (\theta_2 - \theta_1)/\theta_{max} \tag{12}$$

where:

$$\theta_{max} = \tan^{-1}(c/n_e) \tag{13}$$

and for $G_C > G(median)$

$$\theta_1 = \tan^{-1}\left\{(G(w^2-1)-G(w^2+1-n_I))/(n_I-1)\right\} \tag{14}$$

$$\theta_2 = \tan^{-1}\left\{(G(w^2+1-n)-G(w^2+2-2n))/n_I\right\} \tag{15}$$

and $G_C < G(median)$

$$\theta_1 = \tan^{-1}\left\{(G(n_I)-G(2))/(n_I-1)\right\} \tag{16}$$

$$\theta_2 = \tan^{-1}\left\{(G(2n_I)-G(n_I))/n_I\right\} \tag{17}$$

354

where if $n_I > n_e$ then the factor $2n_I$ is replaced by $n_I + n_e$.

The second feature, (b), is formulated so that it tends to suppress signals from general edge pixels and pixels not at the corner position (i.e. pixels having greylevels different to that predicted by the corner model). It is given by:

for $G_C > G$ (median)

$$F_2 = (G(w^2-1)-G(\text{median}))/(1+|G(w^2-n_I)-G_C|/(G(w^2-1)-G(\text{median}))) \qquad (18)$$

and for $G_C < G$ (median)

$$F_2 = (G(\text{median})-G(2))/(1+|G_C-G(n_I)|/(G(\text{median})-G(2))) \qquad (19)$$

Thus the complete algorithm is:

$$S = (G_C - G(\text{median})).N.F_1.F_2 \qquad (20)$$

N = normalisation constant computed from the corner model and the information from the first step of the algorithm. (e.g. $N=S_{90}(0)$ for 90° corners) and

$$F_1.F_2 = 2.c/2 = c \quad \text{at perfect matching.}$$

This algorithm will deal with corners of either sign of contrast, a very necessary ability because objects are likely to contain re-entrant sections in their boundaries.

4. EXPERIMENTAL RESULTS

The image used to examine the performance of the proposed algorithm is shown in figure 4(a). It is similar to one used in [2] and contains a variety of corners of different sharpness and orientation. This image, which had an overall contrast of 35 units of greylevel was quantised into an array of 128x128 pixels. The average width of the edge region was 2 pixels. Image processing was performed using a PDP11/44-CRS4000 system. The experimental results to be discussed were obtained by using the proposed algorithm optimised for 90° corners. The corner map produced using a 5 x 5 mask is shown in figure 4(b). The main corners produced significant signals, typically with a global S/N of 3:1 but locally the S/N=10:1. An example of the output signal as a function of pixel position is shown in figure 5. Two clear peaks are observed. The narrower peak has a signal to noise of S/N=31:1 and for the other peak the S/N = 20:1. In both cases the noise is estimated from a local average over pixels around the peaks. The positions of the largest local maxima, for sixteen signals with the largest output values which should correspond to the sixteen possible corner positions in the input image are shown in Table 1. The corners are numbered in order of decreasing signal strength. The image positions corresponding to this numbering are shown on fig 4(c). In fact only fourteen of the sixteen possible corner points were detected. The missing corners however are the ones with very oblique angles, 135° and 154°. The output signal strength of the two spurious corners gives a measure of the noise level, ie ranging from a S/N of 7:1 to 1:1. However for corners with angles < 90° the signal to noise is 4:1. The corner correspondence is generally within 1 pixel.

The large angle corners can be detected by using a larger window, thus increasing the difference in greylevel between the median and corner pixel. The result from a 9 x 9 window on the same input image was that all 16 corners were found to be in the set of the 16 largest corner signals and the mean deviation was only slightly increased. The range of signal to noise was approximately in the range S/N=22:1 to 2:1.

A comparison of the signal strengths for corners of equal angle but differing orientation is shown in Table 2. Although the signal strengths do vary with orientation, the variation is small, generally between 10% and 30%. The relative strengths of the corner signals were found to exhibit a monotonic decrease with increasing corner angle. This feature is consistent with the corner model used.

CORNER LABEL	RELATIVE SIGNAL STRENGTH	DETECTED CORNER CO-ORDINATES (PIXELS)		MEASURED CORNER CO-ORDINATES (PIXELS)		DEVIATIONS (MEAS.-DET.) (PIXELS)		CORNER ANGLE (DEGREES)
		X	Y	X	Y	X	Y	
1	189	38	44	38	44	0	0	63
2	180	105	21	106	21	+1	0	53
3	174	33	101	33	101	0	0	53
4	156	73	86	72	86	-1	0	63
5	149	17	81	16	81	-1	0	53
6	135	50	62	51	61	-1	-1	90
7	109	56	70	57	69	+1	-1	90
8	84	31	87	31	86	0	-1	106
9	75	40	97	40	97	0	0	135
10	70	21	75	21	76	0	+1	135
11	47	41	84	41	83	0	-1	135
12	44	96	23	96	24	0	+1	154
13	42	93	38	91	40	-2	+2	154
14	33	103	29	102	29	-1	0	154
15	26	SPURIOUS CORNERS						
16	24							

TABLE 1   Position Comparison of measured and detected corners

CORNER ANGLE (DEGREES)	SIGN OF CONTRAST	CHANGES IN ORIENTATION (DEGREES)	RELATIVE SIGNAL STRENGTH			
			5 X 5 MASK		9 X 9 MASK	
			ORIENTATION		ORIENTATION	
			A	B	A	B
53	−	53	149	174	236	223
53	−	154	180	174	249	223
53	−	154	180	149	236	249
63	−	120	135	109	251	207
90	+	90	75	70	159	149
135	−	135	47	26	95	87
135	+	135	−	−	56	55
154	−	154	44	33	49	43
154	+	154	−	−	40	38

TABLE 2   Orientation comparison of corner signal strength

5. CONCLUSIONS AND SUMMARY

This preliminary study of a novel method for corner detection has shown it to be very effective. A hardware implementation using only the difference signal would be extremely easy although it would only be practical for processing images with a small amount of noise. The primary advantage of the method is one of orientation independence and this alone makes it extremely valuable.

In principle, for constant contrast, the output signal is proportional to the sharpness of the corner. However in most practical applications the separation of sharpness and contrast may not be possible and information on sharpness will not be available as it is with other methods. Although, in general, sharpness is very prone to noise from pixels with low contrast. Also the method does not provide information on the orientation of corners.

Thus for images composed of simple shapes with high contrast the method may provide all the corner information required. In the case of more complex images the method may be useful as a fast preprocessing stage which will locate areas containing potential corner points. These areas can then be subjected to detailed edge gradient analysis and more quantitative corner information extracted.

However, even in a complex image with a structured background and object the method maybe applicable in the following situations:

(i) The required corner points produce the largest corner signals, i.e. the object corners have the largest contrast or the object has corners of a specific angle to which the algorithm can be tuned.

(ii) There exists a priori knowledge about the expected configuration of corner points corresponding to the object such that a point matching algorithm can be devised to search the corner map image to locate the object.

ACKNOWLEDGEMENTS

One of us, J.F., gratefully acknowledges the support he has received from the British Council Scholarship Scheme.

REFERENCES

1. L Kitchen & A Rosenfeld. Greylevel Corner Detection. Pattern Recognition Letters 1(1982) 95.
2. Z-Q Wu & A Rosenfeld. Filtered Projections as an Aid to Corner Detection. Pattern Recognition 16(1983) 31.
3. O A Zuniga & R M Haralick. Corner Detection using the Facet Model. Proc. of IEEE Comp. Soc. Conf. on Computer Vision and Pattern Recognition (1983) 30.
4. A Rosenfeld & E Johnston. Angle Detection on Digital Curves. IEE Trans on Comp. C-22(1973) 875.
5. A Rosenfeld & J S Wesjka. An Improved Method of Angle Detection on Digital Curves. IEEE Trans. on Comp. C-24(1975) 940
6. B Kruse & C V Kameswara Rao. A Matched Filtering Technique for Corner Detection. 4th International Joint Conference on Pattern Recognition, Kyoto 1978 p.642.
7. P R Beaudet. Rotationally Invariant image Operators. 4th International Joint Conference on Pattern Recognition, Kyoto, 1978, p.579.
8. H Freeman & L S Davis. A Corner finding Algorithm for Chain Coded Curves. IEEE Trans. on Comp. C-26(1977) 297.
9. W S Rutkowski & A Rosenfeld. A Comparison of Corner Detection Techniques for Chain Coded Curves. Technical Report No. 623, Computer Science Centre, University of Maryland, College Park, MD20742.
10. M L Baird. SIGHT-1: A Computer Vision System for Automated IC Chip Manufacture. IEEE Trans. on Syst., Man & Cybs. SMC-8(1978) 133.
11. S G Tyan. Median Filtering : Deterministic Properties. Topics in Applied Physics, Vol 43, Ed. T S Huang Springer Verlag 1981.
12. C Wang, H Sun, S Yada & A Rosenfeld. Some Experiments in Relaxation Image Matching using Corner Features. Pattern Recognition 16(1983) 167.

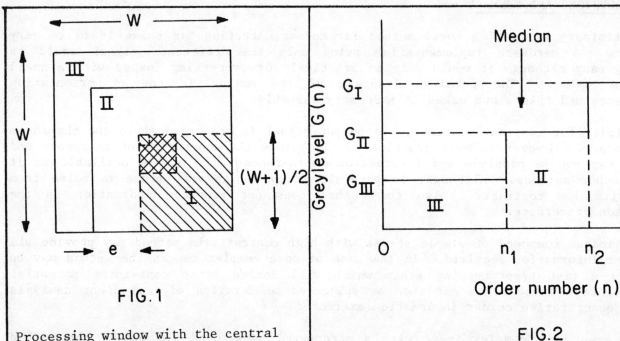

FIG.1

Processing window with the central

pixel located at the corner point.

Regions denoted by:

I = Object, II = Edge,

III = Background.

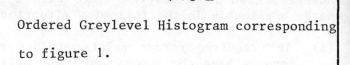

FIG.2

Ordered Greylevel Histogram corresponding

to figure 1.

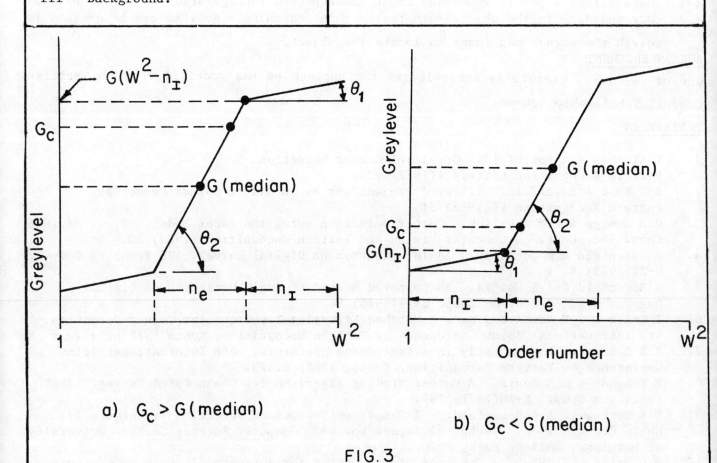

a) $G_c > G$(median)

b) $G_c < G$(median)

FIG.3

Ordered Greylevel Distributions showing Corner Features.

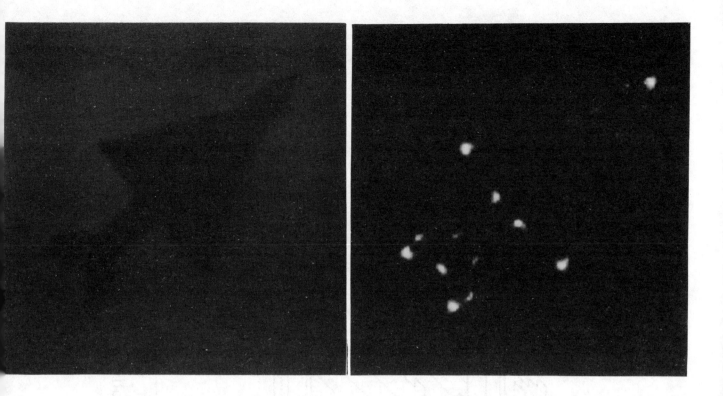

a)   b)

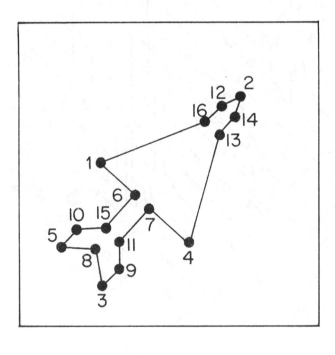

(c)

Fig. 4.   a)   Test Image.

           b)   Corner map corresponding to test image.

           c)   Schematic of test image showing corner labelling corresponding
                 to table 1.

359

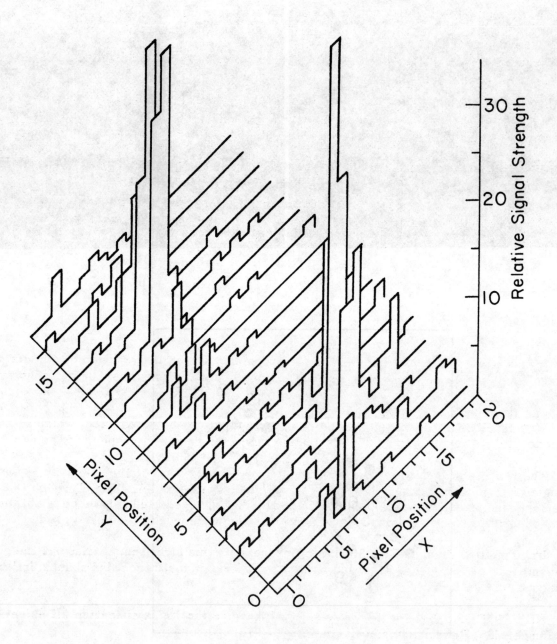

FIG. 5

Examples of corner signals as a function of pixel position.

An Heuristic Method of Classification and Automatic Inspection of Parts: The Recognition System of ANIMA

D. Juvin

and

B. Dupeyrat

Centre D'Etudes Nucleaires de Saclay, France

In this paper, we present a method for automatic pattern recognition of objects with a limited number of stable positions, taking place in an industrial environment (e.g. : conveyor, working area, ...).

We describe first a contour encoding of the pattern to be recognized, which is invariant in translational and rotational modes. This encoding keeps the geometric aspect of the contour.

The proposed algorithm then carries out the identification of the shapes to be recognized by using an Euclidian distance specified in terms of the invariant encoding. Through this distance, an estimation of the shape distorsion and the recorded selected reference is obtained. A dynamic acceleration heuristic using a pre classification technique is then presented.

Implemented on an INTEL 8086 microprocessor , this algorithm carries out the object recognition and localization of pre-recorded points, by reference to 10 recorded shapes, in less than one second.

KEY WORDS : Pattern Recognition, Coding (Invariant), Classification (of objects), Automatic Inspection, Robotics (vision system applied to).

I - INTRODUCTION

The localization and identification of object patterns is an important problem for to-day's robotics.

We present here a recognition method we have developped for the industrial system ANIMA at the Electronic and Nuclear Instrumentation Department. Intended to be the vision system of a robot, ANIMA, starting from a low contrasted image, is able :

- to perform contour extraction fo every pattern located in the image using a dynamic edge tracking algorithm and encode each contour with respect to Freeman's chain ;
- to perform a contour linear segmentation starting from Freeman's chain using an invariant encoding in translational and in rotational mode ;
- to identify the above invariant code by reference to a pattern dictionary that has been built during an interactive learning procedure (training by showing) ;
- to evaluate object distorsion relative to the recorded reference ;
- to indicate localization, orientation and grasping points

in a time compatible with most applications in robotics e.g. one second. Process time is about 50 % for contour extraction, 50 % for recognition.

For cost effectiveness, the ANIMA system is implemented on a microprocessor and hence, its process is sequential.

II - CONTOUR EXTRACTION, COMPRESSION, ENCODING

II.1 Contour extraction

We use a standard video camera, the resulting image is digitized in 256 x 256 x 4 bits.

The CONTOZ module performs the dynamic tracking of every edge located in the picture using a multivariable cost function [3] , [4] .

At the end of this step, detected contour lines are encoded, using Freeman's chain [1] . Starting points cartesian coordinates are also memorized.

II.2 Contour linear segmentation

Starting from Freeman's representation, EXTRA module performs linear extrapolation of contour lines by looking into groups of 4 elements overlapping by 2.

```
___  ____  ____  ____  ____  ____  __
X  X  X  X  X  X  X  X  X  X  X  X  X  X  X  X  X  X  X  X  X  X  X  X  X  X
___  ____  ____  ____  ____  ____
```

The algorithm basically consists in unrolling Freeman's chain and testing the ability of each chain element to belong to a straight line with a given angular error ε . An element which does not fit this condition is considered as associated with a "stopping point" on the contour line. Cartesian coordinates of this extremity are memorized and the algorithm look for a new straight line starting from this point. By doing this, a segment has been created. At the end of the algorithm execution, a segmented picture fitted with straight elements is avaible. The number of points corresponding to a contour line is usually quite low (typically 50) and the compression factor related to the initially digitized picture is up to 1000 (see fig.10).

II.3 Invariant encoding

We intend to recognize an object on the image whatever its orientation and position ; so it is interesting to encode the segments in a rotation and translation invariant mode.

The relative polar encoding fullfills quite perfectly this double purpose ; it consists in memorizing the curvilinear abscissa of each segment tops and the corresponding orientation of the segments.

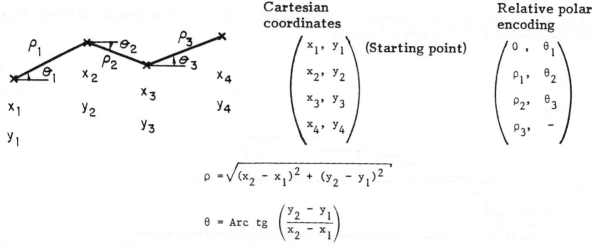

<table>
<tr><td></td><td>Cartesian coordinates</td><td>Relative polar encoding</td></tr>
<tr><td></td><td>$\begin{pmatrix} x_1, \ y_1 \\ x_2, \ y_2 \\ x_3, \ y_3 \\ x_4, \ y_4 \end{pmatrix}$ (Starting point)</td><td>$\begin{pmatrix} 0, \ \theta_1 \\ \rho_1, \ \theta_2 \\ \rho_2, \ \theta_3 \\ \rho_3, \ - \end{pmatrix}$</td></tr>
</table>

$$\rho = \sqrt{(x_2 - x_1)^2 + (y_2 - y_1)^2}$$

$$\theta = \text{Arc tg} \left(\frac{y_2 - y_1}{x_2 - x_1} \right)$$

Figure 2 : The curvilinear polar encoding

For this encoding :

- a translation of the figure is invariant
- a rotation θ on the figure adds a constant θ on each component ; an "offset" is added on all the graph.

II.4 Developped representation, signature

We call <u>developped encoding</u>, the graph of relative polar encoding drawn on Cartesian axes, i.e. :

- on Ox the curvilinear abscissas of the segments
- on Oy the orientations of the considered segments.

Developped encoding of a circle is a ramp ; these of a regular polygon is a stair with regular steps.

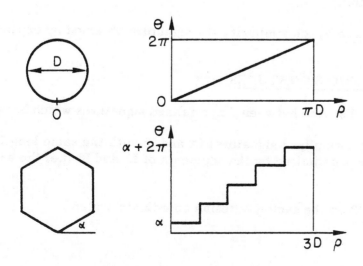

Figure 3 : Encoding of some simple figures

Characteristic : every closed contour has an offset of $\pm$ 2 π between the origin and the end of its developped encoding.

The developped encoding of a closed contour is a functional representation in the picture domain ; there is a full equivalence between both representations. We define the **signature** of a closed contour by its developped encoding, whose curvilinear abscissa is limited between O and the perimeter of the outline.

Perimeter is called **length** of the signature.

We call **normalized signature** with respect to the Cartesian coordinate system IR, the signature whose angular mean value is null in this reference.

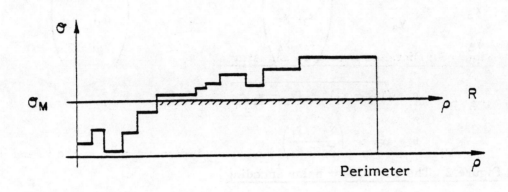

Figure 4 : Normalized signature of a closed contour

We normalize a signature by computing the angular mean value θ_M on the full length, and then we consider the new reference IR, translated on θ axes by θ_M.

We call **extended signature** the developped encoding we would have by running indefinitely around the pattern to encode.

The extended signature is drawn by copying indefinitely the signature, joining the origin of the copied signature to the extremity of the graph and adding, at each time we copy, on offset of $\pm$ 2 π (depending on clockwise or not).

We call ρ_o **translated signature**, the part of the extended signature between the curvilinear abscissa ρ_o and (ρ_o + Perim).

We call **expanded signature** by a k similarity the signature obtained by expanding the abscissa axes in a ratio of k.

III - DISTANCE BETWEEN TWO SIGNATURES

We can only define this distance between 2 normalized signatures which have the same length.

The distance between 2 normalized signatures IA and IB with the same length is equal to the absolute sum of the areas delimited by the segments of IA and IB (i.e. the area between both curves IA and IB).

Let's take back IA and IB on the same Cartesian coordinate system.

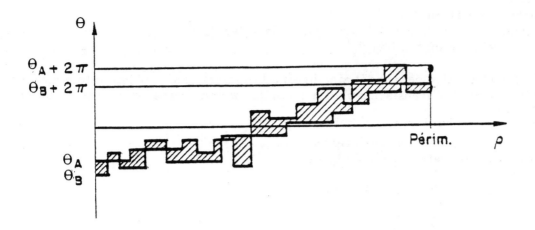

Figure 5 : Distance between 2 signatures

$$d(IA, IB) = \int_{\rho=0}^{Perim} \left| \Theta_A(\rho) - \Theta_B(\rho) \right| \, d\rho$$

This distance is Euclidian [4] . By homomorphism, an Euclidian distance in the image domain has been defined.

The distance between 2 signatures expresses, in the image, the sum of the direction gaps of elementary segments located at the same curvilinear abscissa on both curves.

For 2 closed contour $\mathcal{A}$ and $\mathcal{B}$ which have the same perimeter, to compute this distance means to report to each point of $\mathcal{A}$, with a curvilinear abscissa ρ, a segment whose orientation is the tangent in $\mathcal{B}$ at the same curvilinear abscissa ρ, and to estimate $\Theta_A(\rho) - \Theta_B(\rho)$ x dρ . This is an estimation in each point of

the error made by drawing the elementary segment in the direction forced by the contour $\mathcal{B}$, instead of drawing it in the reference direction given by $\mathcal{A}$.

Remark :   In our case of segmented representation, the integral is easier to compute because $\Theta(\rho)$ remain unchanged on the whole length of the segment.

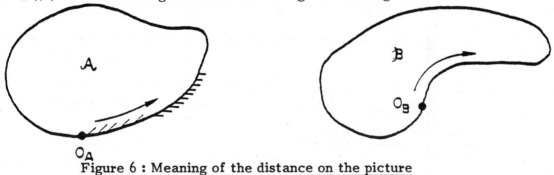

Figure 6 : Meaning of the distance on the picture

IV - RECOGNITION

IV.1 Correlation between 2 signatures

We defined the <u>correlation on a length ρ_L</u> as the operation consisting in searching the minimum distance between a normalized signature IA and the normalized signature IB translated by ρ, for ρ varying continuously between O and ρ_L.

Here is a simple recognition algorithm :

- normalize unknown pattern IB ;
- expand IB signature to IB' so that IB' has a normalized length ℓ;
- compute the correlation on the length ℓ.

This method is attractive because it simplifies a bi-dimentional problem into a uni-dimentional one. On the other hand, the translation vector (ρ_T, θ_T) which corresponds to the minimum distance gives immediatly :

- the curvilinear gap ρ_T between the origins
- the orientation gap θ_T between the 2 objects.

On figure 7, this algorithm gives the best fit :

ρ_T = 66 pixels, θ_T = 112,4°

Figure 7 : Correlation
    a) Reference pattern
    b) Extended developped encoding of
    c) Unknown pattern
    d) Developped encoding
    e) Best fit (correlation)

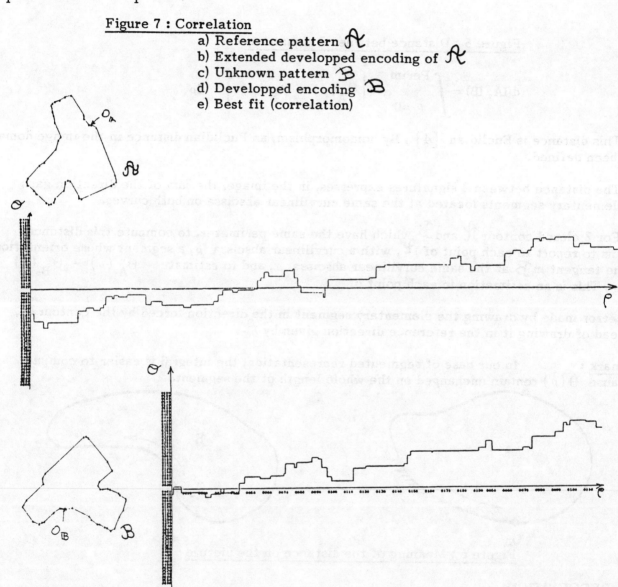

366

The minimal distance corresponds to this fit.

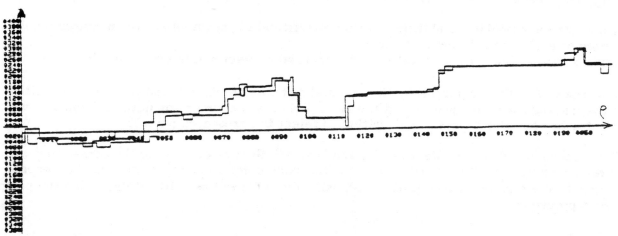

IV.2 Acceleration of the recognition method by a sorting out on the big segments

The correlation method based on object signatures is quite efficient and the associated distance gives a fair estimate of the distorsion between 2 contours. However, it takes a long time to compute systematically this distance (about 10 seconds for a 10 objects dictionary).

Our acceleration method consists in trying to "guide" the correlation to the most probable fit thanks to a sorting out based on global characteristics extracted from the precedent encoding.

We call **big segment** every succession of segments whose developped encoding can be inscribed in a $(\Delta\rho, \overline{\Delta\theta})$ rectangle.

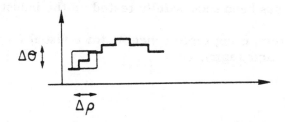

If $\Delta\rho \gg \Delta\theta$ the big segment is named <u>linear characteristic of $\Delta\rho$ length.</u>

If $\Delta\theta \gg \Delta\rho$ the big segment is named <u>angular characteristic $\Delta\theta$</u> .

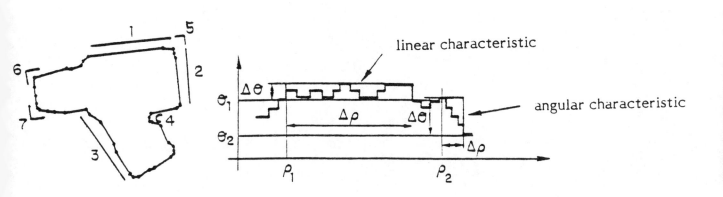

Our acceleration heuristics consist in :

- pointing out the coincidence abilities on characteristical big segments of the unknown object with those of each object in the dictionary ;
- starting a correlation with the most probable translation vector this sorting out has given.

If, at the end of this computation, the minimal distance obtained is too great, the algorithm takes the translation vector just behind the most probable one, and so on untill coincidence vector become too low and this causes the algorithm to reject the presented part.

This method has given excellent results and is as efficient as computing the correlation with every reference objects. Recognition time, for the same object, has fallen to about 0,5 second. We have gained a temporal ratio of about 20, using these heuristics, without degrading the performances in precision.

IV.3 Results

The former algorithms have been successfully implemented on a 8086 Intel microprocessor. Figure 10 shows some experimental results.

V - CONCLUSION

Using a distance between patterns, based on a special encoding of the segmented contour lines allows a very effective recognition and is a good estimator of the distorsion between 2 patterns. Confusion probability between objects having different aspects is theorically impossible.

This method allows, moreover, to locate directly on the object some characteristical points (such as grasping points).

Our recognition algorithm has been successfully tested on the industrial system ANIMA.

An automatic grasping system, using those concepts for a visual feed-back is now studied and the first results are quite encouraging.

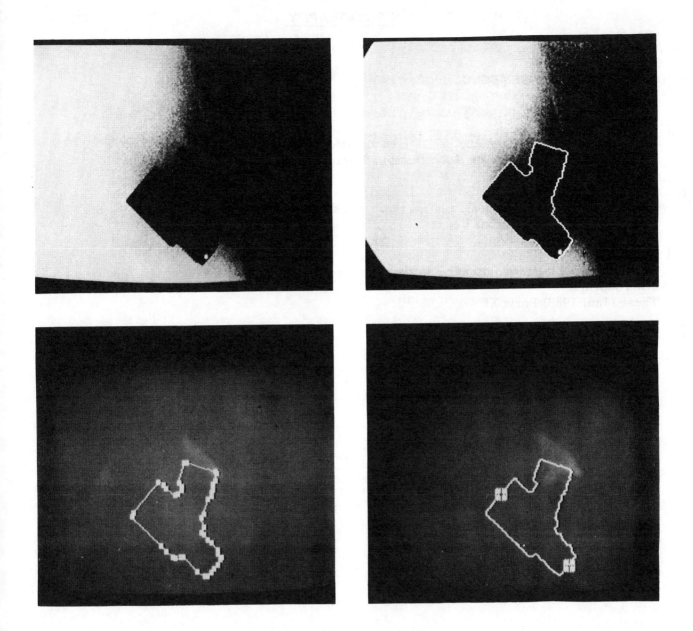

Figure 10 : Recognition

a) Presented picture
b) Detected outline (t = 0.3 s)
c) Segmentation (t = 0.01 s)
d) Correlation and grasping points (t = 0.5 s)

BIBLIOGRAPHY

[1] J.C. SIMON, A. ROSENFELD
Digital Image Processing and Analyses

[2] W. GEISLER
A vision system for shape and position recognition. Proc. of the 2nd
International Conference on Robot Vision, Stuttgard, 1982

[3] B. DUPEYRAT, D. JUVIN
ANIMA (Analyses of Images), a quasi real time system.
Proceeding of PRIP 82, Las Vegas

[4] D. JUVIN
Contribution à la reconnaissance automatique des images appliquée
à la robotique
These (Jan. 1983) Paris XI

[5] T. PAVLIDIS
Structural Pattern Recognition
Springer Verlag, Berlin 1977.

Combining Vision Verification with a High Level Robot Programming Language

A. P. Ambler
and
R. J. Popplestone
Department of Artificial Intelligence, University of Edinburgh, UK
B. Yin
and
Y. Baolin
Beijing Institute of Aeronautics, Peoples Republic of China

Abstract

This paper describes the use of vision verification in the RAPT system. RAPT is an offline robot programming language which uses descriptions of spatial constraints between features of objects to determine the planned positions of objects. By vision verification we mean the use of vision information to determine the difference between the planned positions and the actual positions taken up at run time. We show how vision commands can be used to allow the combination of run time and compile time information in a general way so that all ramifications of the effect of the changed positions are taken account of. This involves symbolic reasoning about spatial constraints at compile time and the construction of a framework indicating the dependance of object positions on each other.

1. Introduction

When using a robot system to perform small batch tasks it is inappropriate to require special purpose work station layouts in order that all workpieces be accurately presented to the robot. Ideally one would have a general purpose work station which can be used for a variety of tasks. However this does mean that parts may not be so accurately positioned: one way of compensating for this is to use sensory information. The problem then becomes one of indicating to the robot system how and when the sensors are to be used and what use is to be made of the data they provide. One point to note is that whatever way the robot system is programmed, whether it be by showing, or by means of a formal programming language, the data will only be obtained from the sensors

----

*Baolin Yin is now at Department of Computer Science, The Beijing Institute of Aeronautics, Beijing, The People's Republic of China.

as the task is being executed and the interpretation of the data and the modification of the robot program therefore have to be done then. This means that it has to be done fast and efficiently. Another point to note is that a robot task is usually a well-ordered one in which there can be firm expectations about the general layout of the workstation and pieces. This means that sensory information will be used to confirm expectations and to make only minor modifications to planned movements, or to choose between a small set of alternative movements.

There are a variety of levels at which one may program a robot system, ranging from teaching by showing, through manipulator level programs and up to task level programs. In a manipulator level program the task is described in terms of movements that the manipulator has to make in order to achieve it. These movements may be described relative to frames of reference, some of which can be attached to work pieces. There is no need for the system to have an understanding of the task or much knowledge of the objects involved. In a task level program, the task is described in terms of the final result required and the devices available, and the robot system is left to decide how to achieve the result. An object level program is intermediate between these - the task is described in terms of how the objects involved are to be moved around and brought into contact. The robot system therefore has more understanding of the task and requires to have knowledge about the objects to be handled - their shape, surface features etc.

If we are to incorporate the use of sensory information into a robot program, then the way in which this can be done depends on the level at which the task is programmed, and the knowledge and understanding that is available in the system. At manipulator level, with relatively little knowledge of what is going on, the programming has to be fairly explicit about how the data is to be collected and interpreted and the conditions in which the sensors are used have to be rigidly defined and adhered to. At the task level the programmer merely describes what sensory systems are available and leaves the system to work out the details. At the object level the collection and interpretation of the sense data can be planned by the system in the light of the knowledge that it has about the objects in the workstation and their expected layout.

In this paper we show how an object level robot programming system (RAPT) can be extended to incorporate the use of vision information to establish the actual position of objects. This includes a description of the commands in the input language, the reasoning about these commands that can be carried out at compile time and a discussion of the requirements of the system at execution time.

2. RAPT

For those readers not familiar with the RAPT system a brief description is given here: more detailed descriptions can be found in refs 1,2 and 3.

2.1 RAPT modelling system

RAPT is a model based object level robot programming system in which the parts to be handled and the robot workstation are described in terms of their surface features. The surface features which can be used are plane faces, spherical faces, cylindrical shafts and holes. Straight edges are represented as cylindrical features and vertices are represented as spherical features, both with very small radius. While the system takes no note of the actual extent of a surface feature (a plane face can be considered to be of infinite extent) it is possible to define edge features in such a way that their extent is known to the system. Parts modelled in the RAPT system are necessarily incomplete, and the system has no notion of space occupancy of objects.

2.2 Assembly description

The assembly program is defined in terms of a sequence of distinct <u>situations</u> that the programmer wants established, and <u>actions</u> occurring between these situations.

The nominal layout of the parts in particular situations may be described in terms of spatial relationships holding betweeen features. For example a paraphrase of some RAPT statements might be "The bottom face of the plate is against the top face of the block and the hole in the plate is aligned with the hole in the block". Actions are described in terms of movements of objects relative to their features (eg rotate the plate about the hole in the plate, move the plate perpendicular to the bottom of the plate by 10). When two objects have become firmly attached to each other, so that movement of one is exactly matched by movement in the other, they are said to be <u>tied</u>. Sometimes a set of objects are loosely attached as in for example the links of a manipulator, in which there can be relative movement of the separate parts, but some spatial relationships are always maintained. Such a set of objects is said to form a <u>sub-assembly</u>.

The spatial relationships which may be specified by the programmer include:

- AGAINST: between two plane faces; a plane face and a spherical face; a plane face and a shaft.

- FITS and ALIGNED: between cylindrical features

As far as this paper is concerned the important ones are AGAINST plane,plane (written as AGPP) and AGAINST plane,shaft (written as AGPC). Informally if two plane faces are against each other then their surfaces lie in the same plane, but with the axes opposed. This relationship, on its own, implies three degrees of freedom between the two bodies involved - two translational and one rotational. If a plane face is against a shaft, then the axis of the shaft is parallel to the surface of the face, and removed from it by a distance equal to the radius of the shaft. There are thus four degrees of freedom allowed by this relationship - two translational and two relational.

A movement of a body relative to a feature of itself can be considered to imply a spatial relationship between features of the instance of the body before movement and the instance after. Thus moving perpendicular to a plane feature an unknown amount can be treated as a LINEAR relationship; if the distance to be moved is known, then the relative position of the instances of the body is known exactly. This is termed a FIX relationship.

2.3 Inference system

If several spatial relationships are to be satisfied simultaneously, then together they form a tighter constraint than implied by each relationship separately. The idea behind the RAPT system is to allow the user to specify sufficient spatial constraints to determine completely the relative positions of the parts involved in each stage of the assembly. The RAPT inference system will then convert the user description into relative positions.

Let us consider two AGPP relationships that are to hold between features of two bodies simultaneously. If the plane features on each body are parallel to each other, and equidistant, then the combination of the two constraints is exactly the same as either one of them. If the two features are at an angle, and at the same angle in each body, then the combination reduces to a LINEAR constraint, and there is only one (translational) degree of freedom between the two bodies. In all other cases it will be impossible to satisfy the two constraints simultaneously.

The RAPT inference system works by considering the relationships two at a time, looking at the relative positions of the features involved in each body, and where possible, rewriting the pair of relationships as a single one. If a FIX relationship is deduced, then the body instances involved can be "merged" - that is they can together be treated as a single body instance with the features and spatial relationships of both suitably transformed. In this way the spatial relationship graph is reduced and eventually relative positions of all objects determined. The inference system also uses the transitive nature of some relationships to further manipulate the graph, but this is irrelevant to this paper.

3. Vision Verification in RAPT

The RAPT system as described above provides the planned (or nominal) relative positions of objects as implied by the spatial relationships stipulated by the user. Now at some stages the programmer may be aware that the actual positions taken up by objects may differ from the planned positions, because of inaccuracies in the movements of devices or in the action of parts feeders. The uncertainties in position may mean that the programmer cannot be certain that the planned actions will always succeed, so he may decide to use sensory information to reduce the uncertainty. The point to note is that the RAPT system has inferred the planned positions, and since the actual positions will differ from the planned ones by only a small amount, the system will be able to use the planned positions to make predictions about the information to be gathered when the program is executed, and to decide how to interpret the information once it has been obtained.

3.1 Vision verification commands

In a RAPT system with vision verification the programmer specifies his planned positions as previously, but he also adds in some commands to use vision information at suitable points. These commands allow him to:

1. specify which features of a particular object are to be looked for with which camera. This is the LOOK command.

2. describe in broad terms the maximum uncertainty expected in the position of the object being viewed. This is the TOLERANCE command.

3. specify limitations on the uncertainty in terms of spatial relationships which are guaranteed to hold whatever the uncertainty. This is the INVIOLATE command. The idea of the inviolate command is to allow the system to take advantage of known restrictions: thus if a flat-bottomed object is roughly placed on a flat-topped table, one can be certain that wherever it ends up, the bottom of the object will be against the top of the table.

The vision verification commands pertaining to one particular state of the world are enclosed in COMBINE ... TERCOM brackets. An example of their use is the following:

```
COMBINE;
        LOOK/edge1 of body1, cameraA;
        LOOK/edge2 of body1, cameraB;
        INVIOLATE/against,bottom of body1, top of table;
TERCOM;
```

3.2 Interpretation of vision verification commands

In the current implementation of RAPT the features which can be LOOKed at are edge features. Now the image of an edge feature will be a straight line in the picture, and

374

we can imagine a plane which contains this straight line and the centre point of the camera. Furthermore we can say that the edge which formed the image lies against that plane. Thus at plan time we know that when the program comes to be executed, and the image of the edge feature is formed, we shall be able to assert that an AGPE has been set up between this plane and the edge feature. (AGPE is the special case of AGPC where the cylindrical feature is actually an edge feature). At plan time we can therefore form a symbolic plane feature and assert a symbolic spatial relationship for each LOOK command in the package.

Now there will also be one or more INVIOLATE commands in the package. Depending on the relative positions of the features in the bodies involved, and AGPP combined with an AGPE can give rise to an AGPP relationship, or a "ROTYLIN" relationship with both a rotational and a linear degree of freedom, or one of two LIN relationships, the first being at 180 degrees to the second. A LIN relationship combined with an AGPE will give rise either to another LIN or to a FIX relationship. Therefore at plan time the RAPT system can look at the inviolate and symbolic relationships involved in a package of vision commands, and, using the relative positions of the features involved, can determine how they are to be combined together. The RAPT system can in general avoid the awkward ROTYLIN relationship by careful ordering of the pairs of relationships to be combined. In the case where an AGPP combines with an AGPE to produce an AGPP the vision information producing the AGPE relationship will be adding nothing to the system's knowledge of the layout of the parts. The RAPT system will be able to point this out to the programmer at plan time and suggest he uses a different camera, or looks at a different feature. Since the RAPT system knows the planned position of the object being examined, and knows that the actual position will differ only slightly, it will be able to decide which of the two possible LIN relationships will be the correct one.

If the programmer has given sufficient LOOK and INVIOLATE commands, the system will be able to deduce symbolic FIX relationships for the body being examined. The RAPT system will therefore be able to produce at plan time a symbolic relative position for the body. If insufficient commands have been given to fix the position of the body during program execution, then the system will be able to warn the programmer of this at plan time and ask him for more commands.

3.3 Image prediction and use of tolerance

Since the RAPT system will have deduced the nominal positions of all bodies, and since the cameras in use are also bodies in the system, it will be able to predict whereabouts in the camera image the image of the specified edge feature will be found. The TOLERANCE command is used to determine the window within which the expected image should be found. If the expected image is outside the field of view of the camera, then the system will be able to report this to the user at plan time. Otherwise, the coordinates of the window are included in the command to the vision system that is added to the run-time system.

As pointed out in section 2.1 the current RAPT does not use complete body models: work is in progress to couple RAPT to a solid body modelling system. In this case the system will be able to show the programmer what the view from the chosen camera will look like with objects in their nominal positions, so that he can judge for himself whether he has made a wise choice of features. It will also be able itself to check on the suitability of the commands.

4. Implications of verified positions

A particular stage in an assembly program cannot be considered in isolation — the constraints on the positions of bodies in one situation will affect the positions of bodies in other situations. This is particularly so in the case of vision verification, the purpose of which is to reduce the uncertainty in positions of objects so that subsequent actions should succeed in their purpose. This means that in the program that is executed, the commanded positions of manipulators and other moveable devices need to be modified in the light of the vision information. A necessary part of a vision verification system is therefore the determination of how positions depend upon each other. Since the program that is executed must run in real time the dependency of positions must be determined at plan time. It is shown in refs 4 and 5 how the dependencies can be represented as <u>modifying factors</u> on nominal positions. The modifying factor of a verified body is determined directly by the vision information. The modifying factors of other bodies can be null or can be expressions involving other modifying factors and positions. They are determined by examination of the text of the RAPT program giving rise to the nominal positions - for example, if the destination of the end effector is specified in terms of spatial relationships with features of a body that has been verified then the commanded position of the end effector depends on the modifying factor of the verified body. TIEs and SUBASSEMBLIES complicate the situation, and a set of linking rules are defined for determining the modifying factor array that is set up by the system at plan time.

5. The run-time program

A normal RAPT program without vision verification commands gives rise to a set of nominal positions for all objects for every distinct situation in the assembly process. These positions can be converted by a post processor into a manipulator executable program. With vision verification the situation is more complicated. Together with the set of nominal positions we now have the modifying factor array, the components of which will either be symbolic positions for verified bodies, or null, or expressions involving pointers to modifying factors and matrix multiplications. The symbolic positions will contain variables which refer to run time commands to the vision system. These symbolic positions will be evaluated once the vision system has been activated and has returned values for the variables. The evaluation of the symbolic positions will involve use of a subset of the RAPT inference system, which therefore has to be available in the run-time system.

The commands to the vision system consist of requests to return the endpoints of the best line found within the given window.

6. Conclusion

In extending RAPT for vision verification we have shown how vision can be used in an object level robot programming language. With the addition of a solid modelling system the programmer can obtain a lot of help as he is writing the program in deciding what to look at, and with what camera. The system does not have to make assumptions about positions of cameras or about the manner of presentation of the objects to the vision systems: this information is naturally included in the object level program as it is written.

The run-time system has to have some knowledge of the geometrical properties of the objects involved, and it has to have a limited inference capability. However, we have demonstrated that it is possible to do much of the computation involved in the use of vision verification at plan time and not during program execution.

7. Acknowledgements

This work was carried out with the support of the SERC (RJP and APA) and the Government of the People's Republic of China and Edinburgh University (BY), to all of whom we extend our thanks.

8. References

1. Popplestone,R.J. and Ambler,A.P. and Bellos,I.M. "An interpreter for a language for describing assemblies". Artificial Intelligence, 14,1980

2. Ambler,A.P., Corner,D.F. and Popplestone,R.J. "Reasoning about the spatial relationships derived from a RAPT program for describing assembly by robot" Proceedings of IJCAI-83, Karlsruhe, 1983.

3. Popplestone,R.J. and Ambler,A.P. "A language for specifying robot manipulations" in Robotic Technology: IEE Control Engineering 23, ed A Pugh, Peter Peregrinus, UK 1983

4. Yin, Baolin "A framework for handling vision data in an object level robot language RAPT" Proceedings IJCAI-83, Karlsruhe, 1983

5. Yin, Baolin "Combining vision verification with a high level robot programming language" Ph D Thesis, University of Edinburgh, 1984

Curved 3D Object Recognition Using 2D Shape Analysis Techniques

W. H. Tsai
National Chiao Tung University, Taiwan
and
J. C. Lin
University of Illinois at Chicago, USA

Abstract

A 3-D object recognition system capable of measuring 3-D object surfaces is described. The system is simple, efficient and relatively inexpensive. It relies on parallel-stripe projections for 3-D surface measurement. Data from four camera views are collected to form reference models which consist of horizontal cross section boundaries. The recognition scheme engages 3-D object registration followed by 3-D similarity measurement. The 3-D registration is accomplished by multiple 2-D registrations on the cross section planes using the generalized 2-D Hough transform. The 3-D similarity measure uses the average of all the 2-D cross-sectional similarity measures which are based on the "distance-weight correlation." Since the recognition scheme involves only pixel-level operations, it therefore is amenable to parallel processing.

I. INTRODUCTION

The first step in 3-D object recognition is the measurement of 3-D object surfaces. Many techniques have been proposed for this purpose (1-10), including contrived-stripe lighting and various range findings from texture, focusing, occlusion, motion, highlights, image brightness (or shading), stereo-disparity, Moire patterns, time of flight, etc. Tsai and Lin (11) recently gave an evaluation of these methods according to speed and cost considerations and propose the following as the desired features of a 3-D surface measurement technique:

(1)   Low-cost illumination;
(2)   Minimal feature-point correspondence or stripe tracing time;
(3)   Minimal data computation time;
(4)   Minimal object position and/or surface property constraints;
(5)   Maximal data collection from each camera view.

Few, if any, methods possess all of these desired features. In this paper, we propose a 3-D surface measurement system which possesses all five desired features. More specifically, we use multiple parallel stripe patterns which are projected onto object surfaces with parallel light beams. All the surface points covered by the stripe projections in a single view can be processed. A low-cost ordinary single-point light source is adequate for illumination. Overhead or slide projectors are acceptable. When placed far away from the object, it can generate approximately parallel light beams. No feature-point correspondence is involved; stripe tracing in the image is easy with known camera position and stripe intervals. Data computation formulas are very simple and do not involve trigonometric functions. No constraints are imposed on object surface properties.

To obtain a complete model for a reference object, all surface points around the object should be included. To accomplish this whole-view measurement, four copies of the above-mentioned single-view system are employed. Because of the use of parallel stripes which are also made parallel to the plane on which the object lies, data measured from each of the four views all belong to the same set of object cross section boundaries at some fixed heights from the plane. Therefore, data from all four views can be merged very easily without extensive data manipulation or transformation. The resulting object model consists of all the boundary points of a set of mutually-parallel object cross sections. Note that model setup or conversion usually is not trivial in other approaches (14, 15, 19, 20).

So far there have been only a few approaches proposed for 3-D object recognition (12, 13, 16, 17). Agin and Binford (12) and Nevatia (13) propose recognition methods for objects described by generalized cylinders or cones. Although their methods are more appropriate for elongated objects, the input object topologically is allowed to be quite complicated, without any restriction on its surface nature and 3-D orientation. The cylinders or cones are derived from one-dimensional slit image points. Oshima and Shirai (16) propose the use of small-region surface elements as object representation units. Raw range data are processed to extract surface elements. The object is described in terms of surface properties and relations. Recognition is performed by matching input object descriptions with model descriptions according to heuristic search. A similar matching scheme is also employed in Nevatia (13).

Watson and Shapiro (17) proposed a method for matching 3-D curved objects to 2-D perspective views. A reference object model consists of a closed space curve representing some characteristic connected curve edges of the object. The input is a 2-D perspective projection of some reference object. No 3-D measurement for input scenes is required. The recognition problem is transformed to matching the input curve with 2-D perspective projections of a model space curve after the model is properly rotated and translated. The curve representations used for matching are Fourier descriptors. Sato and Honda proposed a different method for the description and recognition of curved objects in terms of horizontal cross section boundaries which are similar to the model form used in this study. To perform object recognition, 2-D boundaries are also described by Fourier descriptors from which proper features are derived for matching by pseudo-distance measures.

Note that in all the above approaches, it is required to process and transform row 3-D data into proper forms for matching, including surface elements, generalized cylinders or cones, connected curve edges and Fourier descriptors. This is due to the specific object description methods used in these approaches. In contrast, the proposed approach requires minimal raw data processing. Data points obtained from 3-D measurement are used directly as the model and the input. This is possible because the recognition scheme proposed is based on pixel-level 2-D shape analysis techniques to be described next.

Briefly, 3-D object recognition can be accomplished by registering two objects first and then giving a measure for the goodness of the resulting registration. An input object is recognized to be a reference object if the surface points of the former register well with those of the latter. This is the basic idea of the proposed 3-D recognition method. Since the models are in the form of horizontal cross section boundaries, a 3-D registration can be decomposed into a set of 2-D registrations which are then solved by the generalized Hough transform (21-23). The goodness of a 2-D registration is measured by a simplified version of the so-called distance-weighted correlation proposed by Fan and Tsai (24) for automatic Chinese seal identification. The goodness of an overall 3-D registration is then computed as the average of all 2-D goodness measures on the cross section planes. Since the generalized Hough transform and the distance-weighted correlation computation are performed on the pixel level and can be executed in parallel, the proposed 3-D recognition scheme is amenable to parallel processing. Thus, compared with other approaches, the proposed system possesses the following merits:

(1) uses a low-cost but fast method for 3-D measurement;
(2) imposes no restriction on object surface properties;
(3) requires no extensive processing on raw 3-D data for model set-up and recognition;
(4) adopts well-developed 2-D shape analysis techniques for 3-D recognition;
(5) employs parallelly-executable basic operations.

II. 3-D OBJECT SURFACE MEASUREMENT USING PARALLEL-STRIPE LIGHTING

We use parallel-stripe lighting and a TV camera for the 3-D measurement of object surfaces. The measurement configuration is shown in Fig. 1. A global rectangular coordinate system is established in such a way that the object is located around the system origin O with its bottom on the X-Y plane. The X-Y plane may be chosen to be a large table on which the object lies. A flat transparent pattern plate with black parallel stripes is placed vertically on the X-Y plane in the first quadrant of the coordinate system at a reasonable distance from O in the direction of 45° with respect to the X-axis (i.e., the plate is vertical to the line $y = (\tan 45^{\circ})x$ on the X-Y plane and goes through O). The stripes in the plate, in addition to being mutually parallel, are also parallel to the X-Y plane. The stripes are equally spaced with interval d. The lowest stripe nearest to the X-Y plane is also at distance d from the X-Y plane. Each stripe is indexed by a number i so that the ith stripe is at distance id from the X-Y plane (i.e. with height id) where i > 1. Parallel light is illuminated onto the plate vertically so that the light beams are also parallel to the X-Y plane. Owing to the above special arrangement of stripe lighting, it is not difficult to figure out that the projection of the ith stripe (with height id) on the object surface will also be at height id from the X-Y plane (i.e., the z coordinate of each surface point on the ith stripe projection is id).

The TV camera is located at (a, 0, h), such that the camera lens center is placed at height h from the X-Y plane on the X-axis at distance a from the origin O. The focal length of the camera is f. Therefore the image plane is located at distance a − f from O, and is perpendicular to the X-axis. The rectangular coordinate system within the image plane has its origin I located at global coordinates (a-f, 0, h) with its two coordinate axes, U-axis and V-axis, parallel to the Y-axis and the Z-axis, respectively. In the following, we shall derive the equations for transforming the image coordinates of a given point on a stripe projection into the corresponding global coordinates.

First, imagine that the plane containing the pattern plate divides the global coordinate space into two halves. In the half space containing the object, we can also imagine the existence of a set of parallel "shadow planes," each being formed by a black stripe illuminated by the parallel light beams from the other half space. Each plane extends toward the Z-axis is parallel with the X-Y plane. These planes may be regarded as a set of "flying blades" in space . Any object met by them may be imaged to be "cut" into several slices, each slice including two object cross sections whose boundaries are the projections of two stripes on the object surface. Obviously, the shadow plane S_i corresponding to the ith stripe L_i can be described globally by the following equation:

$$z = id, \qquad (1)$$

because S_i goes through L_i and is parallel to the X-Y plane.

Next, let P with global coordinates (x, y, z) be an object surface point covered by the projection of the ith stripe L_i, and let (u, v) be the image coordinates of the corresponding point P' of P in the image plane, viewed through the TV camera. Since P is also on S_i, by (1) we have the z coordinate of P as:

$$z = id. \qquad (2)$$

On the other hand, by the well-known geometric principle of similar triangles, we get from Fig. 2 the following equalities:

$$\frac{f}{a - x} = \frac{u}{y} = \frac{v}{z - h}. \qquad (3)$$

With z known to be id by (2), we can solve x and y from the above equations in terms of u, v, a, f, and h as follows:

$$x = a - [f (id - h) / v], \qquad (4)$$

$$y = u (id - h) / v. \qquad (5)$$

Note that in the derivation of the global coordinates (x, y, z) using (2), (4) and (5) above, no knowledge about the position of the stripe pattern is required. In particular, the height of the projection of the ith stripe as specified by (2) has nothing to do with the position of the object. This will not be the case if the light beams are not parallel to the X-Y plane. It means that no matter where the object is located on the X-Y plane, as long as the object surface which touches the X-Y plane is always kept unchanged, the projection of each stripe (and so the surface points covered by the projection) will always appear with the same height from the X-Y plane. This characteristic of the proposed 3-D measurement scheme will become a great advantage for 3-D object model setup and 3-D object recognition, as well be shown in the following sections.

Finally, to facilitate tracing the projection curves within the image plane, let the height h of the TV camera be adjusted in such a way that the following inequalities hold for some j:

$$jd < h < (j + 1) d. \qquad (6)$$

In the next paragraph, we will prove that the U-axis falls in between the two projections of the jth and (j+1)th stripes as viewed from the camera. Therefore, tracing of the projections in the image plane to determine the value i in (2), (4), and (5) can be simplified by starting from the projections right above and right below the U-axis and then extending to neighboring ones.

Let P be any object surface point covered by the U-axis in the image plane. Let (x, y, z) and (u, v) be the coordinates of P and P', respectively. From (3), we have:

$$v = \frac{f}{a - x} \cdot (z - h). \qquad (7)$$

Since P is on the U-axis when it is viewed from the camera, we also have:

$$v = 0. \qquad (8)$$

From (7) and (8), we get:

$$z = h. \qquad (9)$$

And from (6) and (9), we see that any object surface point covered by the U-axis in the image plane is with a height just between jd and (j+1)d, or equivalently, that the U-axis falls in between the two projections of the jth and (j+1)th stripes in the image planes.

In summary, after the image of an object is taken, we trace the projections one by one, starting from the one either above or below the U-axis. Let the projection just traced correspond to the ith stripe. Then, for each point on the projection with image coordinates (u, v), its global coordinates (x, y, z) can be determined by Eqs. (2), (4), and (5).

III.   3-D OBJECT MODEL CONSTRUCTION

To recognize an unknown object by its 3-D surface data, we have to match the data with those of reference objects. The purpose of 3-D object model construction is to establish a complete data set of 3-D surface points for each reference object. The 3-D surface measurement method proposed in the last section provides partial 3-D surface data from a single camera view. This is not sufficient for a reference object. To obtain the data for all surfaces, we propose in this section a more complete measurement configuration which, as shown in Fig. 3, includes four identical copies of the single-view measurement equipment shown in Fig. 1. The four identical sets are arranged symmetrically around the global system origin O so that the cameras are located along the four half-axes, and the stripe patterns are located in the directions of 45°, 135°, 225°, and 315° with respect to the X-axis. The width of the patterns are large enough for the four sets of stripe projections to fully cover the whole region where objects may be placed.

As mentioned in the last section, the projection of the ith stripe of each measurement set will appear on the object surface at an identical height z = id from the X-Y plane, no matter where the object is located. This means that the four stripe projections at height z = id together compose the boundary of the object cross section at that height. The object surfaces can be imagined to consist of a set of cross section boundaries formed by all the stripe projections. Such cross section boundaries are all at fixed heights from the X-Y plane, and are equally spaced with interval d. Using Eqs. (2), (4) and (5), we obtain a set of cross section boundary points for each view. The 3-D surface model of a reference object consists of all four sets of data points obtained from the respective views.

Once a reference model object is established in the learning stage, no matter where the object is located in the recognition stage, the same set of cross section boundaries as obtained in the learning stage will again be obtained and used directly for recognition. Additional 3-D data manipulation is not required. The use of cross section boundary points for direct 3-D objection recognition will be discussed later in this paper. Although we propose the above scheme for constructing cross section boundaries as reference object models, we mention here that such models may also be generated from any type of geometric model of the reference objects, if they are available.

In the remainder of this section, we discuss how to combine by coordinate transformation all the data points measured from the four views so that they are consistent under a single global coordinate system. Creation of a so-called cross section image to include object boundary points is also discussed.

As shown in Fig. 3, the first set of measurement equipments (including camera #1) is identical to the one shown in Fig. 1. Thus, 3-D data computation can be performed according to Eqs. (2), (4), and (5). However, for the remaining three sets, Eqs. (2), (4), and (5) cannot be used directly; or if they are to be used, the results must be corrected by coordinate transformation. In general, if (x_k, y_k, z_k) are the coordinates computed for a point according to Eqs. (2), (4), and (5) with the kth set of equipments where k = 1,2,3,4, then the correct cOordinates (x, y, z) should be those obtained after the X-Y plane is rotated through an angle of $-(k-1) \times 90^O$ with respect to the Z-axis, i.e.:

$$x = x_k \cos\theta + y_k \sin\theta , \tag{10}$$

$$y = x_k \sin\theta + y_k \cos\theta , \tag{11}$$

$$z = z_k \tag{12}$$

where $\theta = -(k-1) \times 90^O$ and k=1,2,3,4.

For each cross section boundary at height id, a new image plane is created and is called the ith <u>cross section image.</u> The size of the cross section images are determined by the extremes of the x and y values of object locations. The x and y coordinates computed according to (10),(11) above are rounded into integers, x_I and y_I. For each point with coordinates (x, y, z) computed(10)-(12) above, a cross section boundary point is created by filling the value 1 into the cell at coordinate (x_I, y_I) in the ith cross section image. After all object surface points covered by the four sets of stripe projections are processed in this manner, each cross section image will contain all the cross section boundary points of the object at the corresponding height.

IV. PARTIAL MATCHING OF 2-D SHAPE BOUNDARIES BY GENERALIZED HOUGH TRANSFORM AND A NEW SHAPE SIMILARITY MEASURE

As mentioned previously, the proposed 3-D object recognition approach is based partially on the well-developed 2-D generalized Hough Transform (GHT) technique. Due to the use of planar cross section boundaries as object data, repetitive applications of the 2-D GHT are found adequate for 3-D registration needed in 3-D recognition. In particular, the insensitivity of the GHT to noise, occlusion, and boundary gaps makes the 2-D GHT very suitable for registering the cross section boundary points obtained with the 3-D measurement method described in the last two sections. All the boundary points from the four views, once merged, can be used directly for the GHT without any further conventional image processing like boundary following or segmentation (25). This expedites 3-D recognition. In the following discussion, for simplicity, we will refer to cross section boundaries as shapes although they are not all the same.

The main advantage of the GHT is its usefulness in curve detection. However, for the purpose of object recognition with multiple reference objects, it is desirable to measure the similarity of a given shape to each reference shape instead of just detecting the existence of a reference shape in the given shape. A new shape similarity measure for this purpose is thus also proposed in this section. Its value is computed right after GHT is performed. It can be used for measuring the similarity between any two planar curve-type shapes. It is a modified version of the distance-weighted correlation (DWC) proposed by Fan and Tsai (24) for Chinese seal identification. In the remainder of this section, the GHT is first reviewed, followed by the definition of the new similarity measure which will be called modified DWC (MDWC). Briefly, we use the GHT for 2-D shape registration but the MDWC for shape similarity mesurement.

A. A Review of Generalized Hough Transform

Let B be a set of coplanar points which constitute a reference curve-type shape and S be a set of coplanar points which are extracted from an input curve-type scene. The GHT can be used to determine if any part of B exists in S or vice versa. Assume for the moment that there exists no noise or distortion in the input and that there is no shape rotation or scaling problem, so that if any part of B does exist in S, it can be overlapped on S exactly by a 2-D translation which can be found by the GHT. The GHT involving the two point sets B and S is described in the following.

First, choose a reference origin G for B. The centroid of B is a possible choice. Let (x_o, y_o) be the coordinates of G. From now on, a point P with coordinate (x, y) will be denoted as $P(x, y)$. Next, construct a displacement vector (x_i^d, y_i^d), for each point $P(x_i^B, y_i^B)$ in B such that:

$$x_i^d = x_o - x_i^B, \qquad (13)$$

$$y_i^d = y_o - y_i^B. \qquad (14)$$

Let the set of displacement vectors be denoted as D. Now, create a 2-D accumulator array H which may be thought as a new image. Let $H(i, j)$ denote the cell value of H at coordinate (i, j). All $H(i, j)$ are cleared to zero initially. For each point $P(x_j^S, y_j^S)$ in S, "attached" to it all the displacement vectors in D, find the end position of each vector, and increment the cell value of H at that position by 1. More precisely, this means that for each $P(x_j^S, y_j^S)$ in S, increment $H(x_j^S + x_i^d,$

$y_j^S + y_i^d$) by 1 for each vector (x_i^d, y_i^d) in D for all $1 \leqslant i \leqslant$ #B and $1 \leqslant j$ $\leqslant$ #S where #A means the number of elements in set A. The domain of the accumulator array H is determined by the extreme values of all possible $x_j^S + x_i^d$ and $y_j^S + y_i^d$. This means that a cell value may be incremented at coordinates (i, j) with i or j, or both, negative. If B and S are identical, then at certain coordinates (i_o, j_o), which corresponds to the reference origin G in B, the cell value $H(i_o, j_o)$ will be equal to #B, which is the maximum possible cell value in H. When B and S are partially identical, the maximum cell value in H will be smaller than #B. If a certain part of B can be found in S at several positions, several large values will be found in H. By choosing an appropriate threshold, large cell values can be detected, each of which corresponds to a pair of identical curve portions B' and S' existing in B and S, respectively.

When the rotation problem exists, i.e., when we want to determine whether a reference shape or part of it, possibly rotated, exists in an input scene, the usual way is to incrementally rotate the displacement vectors before they are attached to the input points. Such rotation of displacement vectors may be performed just for possible reference shape orientations. But if no prior information is available about possible reference shape orientations, usually the whole range of 360° is tried. This is proceeded as follows. Let θ_o be the angle increment in degree for reference shape rotation. Then, instead of only attaching the displacement vector specified by Eqs. (13) and (14) to each point P in S, we attach all possible rotated displacement vectors ($x_{i\theta}^d$, $y_{i\theta}^d$) to P for all $\theta = 0^o, \theta_o, 2\theta_o, \ldots \lfloor 360^o/\theta_o \rfloor \theta_o$, where $x_{i\theta}^d$ and $y_{i\theta}^d$ are related to x_i^d and y_i^d in (13) and (14) by the following rotational coordinate transformation:

$$x_{i\theta}^d = x_i^d \cos\theta + y_i^d \sin\theta , \qquad (15)$$

$$y_{i\theta}^d = -x_i^d \sin\theta + y_i^d \cos\theta . \qquad (16)$$

Also, the 2-D accumulator array, H(i,j), now must be extended to 3-D H(i, j, θ), such that the cell value at coordinates ($x_j^S + x_{i\theta}^d$, $y_j^S + y_{i\theta}^d$, θ) is incremented by 1 for each $P(x_j^S, y_j^S)$ in S and each rotated vector ($x_{i\theta}^d$, $y_{i\theta}^d$), specified by Eqs. (13) thru (16). Together, there are #B · #S · ($\lfloor 360^o/\theta_o \rfloor + 1$) rotated displacement vectors attached to all points in S. Since no scaling problem is encountered in the proposed 3-D object recognition approach, extension of the GHT to include scaling is not discussed here.

B. Modified Distance-Weighted Correlation for Partial Shape Matching

We now define the new MDWC similarity measure for matching two sets of curve points. Again, let B be the set of reference points and S the set of input points. After GHT is performed and the modified cell values in the 3-D accumulator array H are computed, we select one of the large cell values, say at (i, j, θ), and then register B and S according to the reference origin G of B and the point in S corresponding to (i, j) after B is rotated through an angle of θ degrees. In the following, whenever we mention B or S, we mean their versions under the above registrtion. If B and S are very similar, then each point in S should be very close to a point in B and vice versa. Therefore, it is feasible to define a measure which indicates overall mutual closeness of all points in B and S. And this is the basic idea of the MDWC to be defined next.

For each point $P(x_i^S, y_i^S)$ in S, let $P(x_{j_i}^B, y_{j_i}^B)$ be the closest point in B to $P(x_i^S, y_i^S)$, i.e., the Euclidean distance d_i from $P(x_i^S, y_i^S)$ to $P(x_{j_i}^B, y_{j_i}^B)$ is smaller than that to any other point in B. We define a weight w_i for each $P(x_i^S, y_i^S)$ in S as:

$$w_i = \frac{1}{d_i^2 + 1} \quad \text{if } 0 \leqslant d_i \leqslant K,$$

$$= 0 \quad \text{otherwise}, \quad (17)$$

where

$$d_i = [(x_i^S - x_{j_i}^B)^2 + (y_i^S - y_{j_i}^B)^2]^{\frac{1}{2}}$$

and K is a constant used as a limit of distances within which the closest point $P(x_{j_i}^B, y_{j_i}^B)$ in B of $P(x_i^S, y_i^S)$ in S should be searched for. That is, if the closest point $P(x_{j_i}^B, y_{j_i}^B)$ is at a distance greater than K (i.e., $d_i > K$), then it will be ignored (with w_i set zero) in computing the final values of MDWC. This in turn means that two points, from the same set, but far away from each other will contribute nothing to the shape similarity measure.

Let S' denote the set of all those points in S whose weights are non-zero, and recall that #S' denotes the number of all points in S', then the MDWC between S and B (with B as the reference shape) is defined as:

$$C(S,B) = (\sum_{P(x_i^{S'}, y_i^{S'}) \epsilon S'} w_i)/ \#S' \quad \text{if } \#S' > N,$$

$$= 0 \quad \text{otherwise}, \quad (18)$$

where N is a threshold for #S'. This means that only when the number of those points in S with corresponding points in B within the limit K is larger than N can we consider the value C(S, B) defined above meaningful. Otherwise, it is set to zero. The purpose of the threshold is to exclude partial matching with too few closest point (within limit K) from being considered as an acceptable matching. Otherwise, in the extreme case, S' with only a single point overlapped by any point in B may give the maximum MDWC value. In other words, only partial matchings involving more than N point from S will be considered "reliable" for shape similarity measurement.

It is easy to verify that $0 \leqslant C \leqslant 1$ and that $C = 1$ when B and S overlaps exactly for more than N points. So, the larger the value C(S, B) is, the more similar S and B are (in the sense of partial matching). The constants K and N are problem-dependent. Their values should be experimentally determined to obtain optimal application results.

On the other hand, we see that a point-to-point correspondence is defined simultaneously when the closest point in B of each point in S is found. Such correspondence defined by the notion of closest point is not 1-to-1, i.e., two points in S may have an identical point in B as their common closest point. Since the search limit K is imposed, we may also say that point P in S has no corresponding point in B if the closest point of P in B is at a distance larger than K from P. Note that the weight for such a point in S is zero. Let B' denote all the points in B which are the closest points of some points in S, or more precisely, of the points in S'. Then, the above-mentioned point-to-point correspondences constitute a many-to-one mapping function from S' to B'. It characterizes the partial matching of S with B. Because of its many-to-one property, the MDWC defined above is in general asymmetric, i.e., $C(S, B) \neq C(B, S)$, unless S and B are identical. Although it can be made symmetric by defining it to be [C(S, B) + C(B, S)]/2 as in (24), it is found that the asymmetric version as defined previously in (18) is adequate for our 3-D recognition purposes. For convenience, S' and and B' will be called the <u>matched portion</u> of S and B, respectively in the remainder of this paper.

V. 3-D OBJECT RECOGNITION BY 2-D SHAPE ANALYSIS

Although we use data from four views to construct a reference object model, recognition will be performed with data from only a single view. In other words, we use only one camera in the recognition stage and try to analyze an input scene by its surface points obtained from a single camera's field of view. Several unknown objects may be presented for recognition. They may also occlude one another. The recognition scheme will identify the objects one by one, each time isolating all the surface points in the input belonging to an object until no more surface points in a camera view can be processed. The recognition scheme is based on the GHT for registration and the MDWC for shape similarity measurement. Due to the use of cross section boundaries, the 2-D GHT is sufficient although attempts have been tried to extend the GHT to its 3-D case (26). In the following, we will illustrate the proposed 3-D recognition scheme with an example.

Suppose that totally m cross section images are used for each reference object model. As mentioned in Sec. III, each cross section image of a reference model is a created image which contains all the cross section boundary points (each filled with value 1) of the reference object at a certain height from the X-Y plane. Let n denote the number of reference objects. For example, Fig. 4 shows two reference objects (Fig. 4(a) and Fig. 4(b)) and their respective cross section images (Fig. 4(d) and Fig. 4(e)), in which n = 2 and m = 3. Also shown in Fig. 4 is an input scene (Fig. 4(c)) and its surface data from a single view. The 3-D recognition procedure, similar to 2-D shape matching discussed in the last section, consists of 3-D registration followed by 3-D shape similarity measurement.

3-D registration of an input scene with the reference models starts from 2-D registration at a certain height, using all the reference cross section images and the input surface data at that height. Assume that B_{ij} is the cross section image of the ith reference model and S_j is the partial input surface data, both at height h = jd from the X-Y plane where d is the interval between the cross sections (originally the interval between every two stripes in the pattern plate). Also assume that the initial 2-D registration starts at height $h_o = j_o d$. h_o may be chosen to the one at which the cross section boundaries of all reference models differ most significantly. This will increase 3-D recognition speed as can be seen later. In the example of Fig. 4, we choose j_o to be 2.

The initial 2-D registration using B_{ij_o} and S_{j_o} goes as follows. First, match S_{j_o} with each of the reference B_{ij_o}, $i = 1,2,...n$. This requires the 2-D GHT for 2-D registration and the MDWC for 2-D shape similarity measurement. After the GHT is performed, an accumulative array H is created. Each cell in H at coordinates (i,j) for some fixed θ is filled with a value which specifies the total number of the displacement vectors whose end points fall at position (i, j). Another modified accumulator array H' is then created from H. The value for each cell in H' at (i, j) for θ is computed as the sum of all the old H values of those cells which are included in an M x M area with (i, j) as its center. The H' array is then searched for large cell values which are above a certain threshold H'_t. Let $L(S_{j_o}, B_{ij_o})$ denote the set of all such cell positions, i.e.,

$$L(S_{j_o}, B_{ij_o}) = \{P = (k, \ell, \theta) | H'(k, \ell, \theta) > H'_t\}.$$

Each element $P = (k, \ell, \theta)$ in L indicates a good position for registering S_{j_o} and B_{ij_o}. More specifically, for each $P=(k, \ell, \theta)$, we register S_{j_o} and a rotated version of B_{ij_o} (through an angle of θ degrees) according to the reference origin G_i of B_{ij_o} and a point in S_{j_o} at (k, ℓ). For convenience, we denote a registration as a 2-tuple $r = (P, G_i)$. Each registration $r = (P, G_i)$ corresponds to a good overlapping of S_{j_o} on B_{ij_o}.

Let $R(S_{j_o})$ denote all registrations defined by the elements in $L(S_{j_o}, B_{ij_o})$ for all i, i.e.,

$$R(S_{j_o}) = \bigcup_{i=1}^{n} \{r=(P,G_i) | P \in L(S_2, B_{i2})\}.$$

Then, $R(S_{j_o})$ includes the good registrations of S_{j_o} with all the reference shapes B_{ij_o} at the height $h_o = j_o d$. Each of these 2-D registrations will then serve as a basis for 3-D registration. All of them should be tried, but we have to determine which registration in

$R(S_{j_o})$ should be tried first. Of course, we hope the one with the largest possibility to yield a successful 3-D registration can be selected first. This will improve recognition speed. One way for this is to use the MDWC as a measure for the goodness of each registration in $R(S_{j_o})$. Therefore, for each $r = (P, G_i)$ in $R(S_{j_o})$ with $P \in L(S_{j_o}, B_{ij_o})$, we compute the corresponding MDWC as :

$$C_{ij_o}^r = C(S_{j_o}, B_{ij_o}),$$

according to Eqs. (17) and (18). And the registration r with the maximum $C_{ij_o}^r$ is then selected for 3-D registration (to be described next). If the resulting 3-D registration is not good, then the registration in $R(S_{j_o})$ with the second largest $C_{ij_o}^r$ value is picked for another trial. The process is repeated until either all the registrations in $R(S_{j_o})$ are claimed no good or some one is found good. For the former case, we determine that the reference objects do not exist in the input scene. For the latter case, if the registration found good for 3-D registration is $r = (P, G_i)$ with $P \in L(S_{j_o}, B_{ij_o})$ then we decide that reference object B_i exists in the input scene. The matched portion of S_{j_o} is then removed from S_{j_o}, resulting in a reduced input shape. The reduced input shape is then searched for another instance of reference shape or part of it by repeating the whole process described above. Thus, the final decision may be one of the following three cases: (i) no reference object is found in the input; (ii) the whole input are found to match reference objects, with possibly more than one instance of a specific reference object; (iii) the input contains some reference objects and some unknown objects. What remains now is how to construct a 3-D registration based on a 2-D registration in $R(S_{j_o})$ and how to determine whether the resulting 3-D registration is good or not.

Let $r = (P, G_i)$ be a 2-D registration with $P = (k, \ell, \theta)$. The 3-D registration based on r is constructed in such a way that while B_{ij_o} and S_{j_o} are being registered according to P and G_i through rotation and translation, all the cross section boundary points in other cross section images are also moved accordingly without changing all their 3-D relative positions. For example, if B_{12} and S_2 from Fig. 4 are registered as shown in Fig. 5(a), then the corresponding 3-D registration of S_1, S_2 and S_3 on all the cross section images of B_1, B_{11}, B_{12}, and B_{13}, will look like Fig. 5(b), which obviously is no good.

In general, the goodness of a 3-D registration is determined by a 3-D shape similarity measure which we define, after 3-D registration, as the average of the 2-D MDWC values computed for all the registered shapes on all the m cross section images. That is, if r is the initial 2-D registration, the corresponding similraity mesure M_i^r is defined as:

$$M_i^r = \frac{1}{m} \sum_{j=1}^{m} C_{ij}^r \text{ where } C_{ij}^r = C(S_j, B_{ij})$$

with C_{ij}^r as the MDWC computed after S_j adn B_{ij} are registered according to the 3-D registration defined previously. And a 3-D registration input S and a reference B_i based on r is said to be good if M_i^r is larger than a certain threshold value M_t.

References

(1)   R. A. Jarvis, "A Perspective on Range Finding Techniques for Computer Vision," IEEE Trans. Patt. Anal. Mach. Intell., Vol. PAMI-5, No. 2, March 1983.

(2)   L. S. Davis. L. Janos, and S. M. Dunn, "Effcient Recovery of Shape from Texture, IEEE Trans. PAMI, Vol. PAMI-5, pp. 485-492, Sept. 1983.

(3)   R. A. Jarvis, "Focus Optimization Criteria for Computer Image Processing," Microscope, Vol. 24, pp. 163-180, 2nd quarter, 1976.

(4)   D. Rosenberg, M.D. Levine, and S. W. Zucker, "Computing Relative Depth Relationships from Occlusion Cues," in Proc. 4th IJCPR, Kyoto, Japan, pp. 765-769, Nov. 1978.

(5)   R. Nevatia, "Depth Measurement by Motion Stereo," Computer Graphics and Image Processing, Vol. 5, No. 2, 1976.

(6)   P. Thrift and C. H. Lee, "Using Highlights to Constrain Object Size and Location," Conf. Record IEEE 1982 Workshop on Industrial Applications of Machine Vision, Research Triangle Park, N.C., May 1982.

(7)   E. N. Coleman adn R. Jain, "Obtaining 3-D Shape of Textured and Specular surfaces using 4-Source Photometry," Computer Graphics and Image Processing, vol. 18, 1982.

(8)   S. Barnard and W. B. Thompson, "Disparity Analysis of Images," IEEE Trans PAMI, Vol. PAMI-2, No. 4, 1980, pp. 333-360.

(9)   M. Idesawa, T. Yuatagai, and T. Soma, "Scanning Moire Method and Automatic Measurement of 3-D Shapes," Appl. Opt., Vol. 16, pp. 2152-2162, Aug. 1977.

(10) S. Nitzan, A. E. Brain, and R. O. Duda, "The Measurement and Use of Registered Reflectance and Range Data in Scene Analysis," Proc. IEEE, Vol. 65, Feb. 1977.

(11) W. H. Tsai and J. C. Lin, "Range Data Driven 3-D Surface Measurement Using a Single View of Multiple-Stripe Projection," submitted.

(12) G. J. Agin and T. O. Binford, "Computer Description of Curved Objects," IEEE Trans. Computers, Vol. c-25, No. 4, April 1976.

(13) R. Nevatia, "Description and Recognition of Curved Objects," Artificial Intelligence, Vol.8, 1977.

(14) B. I. Soroka and R. K. Bajcsy,. "A Program for Describing Complex Three-Dimensional Objects Using Generalized Cylinders as

Primitives," in Proc. IEEE Conf. PRIP, Chicago, Illinois, pp. 331-339, June 1978.

(15) J. K. Aggarwal, L. S. Davis, W. N. Martin, and J. W. Roach, "Representation Methods for Three-Dimensional Objects," in Progress in Pattern Recognition (L. N. Kanal and A. Rosenfeld, Eds.), Vol. 1, North-Holland, New York, 1981.

(16) M. Oshima and Y. Shirai, "Object Recognition Using Three-Dimensional Information," IEEE Trans. PAMI, Vol. PAMI-5, No. 4, pp. 353-361, July 1983.

(17) L. T. Watson and L. G. Shapiro, "Identification of Space Curves from Two-Dimensional Perspective Views," IEEE Trans. PAMI, Vol. PAMI-4, No. 5, PP. 469-475, Sept. 1982.

(18) Y. Sato and I. Honda, "Pseudodistance Measures for Recognition of Curved Objects," IEEE Trans. PAMI, Vol. PAMI-5, No. 4, PP. 362-372, July 1983.

(19) C. Dane and R. Bajcsy, "An Object-Centered 3-D Model Builder," in Proc. 6th Int'l Conf. Patt. Recog., Munich, Germany, Oct. 1982.

(20) R. Bajcsy, "Three-Dimensional Object Representation," in Pattern Recognition Theory and Applications (J. Kittler, K. S. Fu, and L. F. Pau, Eds.), D. Reidel Publishing Co., U. K., 1982.

(21) P. M. Merlin and D. J. Farber, "A Parallel Mechanism for Detecting Curves in Pictures," IEEE Trans. Computers Vol. C-24, No. 1, pp. 94-95, January 1975.

(22) P. H. Ballard, "Generalizing the Hough-Transform to Detect Arbitary Shapes," Pattern Recognition, Vol. 13, No. 2, pp. 111-122, 1981.

(23) S. Yam and L. S. Davis, "Image Registration Using Generalized Hough Transforms," in Proc. IEEE Conf. PRIP, Dallas, Texas, pp. 526, 532, 1981.

(24) T. J. Fan and W. H. Tsai, "Automatic Chinese Seal Identification," Computer Vision, Graphics and Image Processing, Vol. 25, pp. 311-330, March 1984.

(25) A. Rosenfeld and A. C. Kak, Digital Picture Processing, Vol. II, Academic Press, New York, 1982.

(26) D. H. Ballard and D. Sabbah, "Viewer Independent Shape Recognition," IEEE Trans. PAMI, Vol. PAMI-5, No. 6, pp. 653-659, November 1983.

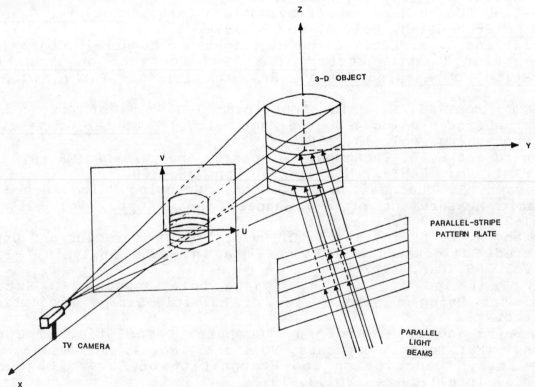

FIG.1 3-D OBJECT SURFACE MEASUREMENT USING
PARALLEL-STRIPE LIGHTING

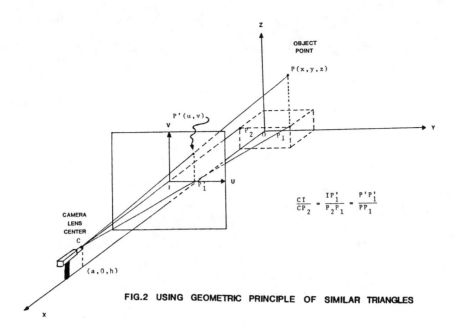

$$\frac{CI}{CP_2} = \frac{IP'_1}{P_2P_1} = \frac{P'P'_1}{PP_1}$$

FIG.2 USING GEOMETRIC PRINCIPLE OF SIMILAR TRIANGLES

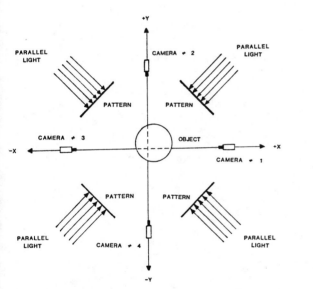

FIG.3 TOP VIEW OF THE 3-D MEASUREMENT CONFIGURATION USED FOR OBJECT MODEL CONSTRUCTION

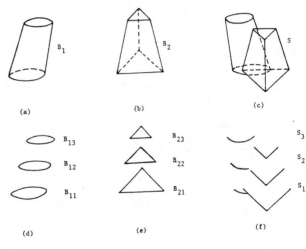

Fig. 4 Illustration of 3-D recognition scheme; (a) and (b) are two reference objects; (d) and (e) are the cross section images of (a) and (b), respectively; and (c) is an input scene with (f) as its surface data.

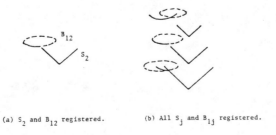

(a) S_2 and B_{12} registered.   (b) All S_j and B_{1j} registered.

Fig. 5 3-D registration.

393

Techniques of Multisensor Signal Processing and Their Application to the Combining of Vision and Acoustical Data

J. M. Richardson, K. A. Marsh

and

J. F. Martin

Rockwell International Science Center, USA

ABSTRACT

We consider first the general problem of single-time state estimation in the case where the measured data are produced by sensors of different types. Attention is devoted to the question of what kind of signal processing is involved in the merging of the sensor outputs. Other questions pertinent to the later stages of the total state estimation problem are discussed.

As an illustrative example, we consider the problem of estimating the three-dimensional geometry of an object using a combination of data from an acoustical scattering measurement system and an optical system. The present treatment is limited to conventional monocular optical systems involving a variety of illumination directions. Results based upon synthetic data are presented and evaluated.

1. INTRODUCTION

It is widely assumed that in the estimation of the state of a system, the use of data produced by a variety of sensor types gives better results than that based upon data from a single type of sensor. In the next section, this plausible assertion is proved rigorously and the proper procedure for combining data from sensors of different types is outlined. In a later section, we present a detailed analysis of an application of these concepts to the combining of acoustical and optical data in the imaging of three-dimensional objects.

2. GENERAL THEORY OF STATE ESTIMATION WITH MULTISENSOR DATA

In this section, we present a brief discussion of the problem of estimating the state of the system (e.g., a static object, a dynamical device, etc.) at a given time from a set of measurements produced by two or more different types of sensors. We will avoid subtle questions dealing with state vectors containing components that are correlated only with subsets of measurements.

Let us consider first an M-dimensional vector x representing the state of a system. The systems of interest here are time independent and, hence, the problem of time evolution of the state does not enter. We consider secondly an N-dimensional vector y representing the set of all measurements. Let us denote an estimate of x, based upon y, to be denoted by the function x(y) which we will henceforth call an estimator. Let us consider the following criterion of optimality

$$S = E\delta(\hat{x}(y)-x) \qquad (2.1)$$

in which E is the operation of averaging on all random variables (including random processes, if present). $\delta(\ )$ is the M-dimensional Dirac δ-function. The criterion S is proportional to the fraction of times the estimator gives approximately the right answer if the ideal δ-function is replaced by a somewhat smeared-out δ-function. For a more detailed discussion of decision theory, the reader is urged to consult the monogram by Ferguson.[1]

The optimal form of $\hat{x}(y)$ can readily be shown to be the most probable value of x given y (assuming that this is unique), i.e., the optimal form $x_{opt}(y)$ is the value of x that maximizes $P(x|y)$, the probability density (p.d.) of x given y. This p.d. (often called the a posteriori p.d.) is given by

$$P(x|y) = P(y|x)P(x)/P(y) \qquad (2.2)$$

where

$$P(y) = \int dx P(y|x)P(x) \quad . \qquad (2.3)$$

In the above expressions, $P(y|x)$ is the p.d. of y given x, and $P(x)$ is the a priori p.d. of x. In the problem of maximizing on x, the factor $1/P(y)$ is a constant, since y is given, and hence it can be ignored.

Let us consider two measurement vectors, y_1 and y_2, corresponding to measurements of types 1 and 2, respectively. Each type of measurement may involve many sensors of the same type. Let us now assume that the total measurement vector y is formed by a concatenation of y_1 and y_2, namely

$$y = \begin{pmatrix} y_1 \\ y_2 \end{pmatrix} \qquad (2.4)$$

in which the dimensionality of y is given by

$$N = N_1 + N_2 \qquad (2.5)$$

where N_1 and N_2 are the dimensionalities of y_1 and y_2, respectively. From the results of the last section, it follows trivially that

$$P(x|y) \equiv P(x|y_1,y_2) = P(y_1,y_2|x) \ P(x)/P(y_1,y_2) \quad . \qquad (2.6)$$

We now require the state x to have the property that, if x is known, y_1 and y_2 are statistically independent, namely

$$P(y_1,y_2|x) = P(y_1|x)P(y_2|x) \quad . \qquad (2.7)$$

This relation can be rewritten in the form

$$P(y_1|x,y_2) = P(y_1|x) \qquad (2.8)$$

We can also write a similar expression with y_1 and y_2 interchanged. The above relation says that the p.d. of y_1, given x and y_2, is unaffected by the knowledge of y_2, since x carries all of the information required for specifying the statistical properties of y_1 independently of y_2. If x is unknown, then y_1 and y_2 are dependent, except in sterile trivial cases of no interest to us. In this case, we must write

$$P(y_1,y_2) \neq P(y_1)P(y_2) \quad . \qquad (2.9)$$

This lack of factorizability is of no consequence since $P(y_1,y_2)$ plays the role of an ignorable normalization factor in the problem of maximizing the a posteriori p.d. on x. Thus, the latter problem reduces to the maximization of

$$\phi(x|y_1,y_2) \equiv \log P(y_1|x) + \log P(y_2|x) + \log P(x) \qquad (2.10)$$

with respect to the state x. The point at which the maximum occurs can be written in the form

$$x = \hat{x}_{opt}(y_1,y_2) \qquad (2.11)$$

where the right hand side is the optimal estimator using the optimality criterion (Eq. 2.1)). Equation (2.10) provides the basis (at least in somewhat abstract form) for our subsequent investigation of the problem of combining acoustical and vision measurement.

We turn briefly to the proof of the assertion that, with an optimal estimator, the more measurements the better. In particular, if we have a restricted optimal estimator, $\hat{x}_{opt}(y_1)$, why is the more general optimal estimator, $\hat{x}_{opt}(y_1,y_2)$, better?

Let us consider this equation using the optimality criterion (Eq. 2.1)), although this assertion is valid for any optimality criterion whatsoever. We will assume that the averaging operator E is the same in both cases, an assumption that is justified by the simple device of using the single most general statistical model under the constraints pertaining to each case. Now, the differences between different cases resides only in the various constraints applied to the candidate estimator, $x(y_1,y_2)$. Thus, in the maximization of the optimality criterion, S (defined by Eq. (2.1)) with respect to the _functional form_ of the candidate estimator, the candidate estimator with no constraints will almost always give a higher maximum of S than the candidate estimator with constraints. Thus, if, for example, $x(y_1,y_2)$ is unconstrained, its optimal form will almost always be a better estimator than the optimal form obtained when $x(y_1,y_2)$ is constrained to be independent of y_2.

If the optimization procedure for the estimator is suboptimal, the addition of more measurements does not necessarily improve the estimator. It is hard to imagine a degree of suboptimality that would obtain no improvement from additions of the types of measurements that are exactly the same as some of the previously existing ones. However, with the addition of distinctly different measurements, it is not very unlikely that suboptimal procedures will lead to worse results.

3. APPLICATION TO THE COMBINING OF ACOUSTICAL AND OPTICAL MEASUREMENTS

In the following sections, we present a detailed formulation of the problem of estimating the deviation functions of an object from both acoustical and optical measurements. In the section Represenation of the Object, we discuss certain preliminary mathematical matters and introduce some special notation that will significantly simplify the writing of formulas. In the section Acoustical Measurement model, we discuss the representation of the external geometry of the object by a single-valued elevation function. In the sections Optical Measurement Model and Procedure for Estimating the Elevation Function, the measurement models for acoustics and optics are formulated. Finally, in section 4, the procedure for estimating the elevation function from both kinds of measurements are outlined.

Mathematical Preliminaries and Notation. The writing of equations will be simplified significantly by the introduction of a special notation that differentiates between two- and three-dimensional vectors. The desirability of this arises from the fact that the horizontal plane passing through the origin (i.e., xy-space) is a preferred geometrical entity, as we shall see in the later analysis.

The three-dimensional vector $\vec{r}$ in xyz-space is defined by the expression

$$\vec{r} = \vec{e}_x x + \vec{e}_y y + \vec{e}_z z \qquad (3.1)$$

where x, y and z are the usual Cartesian coordinates and $\vec{e}_x$, $\vec{e}_y$ and $\vec{e}_z$ are unit vectors pointing in the coordinate directions. On the other hand, the two-dimensional position vector $\underline{r}$ in xy-space (i.e., in the xy-plane) is defined by

$$\underline{r} = \vec{e}_x x + \vec{e}_y y \quad . \tag{3.2}$$

In general, we will denote any three-dimensional vector in xyz-space by an arrow over the symbol and any two-dimensional vector in xy-space by an underline. Clearly, $\vec{e}_x$ and $\vec{e}_y$ are equal to $\underline{e}_x$ and $\underline{e}_y$, but we will adhere to the super arrow notation for the sake of definiteness. The vectors $\vec{r}$ and $\underline{r}$ are obviously related by the expression

$$\vec{r} = \underline{r} + \vec{e}_z z \quad . \tag{3.3}$$

In the case of a general three-dimensional vector $\vec{u}$, we can write the relations

$$\vec{u} = \vec{e}_x u_x + \vec{e}_y u_y + \vec{e}_z u_z \quad , \tag{3.4}$$

$$\underline{u} = \vec{e}_x u_x + \vec{e}_y u_y \quad , \tag{3.5}$$

$$\vec{u} = \underline{u} + \vec{e}_z u_z \quad . \tag{3.6}$$

We will use the vector magnitude operation $|\ |$ for both two- and three-dimensional vectors, e.g.,

$$|\underline{u}| = (u_x^2 + u_y^2)^{1/2} \tag{3.7}$$

$$|\vec{u}| = (u_x^2 + u_y^2 + u_2^2)^{1/2} \quad . \tag{3.8}$$

It is also useful to consider the two- and three-dimensional unit tensors defined by

$$\underline{1} = \vec{e}_x \vec{e}_x + \vec{e}_y \vec{e}_y \tag{3.9}$$

$$\vec{1} = \vec{e}_x \vec{e}_x + \vec{e}_y \vec{e}_y + \vec{e}_z \vec{e}_z \tag{3.10}$$

in which the terms $\vec{e}_x \vec{e}_x$, etc., are dyadic.* The two-dimensional unit tensor $\underline{1}$, along with $\vec{e}_z$, is useful in the decomposition of a three-dimensional vector into a two-dimensional vector and the z-compound, namely

$$\underline{u} = \underline{1} \cdot \vec{u} \tag{3.11}$$

---

* In indicial notation, a general dyadic $\vec{a}\,\vec{b}$ has the definition $(\vec{a}\,\vec{b})_{ij} = a_i b_j$, where i,j = 1,2,3 corresond to the Cartesian coordinates x,y,z. In some notational systems, $\vec{a}$ would be regarded as a column vector and $\vec{b}$, a row vector and, thus, the dyadic $\vec{a}\,\vec{b}$ might be represented by the expression ab^T. This is in contrast with the scalar (or dot product) $\vec{a} \cdot \vec{b}$ being represented by $a^T b$.

$$u_z = \vec{e}_z \cdot \vec{u} \quad . \tag{3.12}$$

This results give the terms on the right hand side of Eq. (3.6) in terms of the general three-dimensional vector on the left hand side.

We close this subsection with a brief consideration of two- and three-dimensional gradiant operators defined by

$$\underline{\nabla} = \vec{e}_x \frac{\partial}{\partial x} + \vec{e}_z \frac{\partial}{\partial y} \tag{3.13}$$

$$\vec{\nabla} = \vec{e}_x \frac{\partial}{\partial x} + \vec{e}_y \frac{\partial}{\partial y} + \vec{e}_z \frac{\partial}{\partial z} \quad . \tag{3.14}$$

Other operators, e.g., the divergence and the curl, are related in an obvious way to the above definitions.

Representation of the Object. We assume that the object of interest (combined with its surroundings) is represented by a single valued elevation function

$$z = Z(\underline{r}) \tag{3.15}$$

where z if the vertical coordinates and $\underline{r}$ is a two-dimensional position vector in the xy-plane. The points $\underline{r}$ are limited to the localization domain, D_L, defined by the inequalities

$$-\frac{1}{2} L_x \leqslant x < \frac{1}{2} L_x$$

$$-\frac{1}{2} L_y \leqslant y < \frac{1}{2} L_y \quad . \tag{3.16}$$

Outside of the localization domain, we assume that the elevation is, in effect, given by $z = -\infty$, or at least a negative number sufficiently large that its contribution to the acoustical measurement can be windowed out. In the optical model, the surface outside of D_L is, of course, at least partially visible, no matter how large and negative the elevation function is. We can handle this by requiring this part of the surface to be perfectly black (i.e., totally nonreflective). This assumption makes the shadow of the pedastal (associated with D_L) thrown onto the low-lying surface outside of D_L to be totally invisible. It is understood that the object of interest need not fill all of D_L; the elevation function between the object and the boundaries of D_L could, for example, be part of the presentation surface. A typical set-up is illustrated in Fig. 1a and 1b.

The single-valuedness of the elevation function requires further comment. Clearly, most objects involved in robotic assembly are not described faithfully by single-valued elevation functions. We can take two alternative points of view. The first is that we will, for the sake of simplicity, limit our investigation to objects that are actually described by single-valued elevation functions. The second is that, although many likely objects are described by multiple-valued elevation functions, the single-valued elevation function considered here corresponds to the top elevation function. It may not be too erroneous to confine attention to the top elevation function in the model of possible objects, since the lower branches of the elevation function will contribute weakly to the measured data.

In general, the way in which the object scatters incident light should be regarded as part of the state of the object, along with the elevation function. However, to simplify the present problem as much as possible, we make the assumptions that the surface has a mat finish that scatters light diffusely in accordance with Lambert's law. With a sufficient number of illumination directions, one can, under the above assumption, derive a combination of directly measurable quantities that are independent of the local reflectivity. Thus, this combination is determined entirely by the elevation function, $Z(\underline{r})$, and thus the latter entity suffices to define the state of the object in the context of optics.

Acoustical Measurement Model. We consider a single pulse-echo measurement of the scattering of acoustical waves from the object of interest. We, of course, assume that the host medium is ordinary air. We further assume that the transducer and the object are in the far field of each other. The experimental set-up is depicted schematically in Fig. 1a. The object is assumed to have a very high acoustial impedance and a very high density compared with the corresponding properties in air and thus the surface of the object may be regarded as a rigid immovable boundary.

The appropriate measurement model is givn by the expression

$$f(t) = \alpha \int_{D_L} d\underline{r} \ p'(t + 2c^{-1} Z(\underline{r})) + \mu(t) \ , \tag{3.17}$$

where the integration on $d\underline{r}$ spans the localization domain D_L in the xy-plane. The other symbols are listed below:

$f(t)$ = a possible measured waveform.
$\mu(t)$ = experimental error and noise.
$p'(t)$ = time-derivative of $p(t)$, the measurement system response function. The latter is defined as the waveform produced by the measurement system if a fictitious scatterer with an impulse response function given by $R(t) = \delta(t)$ is positioned at the origin.
$Z(\underline{r})$ = elevation function discussed in the last section.
α = constant dependent upon the acoustical properties of air.
c = propagation velocity of acoustical waves in air.

It is appropriate to consider a discrete version of the above measurement model, namely

$$f(t) = \alpha \sum_{\underline{r}} \delta\underline{r} \ p'(t + 2c^{-1}Z(\underline{r})) + \mu(t) \tag{3.18}$$

where $\underline{r}$ takes vector values on a suitable grid of points in the xy-plane spanning the localization domain D_L, $\delta\underline{r}$ is the area of one cell or pixel in the grid, and where the time t is now assumed to take a discrete set of values. The grid can be defined by setting $x = aq_x$, $y = aq_y$, where q_x and q_y are integers and where now $\delta\underline{r} = a^2$. In order that the grid fill up, the localization domain is defined by setting $L_x = Q_x a$ and $L_y = Q_y a$ a concordance with the assumptions that $q_x, q_y = -\frac{1}{2} Q + 1, \ldots, \frac{1}{2} Q$. $q_x = -\frac{1}{2} Q_x + 1, \ldots, \frac{1}{2} Q_x$ and $q_y = -\frac{1}{2} Q_y + 1, \ldots, \frac{1}{2} Q_y$ in which Q_x and Q_y are even positive integers.

We turn now to a discussion of the a priori statistical aspects of the model. The random quantities $\mu(t)$ and $Z(\underline{r})$ are assumed statistically independent of each other. The noise $\mu(t)$ is assumed to be Gaussian with zero mean and with a covariance matrix given by

$$E \ \mu(t) \ \mu(t') = \delta_{tt'} \ \sigma_\mu^2 \tag{3.19}$$

where $\delta_{tt'}$ is a Kronecker delta generalized for the case of noninteger subscripts ($\delta_{tt'} = 1$ if $t = t'$ and $= 0$ if $t \neq t'$), and where σ_μ is the standard deviation of

$\mu(t)$. The above result is valid if the covariance function $C_\mu(t-t')$ associated with $\mu(t)$, temporarily regarded as a random process defined on a continuous time t, has a negligible amplitude when $|t-t'|$ is equal to or larger than the time-sampling interval pertaining to the discrete time case. Clearly, the standard deviation σ_μ is then given by $\sigma_\mu^2 = C_\mu(0)$.

In the discussion of the a priori statistical properties of the elevation function $Z(\underline{r})$, it is expedient to decompose it into two parts: 1) a part $\overline{Z}$ that is the area average on the localization domain D_L, and 2) a part $\delta Z(\underline{r})$ that is the deviation from the above area average. To be explicit, we write

$$Z(\underline{r}) = \overline{Z} + \delta Z(\underline{r}) \tag{3.20}$$

where

$$\overline{Z} = Q^{-1} \sum_{\underline{r}} Z(\underline{r}) \tag{3.21}$$

where, in turn, $Q = Q_x Q_y$ is the number of points on the grid in the xy-plane, and where the summation on $\underline{r}$ spans the points on this grid. It follows from the above equations that

$$\sum_{\underline{r}} \delta Z(\underline{r}) = 0 \quad . \tag{3.22}$$

We now assume that the set of quantities $\{\delta Z(\underline{r})\}$ is a Gaussian random vector with the properties

$$E \, \delta Z(\underline{r}) = 0 \tag{3.23}$$

$$E \, \delta Z(\underline{r})\delta Z(\underline{r}') = C_{\delta Z}(\underline{r}-\underline{r}') \quad . \tag{3.24}$$

At this point, we introduce the assumption of periodic (or cyclic) boundary conditions, an artifact that gives meaning to nonlocal quantities that depend, at least formally, on points beyond the boundary of D_L. This assumption necessitates the further assumption that $C_{\delta Z}(\underline{r})$ is doubly periodic with the period D_L. Because of the relation (3.22), we must impose the requirement

$$\sum_{\underline{r}} C_{\delta Z}(\underline{r}-\underline{r}') = \sum_{\underline{r}} C_{\delta Z}(\underline{r}) = 0 \quad . \tag{3.25}$$

The area average $\overline{Z}$ is assumed to be statistically independent of the $\delta Z(\underline{r})$. It is not unreasonable to assume that $\overline{Z}$ is, a priori, a Gaussian random vector with zero mean and variance $C_{\overline{Z}}$.

Optical Measurement Model. In this case, we assume that a TV camera (or its equivalent) is situated at a high altitude (compared with the a priori range of the elevation function of the object) and is pointed straight down. As already noted, we assume that the surface is a diffuse Lambertian scatterer. We make the following additional assumptions: 1) each light source is assumed to be equivalent to a point source at infinity, 2) each element of surface is illuminated by each point source of light, and 3) multiple scattering of light between different elements of surface can be neglected. This model is essentially identical to that treated by Ray, Birk and Kelley.[2] The geometry of the experimental set-up is depicted in Fig. 1b. We will first consider a noiseless model of the directly measurable quantities (i.e., the set of image intensities for each illumination direction) from which we ultimately obtain a set of derived quantities that depend only on the local slopes. A noisy model is then formulated for these derived quantities.

The noiseless optical measurement model can be expressed in the form

$$\rho(\underline{r}, \vec{e}_m) = \beta(\underline{r}) \vec{e}_m \cdot \vec{n}(\underline{r}) \qquad (3.26)$$

where

$\rho(\underline{r}, \vec{e}_m)$ = possible image intensity at the position $\underline{r}$ in the xy-plane* with the illumination direction $\vec{e}_m$.

$\underline{r}$ = xy-position of a point on the object and the corresponding point in the image planes.

$\vec{e}_m$ = unit vector pointing from an element of surface on the object toward illumination source (since the distance from any part of the object to each light source is assumed to be very large compared with the size of the object and the localization domain, $\vec{e}_m$ is independent of $\underline{r}$).

$\vec{n}(\underline{r})$ = outward pointing normal vector for an element of surface above the position $\underline{r}$ in the xy-plane.

$\beta(\underline{r})$ = factor associated with the surface reflectivity, characteristics of the imaging system and illumination intensity.

The physical background of Eq. (3.26) is discussed in the book edited by Kingslake.[3]

The local normal is easly shown to be

$$\vec{n}(\underline{r}) = \vec{\nabla}(z - Z(\underline{r})) / |\vec{\nabla}(z - Z(\underline{r}))|$$

$$= (\vec{e}_z - \underline{\nabla} Z)(1 + |\underline{\nabla} Z|^2)^{-1/2} \qquad (3.27)$$

In this model, we assume that the illumination direction, $\vec{e}_m$, can assume several vector values.

The above model is suitable for continuous $\underline{r}$. For the case of discrete $\underline{r}$, defined on the xy-grid defined in the last section, certain modifications must be made. The above measurement model can stand as written, except that $\underline{r}$ is discrete. The most satisfactory approximation is obtained from an elevation function for continuous $\underline{r}$ obtained by Fourier interpolation applied to the discrete set of elevations defined on the above xy-grid. This procedure leads to the result

$$\underline{\nabla} Z(\underline{r}) = - \sum_{\underline{r}'} \underline{A}(\underline{r} - \underline{r}') \, Z(\underline{r}') \qquad (3.28)$$

where

---

* Here, we assume that the two-dimensional coordinate system on the image plane and the two-dimensional coordinate systems used in the definition of the elevation function was are in one-to-one correspondence.

$$-\underline{A}(\underline{r}) = Q^{-1} \sum_{\underline{r}} i\underline{k} \exp(i\underline{k} \cdot \underline{r}) \quad . \tag{3.29}$$

In the above summation on $\underline{k}$, $\underline{k}$ takes the discrete set of values defined by

$$\underline{k} = 2\pi \left(\frac{P_x}{L_x} \vec{e}_x + \frac{P_y}{L_y} \vec{e}_y \right) \tag{3.30}$$

where P_x and P_y take the same integral values as q_x and q_y, respectively; namely, $p_x = \frac{1}{2} Q_x + 1, \ldots, \frac{1}{2} Q_x$ and $p_y = -\frac{1}{2} Q_y + 1, \ldots, \frac{1}{2} Q_y$. It should be stressed that the above results contain the implicit assumption that $\underline{A}(\underline{r})$ is periodic with a two-dimensional period D_L. This is consistent with the assumption of periodic boundary conditions or toroidal topology.

We turn next to the determination of $\nabla Z(\underline{r})$ in the noiseless case. We assume that the illumination directions $\vec{e}_1$, $\vec{e}_2$ and $\vec{e}_3$ are noncoplanar and define a set of reciprocal vectors $\vec{e}_1^R$, $\vec{e}_2^R$ and $\vec{e}_3^R$ by the relations

$$\vec{e}_m^R \cdot \vec{e}_{m'} = \delta_{mm'} \; , \quad m, \; m' = 1,2,3 \quad . \tag{3.31}$$

It is easily shown that

$$\sum_m \vec{e}_m^R \vec{e}_m = \overset{\leftrightarrow}{1} \tag{3.32}$$

where $\overset{\leftrightarrow}{1}$ is the three-dimensional unit tensor defined by Eq. (3.30). We next define a new observable quantity

$$\vec{\tau} = \sum_{m=1}^{3} \vec{e}_m^R \rho(\underline{r}, \vec{e}_m) \quad . \tag{3.33}$$

Using Eq. (3.26), we obtain

$$\vec{\tau} = \beta(\underline{r}) \sum_{m=1}^{3} \vec{e}_m^R \vec{e}_m \cdot \vec{n}(\underline{r}) = \beta(\underline{r}) \; \overset{\leftrightarrow}{1} \cdot \vec{n}(\underline{r})$$

$$= \beta \; \vec{n}(\underline{r}) \quad . \tag{3.34}$$

We finally consider the ratio of the horizontal and vertical parts of $\vec{z}$ to obtain

$$\underline{\sigma}(\underline{r}) = \underline{\tau}(\underline{r})/\tau_z(\underline{r}) = \underline{n}(\underline{r})/n_z(\underline{r})$$

$$= -\underline{\nabla}\delta Z(\underline{r}) = \sum_{\underline{r}} \underline{A}(\underline{r}-\underline{r}')\delta Z(\underline{r}') \tag{3.35}$$

where Z has been replaced by δZ since $\overline{\nabla Z}$ vanishes. We have thus found a combination of observable quantities that are independent of the local reflectivity (i.e., albedo) and is dependent only on the local slope. It should be reemphasized that, as stated earlier, the object and the illumination directions must be such that no part of the exposed body is in shadow.

We make the approximation that experimental error (the combined effects of errors in the $\rho(\underline{r}, e_m)$) can be represented by Gaussian additive noise, and thus the complete stochastic measurement model takes the form

$$\underline{\sigma}(\underline{r}) = \sum_{\underline{r}'} \underline{A}(\underline{r}-\underline{r}')\delta Z(\underline{r}') + \underline{\nu}(\underline{r}) \quad . \tag{3.36}$$

The a priori statistical behavior of $\delta Z(\underline{r})$ has been described in the last subsection. The measurement error—(or noise) $\underline{\nu}(\underline{r})$ is assumed a priori to be statistically independent of $\delta Z(\underline{r})$ and Z. Also, it is assumed to be Gaussian with the properties

$$E \underline{\nu}(\underline{r}) = 0 \tag{3.37a}$$

$$E \underline{\nu}(\underline{r})\underline{\nu}(\underline{r}') = \underline{1} \delta_{\underline{rr}'} \sigma_\nu^2 \quad . \tag{3.37}$$

The appearance of the two-dimensional unit tensor 1 on the right hand side of the last expression reflects an assumption of statistical isotropy in the xy-plane.

Procedure for Estimating the Elevation Function. Here, we treat the problem of combining the two types of measurements, acoustical and optical, in the estimation of the elevation function using the general formula (2.10) particularized for the present case. We make the substitutions $x \rightarrow \{Z(\underline{r})\}$, $y_1 \rightarrow \{f(t)\}$ and $y_2 \rightarrow \{\sigma(\underline{r})\}$ in Eq. (2.10), with the results

$$\phi(Z|f,\sigma) = \log P(f|Z) + \log P(\sigma|Z) + \log P(Z) \tag{3.38}$$

in which Z, f and σ are shorthand notations for the sets given above. In the above expression, an additive constant related to normalization has been ignored. We will use this practice in all of the subsequent equations. The conditional probability densities are given by

$$\log P(f|Z) = -\frac{1}{2\sigma_\mu^2} \sum_t [f(t) - \alpha \sum_{\underline{r}} \delta \underline{r}p'(t + 2e^{-1}Z(\underline{r}))]^2 \tag{3.39}$$

$$\log P(\underline{\sigma}|Z) = -\frac{1}{2\sigma_\nu^2} \sum_{\underline{r}} |\underline{\sigma}(\underline{r}) - \sum_{\underline{r}'} \underline{A}(\underline{r}-\underline{r}')\delta Z(\underline{r}')|^2 \tag{3.40}$$

The a priori p.d. of $\overline{Z}$ and the $\delta Z(\underline{r})$ is given by the expression

$$\log P(Z) = -\frac{1}{2} \sum_{\underline{r},\underline{r}'} \delta Z(\underline{r}) C_{\delta Z}(\underline{r}-\underline{r}')^+ \delta Z(\underline{r}') - \frac{1}{2} C_{\overline{Z}}^{-1} \overline{Z}^2 \tag{3.41}$$

where $C_{\delta Z}(\underline{r}-\underline{r}')^+$ is the pseudo inverse of $C_{\delta Z}(\underline{r}-\underline{r}')$. The latter quantity is a singular matrix because of the constraint (3.25).

Our general task of finding the most probable elevation function $Z(\underline{r})$, given the measured values $\{f(t)\}$ and $\{\underline{\sigma}(\underline{r})\}$, is tantamount to the maximization of $\phi(Z|f,\sigma)$ with respect to the set of elevations $Z(\underline{r})$ at the grid points $\underline{r}$.

It is of interest to consider the estimate of $\delta Z(\underline{r})$ based upon optical measurements alone. Using the second form of the linear estimator in two-dimensional spatial frequency space, we obtain

$$\hat{\delta Z}(\underline{k}) = [\,|\underline{k}|^2 + \sigma_\nu^2 \; C_{\delta Z}(\underline{k})^{-1}]\; i\underline{k} \cdot \underline{\sigma}(\underline{k}) \quad \underline{k} \neq 0 \quad ,$$

$$= 0 \; , \; \underline{k} = 0 \quad . \tag{3.42}$$

In the above expression, all of the functions originally defined in $\underline{r}$- space have been transformed over to $\underline{k}$- space (the two-dimensional frequency space). The relations between the two representations are

$$\delta Z(\underline{k}) = \sum_{\underline{r}} \exp(-i\underline{k} \cdot \underline{r})\delta Z(\underline{r}) \quad , \text{ etc.,} \tag{3.43}$$

where it is understood that $\delta Z(\underline{k})$ is a different function (i.e., the function defined above), not simply $\delta Z(\underline{r})$ with $\underline{k}$ substituted for $\underline{r}$. It is easily shown from Eq. (3.29) that

$$\underline{A}(\underline{k}) = -i\underline{k} \quad , \tag{3.44}$$

a relation that was used in the derivation of Eq. (3.44). In the above expression, $\underline{k}$ takes the value given by Eq. (3.30).

4. COMPUTATIONAL EXAMPLE

As an illustrative example of the technique for estimating a three-dimensional surface using a combination of acoustical and optical data, synthetic data were generated for a tetrahedron sitting on a flat table. A hidden line representation of the assumed surface is shown in Fig. 2. The surface was defined on a square grid of 16 x 16 points, of dimension 20 mm on a side. The length of one side of the tetrahedron was 15 mm. The coordinate system was such that the table was in the xy-plane at z = 0. Optical and acoustical data were generated in accordance with the physical models outlined in the previous section. In the case of the optical data, the two-dimensional intensity distribution, as seen by a camera looking straight down, was calculated for three illumination directions, each having a polar angle of 15°, but with azimuthal angles of 0°, 90° and 270°, respectively. Figure 3 shows greyscale representations of the three synthetic optical images. In the case of the acoustic data, the assumed transducer response was in the shape of a Hanning window in the frequency domain, whose bandpass between the zero points was 0-20 kHz.

In the inversion procedure, the standard deviations of the measurement noise were assumed to be 0.01 and 0.1 for the acoustical and optical data, respectively, where the above quantities are relative to the peak absolute value of the measured data in both cases. The mean elevation $\overline{Z}$ was assumed to have an ensemble average of zero, and a covariance $C_{\overline{Z}}$ of the square of half the width of the localization domain, i.e., 10 mm, thus providing a slight bias towards a spatial average height of zero in the reconstruction. The covariance of δZ was assumed here to be infinite.

The inversion was performed in three stages:

1. Determine $\delta Z(x,y)$ using the optical data alone.
2. Assuming $\delta Z(x,y)$ to be fixed at the above values, determine $\overline{Z}$ from the acoustic data alone.
3. Perform the final optimization using both acoustic and optical data by maximizing $P(Z|\sigma,f)$ with respect to δZ and $\overline{Z}$.

The results for the three stages are shown in Fig. 4. In the case of stage 1 (Fig. 4a), the reconstructed surface has a spatial mean elevation of zero, in accordance with our definition of δZ, and hence contains no absolute information on vertical position. In the case of stage 2 (Fig. 4b), the acoustic data have brought the surface up to approximately the correct height. For the final image (Fig. 4c), the vertical offset has been improved slightly, bringing the edges of the table into alignment with the x and y axes. This represents the best estimate of the surface on the basis of the optical and acoustic data, and comparison with Fig. 2 shows that the reconstruction is faithful. The height of the apex of the reconstructed pyramid above the table is 11.4 mm compared with the true height of 12.2 mm. This discrepancy is probably due to the approximations involved in calculating gradients from measurements on a finite grid.

5. SUMMARY

The methodology for the probabilistic estimation of the state of a system from multisensor data has been outlined. Application was made to the estimation of a single-valued elevation function of a rigid object in air based on a combination of acoustical and optical measurements, the nature of which we discussed in detail in earlier sections. The optical measurement system described here has the property that it provides complete information about the shape of the elevation function, but no information regarding the absolute elevation or vertical offset for the object and its immediate surroundings. On the other hand, the acoustical measurement system provides significant information about the vertical offset, but only incomplete information about the shape.

In the computational example, the shape of the elevation function was determined with high fidelity from optical measurements alone. The vertical offset was not available and was replaced by a condition of zero average (area) elevation for the quantity $\delta Z(\underline{r})$. The next stage in the procedure was to determine the vertical offset Z from acoustical measurements keeping the shape fixed. The third stage was to obtain an optimal estimate in which simultaneous variations of both Z and $\delta Z(\underline{r})$ were considered with both types of measurements. A somewhat better value of the vertical offset Z was obtained and the shape, represented by $\delta Z(\underline{r})$, was improved quantitatively without striking changes in its qualitative nature.

ACKNOWLEDGEMENTS

This work was supported by Rockwell International Internal Research and Development Funds.

REFERENCES

1. T.S. Ferguson, "Mathematical Statistics - A Decision Theoretic Approach," Academic Press, NY (1967).

2. R. Ray, J. Birk and R.B. Kelley, "Error Analysis of Surface Normals Determined by Radiometry," IEEE Transactions on Pattern Analysis and Machine Intelligence, Vol. PAMI-5, No. 6 (Nov. 1983).

3. R. Kingslake, Ed., "Applied Optics and Optical Engineering," Vols. I and II, Academic Press, NY (1965).

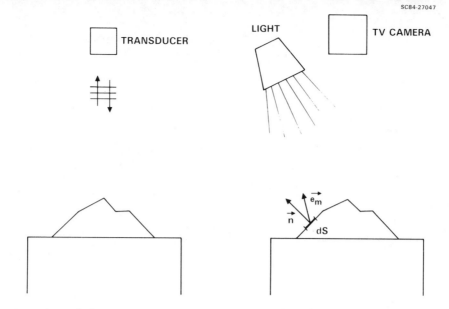

Fig. 1. (a) Acoustic measurement; (b) vision measurement.

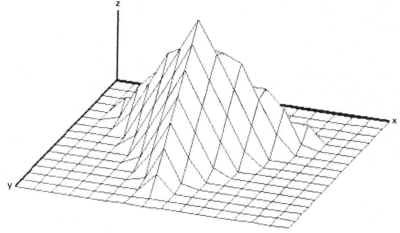

Fig. 2. Hidden line representation of the assumed surface, corresponding to a tetrahedron whose sides are of length 15 mm placed on a flat table of width 20 mm.

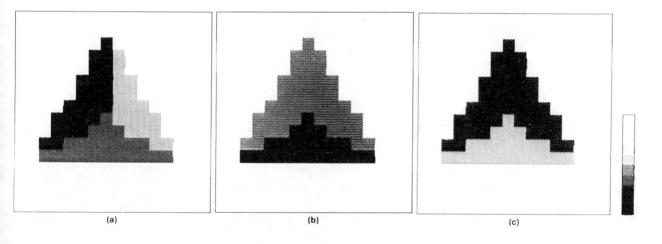

(a)   (b)   (c)

Fig. 3. Synthetically generated optical data representing the intensity distribution which would be seen by a camera looking straight down. The illumination directions correspond to a polar angle of 15° in each case, and azimuth angles of (a) 0°, (b) 90° and (c) 270°.

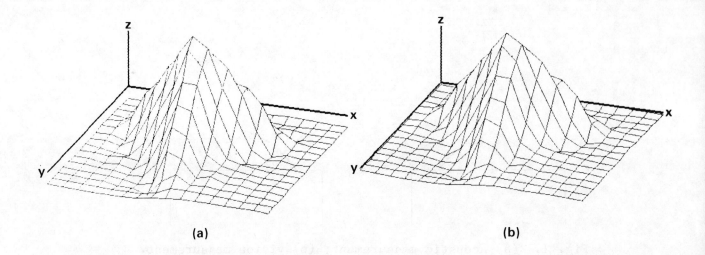

(a)

(b)

Fig. 4. The results of the three
stages of the inversion technique
discussed in the text: (a) the relative profile $\delta Z(x,y)$ deduced from
optical data only, (b) an estimate
of $Z(x,y)$ in which the acoustic data
has been used only to adjust the
vertical position offset, (c) the
optimal image used from the combined
optical and acoustical data.

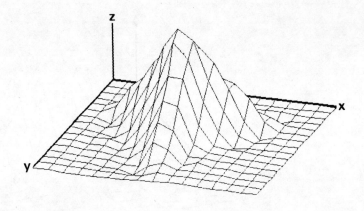

(c)

SENSOR BASED MANUFACTURING II

A Sensory Gripper for Composite Handling

D. G. Johnson
and
J. J. Hill
University of Hull, UK

Assembly of aerospace structures from composites can be carried out more efficiently through the use of flexible automation. The lay-up of carbon-fibre profiles is a major problem and the development of automated techniques is complicated by the flexible nature of the material and the high degree of precision required in the assembly. A sensory gripper is described which provides visual feedback from two gripper-mounted linear array cameras to permit accurate alignment and lay-up of carbon fibre composites.

1. INTRODUCTION

Flexible automation should make the manufacture of composite aerospace structures cost effective by reducing some of the high labour costs traditionally associated with this technology. In collaboration with British Aerospace, this project seeks to develop a flexible manufacturing environment for the assembly of satellite antenna dishes in small batch sizes. This particular assembly encompasses many of the problems associated with automation in the industry.

British Aerospace manufacture a large range of aerospace structures, an increasing number of which involve carbon fibre composites. Supplied in either rolls or sheets, the composite must first be cut to the desired shape and subsequently laid-up on a mould tool. Backing paper protects the composite prior to use and this must be removed to allow the material to be fixed in place. Within each manufacturing process, a number of tasks can potentially be automated. These include cutting and laying-up the profiles, drilling the completed structures and testing with a sonic probe. Ideally, each of these would be integrated in a flexible manufacturing cell using data from existing CADAM sources.

Many aerospace assembly tasks involve small batch production, with batch sizes of 5 not being uncommon. This necessitates rapid reprogramming of the robot and ideally an ability to adapt to variations in surroundings, and to cope with positional deviations; this is of particular significance where flexible materials are involved.

2.   THE ASSEMBLY PROBLEM

The lay-up problem is itself significant and accounts for the majority of the assembly time. Pre-cut pieces of composite must be laid precisely on the mould tool, ensuring accurate butt-jointing of adjacent profiles. Further complications are introduced by the curved nature of the structures. One solution is to have a gripper capable of moulding itself to the shape of the structure, but this has been rejected on the grounds of complexity. Alternatively, the gripper must be able to follow the contours of the surface, applying the profile during the course of its motion. This paper will concentrate on the lay-up of an antenna dish, but the gripper described is intended as general purpose and references to the aforementioned assembly serve only to illustrate the functional role of the gripper. The antenna dish assembly involves laying down four layers of composite, each layer consisting of a number of pie segments of about 500 mm in length. A central crown section is first laid using reference points on the side of the mould tool as a guide. The pie profiles of the first layer are then fitted ensuring that adjacent profiles constituting a layer butt against each other to an accuracy of less than 1mm down the length of the joint. Upon completion of the first layer, the subsequent layers are applied in a similar manner to complete the lay-up.

One major problem in automating composite manufacture is the wide variety of shapes and sizes of profiles typically encountered. To some extent, this problem could be alleviated through a 'Design for Assembly' approach[1] but problems of handling large and small pieces are likely to remain. Solutions to the variable-size problem have included adjustable span grippers[2] or very large grippers with a number of individually controlled air chambers. This particular problem is not considered in this paper because a single size of profile is used, allowing attention to be directed at the more fundamental problem of sensory assembly.

3.   GRIPPER CONSTRUCTION

The construction of the gripper is illustrated in Figures 1 and 2. Six suction cups on the underside face are connected through rubber tubing to a vacuum pump to provide the means of supporting the pre-cut profiles. Sufficient vacuum is generated so that with five cups open to the atmosphere the single remaining cup can still hold the profile. Furthermore, the profiles can slide across the cups without too much trouble.

Visual sensing is provided to monitor profile position and provide a quantitative check on the final butt-joint. This is achieved by two 256-element CCD (charge-coupled device) linear array cameras mounted on opposite ends of the gripper. The use of gripper-mounted linear array cameras for robot vision has been previously reported[3]. The marginal amount of extra information obtainable from area array devices does not justify the (considerable) extra expense. A single line of pixels provides all the necessary information to determine the edge position of the profile. A mirror is used in conjunction with each camera to enable two slits, each of width 25 mm, to be viewed. The available resolution is therefore 10 pixels/mm which is considered adequate for this application. Each slit requires its own source of illumination and this is provided by high power light emitting diodes (LEDs), two for each slit. The peak sensitivity of the CCD devices corresponds with the peak emitted wavelength of the LED and hence sensor exposure time is minimized. The LEDs are only turned on during the exposure period, allowing high intensities to be used and minimising power consumption when the vision system is not required. When both LEDs are illuminated, the observed intensity distribution across the scene is quite uniform and the varying sensitivity of the CCD sensing elements provides the major cause of non-uniformity. Each LED can, however, be controlled individually to achieve side illuminated images. This is important in the derivation of surface profile information to detect overlap and is discussed later.

Prior to use, the vision system runs through a calibration phase in which the optimum exposure time of the CCD sensor is found. The average pixel intensity of a captured image is determined and used to increase or decrease the exposure period, producing an average intensity lying within a specified band. A maximum of eight images are captured to achieve this and the whole process takes less than half a second. The optimisation of the exposure time is necessary to accommodate different coloured backing paper.

The video data from the CCD sensors is digitized into a 6 bit word by the vision system and stored in memory for subsequent image processing. The high contrast images are easily processed to determine edge positions and gap size.

The robot mounting pod is attached to the active surface of the gripper through a pivot, allowing the gripper to rotate about the pod and so providing the basic feedback mechanism to achieve surface following. A potentiometer mounted on the shaft provides the position feedback information which can be related to the pod angle; changes in this angle are used to measure changes in surface height over known linear motions of the gripper. This allows the gripper to 'follow' the surface contours of the mould tool with no a priori knowledge of the surface. For regular surface shapes, e.g. the parabolic section of the antenna dish, a world model can be derived quite rapidly, with corresponding improvements in lay-up time as the system learns the shape of the required trajectory.

The architecture of the system, as illustrated in Figure 3, uses a master-slave protocol, communicating through an in-house bus, Robus[4,5]. This modular approach has been found to be the one most conducive to rapid reconfiguration and development of the hardware. Modularity and portability have been maintained into the software structure through the use of the C programming language for all procedures.

4. THE GRIPPER IN USE

The pre-cut composite profiles are, at present, pre-stacked and only minor positional changes (using visual feedback) are necessary to align the profile on the gripper. The more general problem of selecting the desired profile from a dis-ordered array has not been considered. However, it is anticipated that if the robot could be used to do the cutting then such a bin-picking problem would not arise. Once on the gripper, the lower backing paper is removed and the profile offered to the mould tool. Sufficient pressure is applied to ensure a bond between one end of the profile and the tool. One problem here is the inconsistent tac (stickiness) of the composite which can present difficulties. The gripper is moved along the tool surface using information from the shaft resolver to move the gripper up or down, hence following the surface. Because the profile is now fastened at one end, it slides across the surface of the gripper during the lay-up. The rubber suction cups maintain sufficient vacuum to hold the profile but permit this sliding motion to occur.

The role of the vision sensor in the lay-up is in determining the required path of the gripper to take into account deviations from the predicted model. During the lay-up, the quality of the butt-joint can be monitored using the forward vision slit. However, corrective action is difficult to apply because of the possibility of creasing the material. Current work is directed at quantifying this to ascertain exactly what magnitude of correction can be applied. By using a rotary table, the butt-joint occurs down the same space trajectory each time, although cummulative errors in alignment of previous profiles necessitate fine tuning of this trajectory. A second role of the vision sensor is in inspecting the quality of the completed joint; this amounts to noting the gap size and detecting overlap. Gap size is easy to monitor, but the detection of overlapping profiles presents more significant problems since no visual contrast is apparent in the image.

4.1 Analysis of Surface Profiles

The solution adopted in examining the surface structure is to use glancing illumination to create shadows of the surface discontinuities. Since each slit has two

LEDs associated with it, the scene can be illuminated from the left or the right under software control. To extract a surface representation, two images are taken. The first one with the scene illuminated from the left and the second one with the scene illuminated from the right. These two images are normalized by dividing each pixel intensity by the corresponding 'no-object' intensity obtained from reference images. Any discontinuities on the surface manifest themselves as a unique combination of shadows in the normalized images. Figures 4 through to 10 show the intermediate images during the extraction of the surface profile for an overlap situation. The two reference images shown in Figures 4 and 5 are obtained by illuminating separately from the left and right with the camera positioned over a flat white object. Figures 6 and 7 show the images obtained when an edge is present in the field of view. Depending on the direction of the edge, only one direction of illumination produces a shadow. After normalization, the images shown in Figures 8 and 9 result. The width of each shadow is proportional to the height of the discontinuity and a simple geometrical transformation can map the width to the corresponding height. Figure 10 shows the final reconstructed surface profile.

This technique allows the presence of an overlap to be ascertained but does not allow its extent to be measured. This must be inferred by knowledge of the gap size at the previous points and corresponding interpolation to reproduce the joint structure. Alternatively, a comparison can be made of the position of a second edge of the profile compared to its theoretical position, inferring the extend of overlap on the first edge.

4.2 Data Collection

Once the complete profile has been rolled on to the surface it is smoothed over using three passes, indexing the rotary table each time so that the smoothing action occurs down the same trajectory. The first smoothing action also serves as an inspection pass on the most recently made joint. Data is collected and processed to ensure no significant errors have occurred. The third smoothing action provides information on the position of the profile edge and hence the corresponding position of the next joint is accurately determined.

5. CONCLUSIONS

Prototype testing of the gripper is currently in progress and at the time of writing no quantitative results on the effectiveness of this automated lay-up technique are available. Several significant problems have become apparent, notably size/shape variation and effective removal of the backing paper, which will necessitate some further investigation and possible refinement of the gripper structure. The sensing mechanisms of the gripper appear to be sound and the technique of analysing surface profiles works well. Some limiting factors of the PUMA 560 and its associated software have made control difficult, particularly where motion along curved trajectories is concerned. The discrete updating of position reduces the overall continuity of motion although speed is not of primary concern at this stage of the project.

ACKNOWLEDGEMENTS

The partial support of the Science and Engineering Research Council in providing CASE support for this project is gratefully acknowledged. The help of British Aerospace in providing raw materials, basic equipment and financial support is also gratefully appreciated. The authors acknowledge helpful discussions with Mr. J. Dickinson, M. Higgs and B. Barton of British Aerospace and Professor Alan Pugh of the Department of Electronic Engineering, The University of Hull. The help of John Hodgson in constructing the gripper is also appreciated.

REFERENCES

1.   Boothroyd, G., and Dewhurst, P., "Design for assembly Handbook", Department of Mechanical Engineering, University of Massachusetts, Amherst, Massachusetts.

2. Johnson, D.G., and Hill, J.J., "A Modular Linear Array Camera for Robot Vision", Digital Systems for Industrial Automation, Vol.2.3, June 1984.

3. Bubeck, K.B., "Advanced Composite Material Handling Application", 13th International Symposium in Industrial Robots, Chicago, April 17-21 1983.

4. Stubbings, C.A., "Robus - A Cheap Multiprocessor Robot and Sensor Control Bus", Internal Report, Department of Electronic Engineering, University of Hull, May 1983.

5. Mitchell, I., and Whitehead, D.G., and Pugh, A., "A Multiprocessor System for Sensory Robotic Assembly", Sensor Review, April 1983.

Fig.1 : The Gripper on the PUMA 560 robot

413

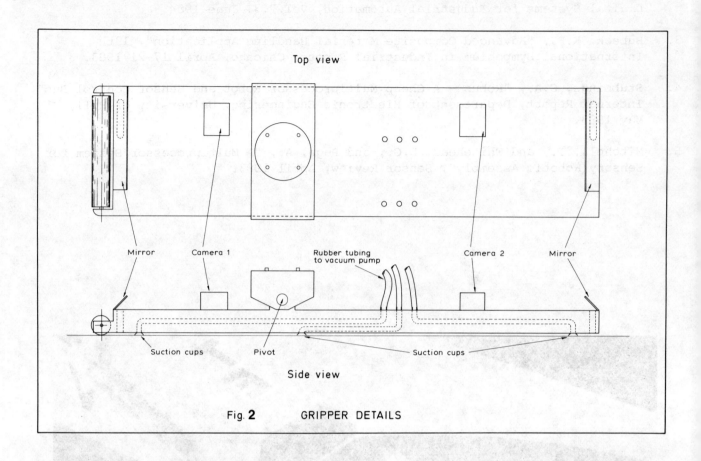

Fig. **2**   GRIPPER DETAILS

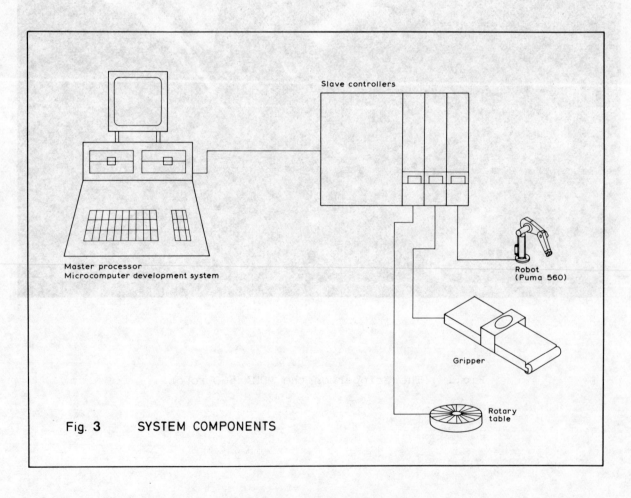

Fig. **3**   SYSTEM COMPONENTS

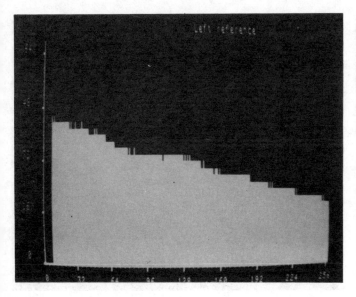

Fig.4 : The left-illuminated reference
image

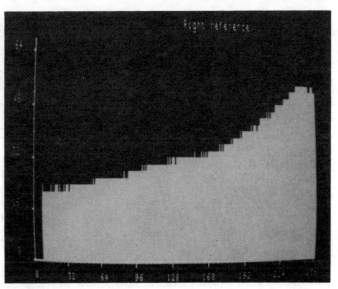

Fig.5 : The right-illuminated reference
image

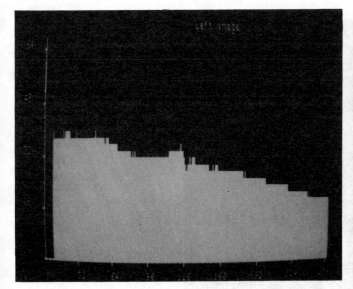

Fig.6 : Test-scene illuminated from left

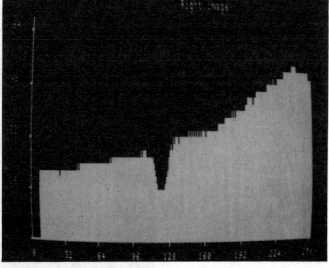

Fig.7 : Test-scene illuminated from right

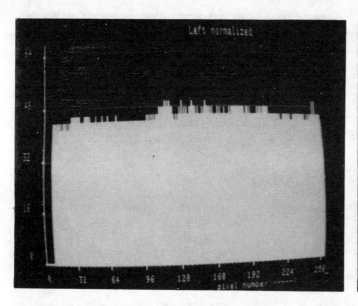 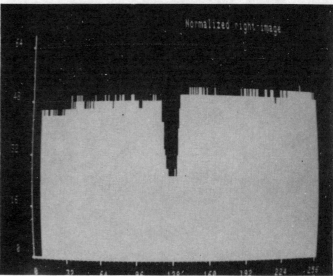

Fig.8 : Normalized left-illuminated image   Fig.9 : Normalized right-illuminated image

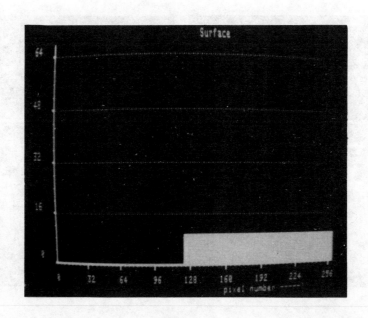

Fig.10 : The Reconstructed surface profile

Computer Vision in a Manufacturing Process

A. A. Hashim

and

P. E. Clements

Leicester Polytechnic, UK

ABSTRACT

This paper describes an automated system for visual
detection of tool breakages. The system, consisting
of a monochrome TV camera, a digitiser, and a frame
store holds the digitised information. Image
processing algorithms were developed to extract the
appropriate features from the captured images before
and after the tooling operation. A statistical
maximum likelihood decision is then taken as to
whether or not a tool breakage has occurred. It has
been shown that the system can detect tool breakages
with a typical reliability exceeding 90% under
laboratory conditions and 80% in a production
environment.

1. INTRODUCTION

Recent advances in both hardware and software technology during the last
few years has meant that many of the tasks currently being performed by
humans in the factory environment due to,

(i) the decision nature of the job and

(ii) the visual aspect of the job eg checking for

defects, orientation, sorting into various classes etc., can be
performed by computer-controlled machinery.

One of the major advances has been the development of many devices
capable of processing visual data received via some type of camera. This
has been very important, as it has meant that the visual sense of the
operator could now possibly be replaced by a computer with a camera
input, and it has opened a whole new area of possible applications. Chin
and Harlow (1) provide an excellent summary of the work currently being
undertaken in the field of computer vision and (2-4) provide some of the
work that has already been achieved in the area of applying vision to
industry.

Already the Machine Tool industry has become aware of the need, as Tom

417

Williams (Director of the GTMA) said, 'Looking to the future, members underline the importance of optical sensing techniques. The cost-effective application of optical metrology including machine monitoring and in-process gauging could substantially benefit the efficiency of UK industry' (5).

This paper details the work performed at Leicester Polytechnic in collaboration with Kearney and Trecker Marwin (KTM), involving the application of computer vision to a particular industrial problem. It also details the way in which Leicester Polytechnic has developed a system to assist in the determination of the feasibility of applying vision to a given industrial problem.

2. BACKGROUND

At the 1981 EMO Exhibition, KTM exhibited a new concept in machine tools, the CNC Multihead Changer. This machine represented a bridge between CNC machine tool technology and the reduced production time associated with multi-spindle transfer lines, traditionally used in medium to high volume production. On such a machine tool machine, designed for low level manning or no manning at all, one area of technology requiring investigation is that of tool breakage detection. Several basic devices, available commercially, aim at detecting tool wear or tool breakage on a single point tooling. Whilst the fact that the simplicity of these devices is, in some aspects advantageous, the simplicity debars them from use on the CNC Multihead Changer. Until now, tool breakages on multiheads has been detected by purpose built electro-mechanical devices (probes) unique to the testing of a particular tool head configuration. This method, if used, would cut the flexibility of the machine by half and the time per job would be increased drastically. Hence due to these factors it is unrealistic to use probes and so a totally new approach had to be found. A possible solution to the problem, is to use a camera(s) instead of an operator to check each time for tool breakages, and to model the operator by computer.

The advantages of this method are :-

(i) it would be a non-contact sensing device and would not interfere with the operation of the machine,

(ii) it would be possible to have any reasonable configuration of tools on a given tool head and be able to detect them. This is useful as the customer may at some stage decide to change a tool head and there would be no extra cost incurred,

(iii) it need not be just attached to the CNC multihead changer, but could be put on other tooling machines, and

(iv) cost.

3. DEVELOPMENT SYSTEM

Early on in the project an image processing system was bought for the initial development of the system, see figure (1).

This developmental system was more powerful than the hardware that would finally be used, but allowed the researcher to determine the optimum system specification such as numbers of grey levels, resolution, etc.

4. THE PROBLEM

4.1 Description of the problem

The first factor that had to be considered was the operation cycle of the CNC Multihead Changer machine. The philosophy behind the design was to incorporate the software into the machine operation cycle rather than to halt the machine while the system checked whether a tool on the tool head had been broken during its last operation.

For a time plan of the machine operation cycle, see figure (2).

The following design constraints were noted after an initial look at the machine operation cycle :-

(i) the prototype had to have a real time capture in order to assure continued running whilst the image was being created.

(ii) the image had to be from a fairly high resolution to obtain pixel-to-pixel correspondence between the before and after image. This was important, as the majority of the processing had to be performed whilst the machine is performing its tooling operation and therefore processing would be minimised by utilisation of information found on the before image to the after image,

(iii) the system took approximately four seconds to decide whether a breakage had or had not occurred during the last tooling operation. Clearly the decision had to be of a fairly simple nature,

(iv) it would have to work for any tool head.

The algorithm that was developed was built with all these constraints in mind and is described in the next section.

4.2 Prototype

The initial design for the prototype is shown in figure 3, as seen in this diagram, the prototype can be conveniently split into four stages of processing which are :-

(i) transform image from 64 grey levels to a binary image,

(ii) locate the tool tips in the binary image,

(iii) generate a statistical parameter for each drill, and

(iv) decide if tool breakage has occurred during operation.

Each of these four separate stages are discussed in sections 4.2.1 to 2.2.5.

4.2.1 Transform multi-grey level image to binary

This part of the prototype proved the most difficult to solve. Solving this problem, the Leicester Polytechnic Image Processing Library developed by the authors, was used to determine the types of algorithm suitable for final use.

Many well known techniques were tested on a trial tool head and the results examined for accepability. After several attempts at a solution, a method was found which provided a relatively good contrast between the background and the tool head. The method used involved computing an average pixel value for the image and use this as a theshold for the image.

Acceptable results were produced, but further work is continuing on this part of the prototype as it is a fairly long process due to the image being read in twice,

(i) the first reading produces the average pixel value, and

(ii) the second transforms the image using the threshold to a binary image.

If the captured image is noisy, then a cleaning up operation is performed on the image before continuing with the next stage.

4.2.2 Locating the tool tips in the binary image

After successfully transforming the image into a binary image, the next stage was to develop an algorithm to search the binary image and locate probable tool tips.

This process involved the binary image and the characteristics of a tool head, and to determine which parts of the image contain tool tips and which parts do not.

The method finally developed was able to locate the perimeter of the current tool head. After the perimeter had been located, it was scanned to establish which regions of the perimeter contained tool tips and which did not. The following rules were used in determining whether or not a region on the perimeter contained a tool :-

(i) if two adjacent pixels differed in height by 5 pixels, then the two pixels belonged to separate regions,

(ii) if the length of the region was greater than 3 pixels in length, then that was probably a tool region.

These rules are fairly simple (processing time is the limitation) but unfortunately, due to their inherent simplicity, they produce a number of false positives, see figure 4. This is considered reasonably acceptable whereas false negatives are not acceptable. The system must identify all the tools in the image.

4.2.3 Generation of statistical parameter for each tool tip

Time has been spent on trying out various methods for obtaining a good tool tip descriptor such as a t-test, chi-sq etc. But again as processing time is always a constraint, the average grey-level of the suspected area is taken as the tool tip descriptor.

4.2.4 Decision Making

When the after image has been captured, the system utilises the information previously found, ie the position of the tools in the image can determine in a short space of time, a new statistical parameter for for the tool. These two measures are then compared and if the difference is greater than some predefined threshold, the machine is flagged to indicate that a tool breakage has probably occurred during the last tooling operation.

5. RESULTS AND DISCUSSION

Whilst developing the prototype, two major problems were encountered :-

(i) Two captured images of the same scene taken in quick succession did not give the same grey-level values. This meant that this variation had to be considered when choosing the threshold value for deciding whether the tool tip had broken.

In order to assess the extent of this variability and its effect on the prototype the following test was performed -

The prototype was run on the tool head in the laboratory and the difference between the 'after' and 'before' descriptors were stored and plotted in the form of a graph, see Graph (1)

It can be seen that the extent of the variability is fairly wide, ranging from - 5.6 to 7.4, and hence this needs to be considered when deciding on the threshold,

(ii) After the tooling operation has been finished, the tools may not be in the same orientation i.e, they may have rotated. This is not a major problem for taps but for drills this can make a significant difference. In order to assess the extent of this variability and its effect on the prototype (one doesn't want to stop the machine when a drill has only rotated into a different orientation), the following test was performed -

The prototype was run on the tool head in the laboratory and the differnce between the 'after' and 'before' descriptors were stored but in this case before the 'after' image had been captured the drills on the tool head were randomly rotated. The results were again plotted in the form of a graph, see Graph (2).

Once these two tests had been performed then a simulation program was written that simulated a tool being broken during operation. This simulation was such that the breakage was fifty pixels in area. The prototype was then run as in the previous two tests the difference between the 'after' and 'before' descriptors were stored and plotted in the form of a graph, see Graph (3).

Plotting these three graphs onto one graph, see Graph (4), it can be seen that by choosing a certain threshold, this will determine the false negative and false positive rates. After discussion with KTM it was felt that it is necessary to pick out every breakage that occurs. The cost of missing a breakage and the wrecking of the work piece by continued tooling operations could be in the region of thousands, but the cost of calling the operator over to check the decision is minimal and hence the threshold was selected with this in mind.

Two tests were carried out, the first was conducted under laboratory conditions, and demonstrated that tool breakages in a Multihead can be detected with a typical reliability exceeding 90%.

The second test was carried out in a production environment, results indicate that under the more demanding conditions of the shop floor, the system can detect breakages with an average reliability of more than 80%.

ACKNOWLEDGEMENTS

The authors gratefully acknowledge the technical assistance of members of the image processing group at Leicester Polytechnic, Mr M Wheaton, Mr P Worley of KTM, and the sponsorship and active support of KTM.

References

1. Chin, R. T and Harlow, C. A.,IEEE Trans. on PAMI., 1982, 4, (6), 557.

2. "Vision systems orientates SCAMP'S parts"
   Sensor Review, Vol 3, No.1, pp 8-11 (January 1983).

3. "Visual sensing is finding its way into industrial applications",
   Sensor Review, Vol 2, No.2, pp 95-96 (April 1982).

4. Rossol, L., "Sight is the key to solving problems in industrial inspection and robotics"
   Sensor Review, Vol 1, No.4, pp 174-177 (October 1981)

5. Williams, T., "Opportunities for the gauge and tool industry"
   Sensor Review, Vol 3, No.3, pp 123-124 (July 1983)

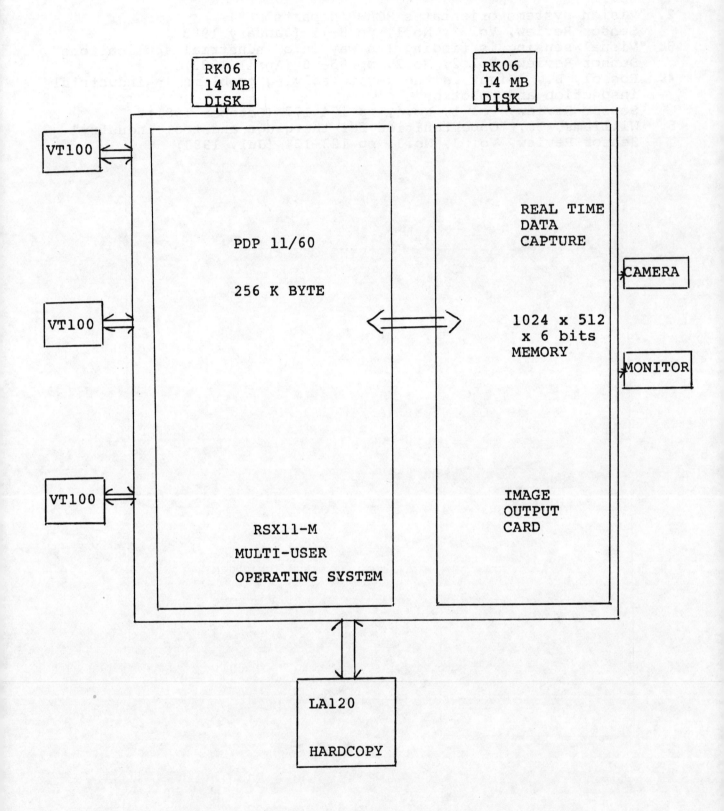

Figure 1.   Overview of Computer System Used

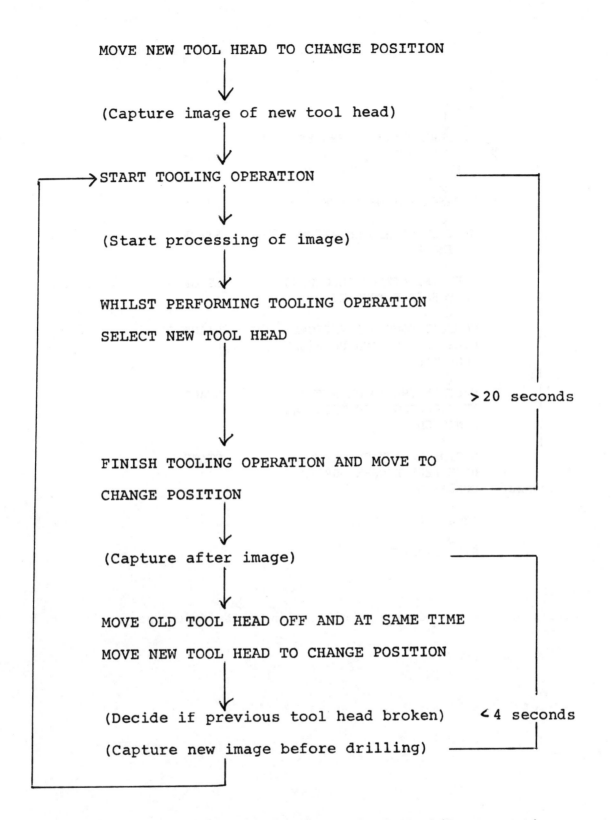

Figure 2.   Time Plan of CNC Multihead Machine Operation.

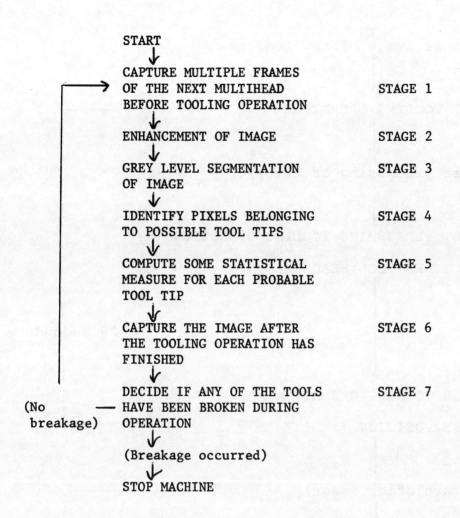

```
         START
            ↓
         CAPTURE MULTIPLE FRAMES
         OF THE NEXT MULTIHEAD              STAGE 1
         BEFORE TOOLING OPERATION
            ↓
         ENHANCEMENT OF IMAGE              STAGE 2
            ↓
         GREY LEVEL SEGMENTATION           STAGE 3
         OF IMAGE
            ↓
         IDENTIFY PIXELS BELONGING         STAGE 4
         TO POSSIBLE TOOL TIPS
            ↓
         COMPUTE SOME STATISTICAL          STAGE 5
         MEASURE FOR EACH PROBABLE
         TOOL TIP
            ↓
         CAPTURE THE IMAGE AFTER           STAGE 6
         THE TOOLING OPERATION HAS
         FINISHED
            ↓
(No      — DECIDE IF ANY OF THE TOOLS      STAGE 7
breakage)  HAVE BEEN BROKEN DURING
           OPERATION
            ↓
         (Breakage occurred)
            ↓
         STOP MACHINE
```

Figure 3.   Initial design specification of the prototype

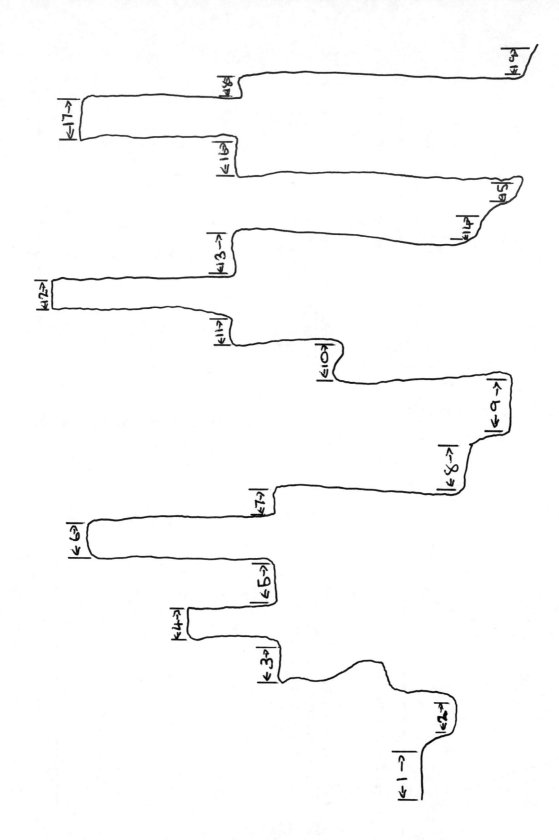

Figure 4. Diagram to show the areas the semantic will choose as
as probable tool tips. Infact only regions 4,6,12,17
are tool tips the other regions are part ot the tool
head

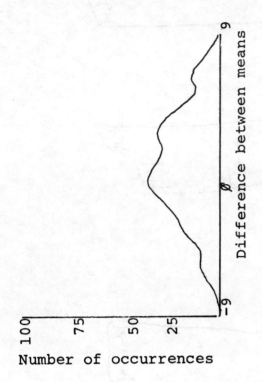

Graph (1). shows the difference between the 'after' and 'before' descriptor values on identical images.

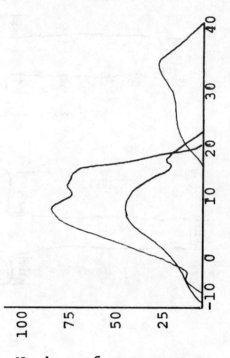

Graph (2) shows the difference between the 'after' and 'before' descriptors where the drills have randomly been rotated.

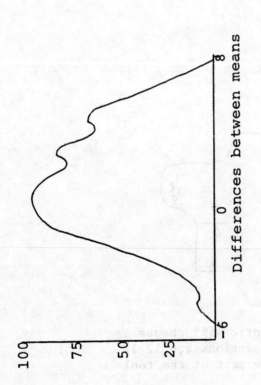

Graph (3). shows the difference between the 'after' and 'before' when the simulation program has been run.

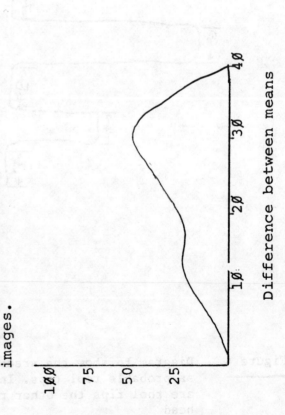

Graph (4). shows choice of threshold.

426

Visual Recognition of Engine Bearing Caps

F. Maali

and

C. B. Besant

Imperial College, UK

and

C. D. Backhouse

R. D. Projects Limited, UK

ABSTRACT

This paper describes a vision system which is currently used in an engine assembly line. It recognises engine component sets and provides real-time information to a gantry robot during component palletisation. The components are non-planar and of variable reflectance. Moreover, they appear against an equally uncomprimising background.

Despite its inexpensive hardware comprising a line scan camera and a light source, which are fitted to the moving gantry arm, and a microprocessor based vision controller, the system is able to perform its tasks reliably.

---

*Dr. Maali is now with Joyce-loebl Div.(Vickers plc), Gateshead NE11 0QW, ENGLAND.

INTRODUCTION

Design of a vision system for automated handling of engine components was reported earlier in [1]. The purpose of this article is to report on a version of the former system which has been installed and is currently operational in an engine assembly plant.

Overview of the handling process

In an engine assembly plant a 3-axis transfer gantry loader (Fig.1) unloads a machine tool and delivers engine bearing caps to the following production stage.

(1) During the normal cycle- from the machine tool ("initial site") to a washing machine ("terminal site") approximately 4 m apart.

(2) During the emergency cycle (i.e. malfunction of the washing machine)- from the initial site to a pallet which lies between the initial and terminal sites.

(3) During recovery from an emergency cycle- from the filled pallet to the terminal site.

The components are treated in sets (of five) as indicated in Fig.2. Each component set sits in a corresponding slot-set. be it in the initial site. terminal site, or anywhere in the pallet. The pallets are stacked up on a transverse carriage which supports a maximum of 4 pallets. Appropriate pallet location is reached through automatic motion of the base carriage on a series of rollers.

The vision system is used to verify the presence of fully aligned component sets during the pallet unloading and clear slot-sets in the pallet loading cycle. The components are non-planar and vary in reflectance as the supporting pallets.

The vision system may also be used to check the position of the pallet itself.

SYSTEM CONFIGURATION

The vision system employs a linear photodiode array camera (IPL 512) as the sensor. The illumination source projects a strip of light and they are both fitted to the moving arm, as shown in Fig.3. To avoid generation of excessive heat, and preserving the life of the bulb, the light is switched on under program control only during the site inspection periods.

A motorola 68B09 based microcontroller, controls the vision system, as shown in Fig.4.

the vision controller interacts with the following equipment:

- The transfer gantry arm controller, namely a Bosch PC400 programmable controller

- The Camera Control Unit (CCU)

- Its own front panel

Communication with the above peripherals is arranged through four 8-bit ports. provided by data lines of the two PIA's, and a serial RS232c port.

The video output from the camera is passed through a comparator where it is thresholded according to a preset level. The binary image data are passed to the CCU where the location of successive transitions are extracted and stacked in a FIFO. the stacked data may be viewed as run-length coded data, but due to the nature of processing in this system the image is reconstructed at the designated buffer in the vision controller.

Fig.5 shows the front panel of the vision controller from which sequeces are initiated except those in the actual run, which remain dependent upon signals from the arm controller.

Formulation of algorithms

The overall problem in the vision system is divided into two distinct subproblems:

a. Recognition of the state of slot-sets/component sets

b. Location of the pallets

Each of the above demands a separate algorithm which will be addressed in the same order below.

Recognition of slot/component sets

Notation:

A slot-set and a site are synonymous. so in the forthcoming discussion they will be used interchagebly.

x   - a slot-set / component set

a   - the ith pixel in the jth frame

L   - the jth frame

N   -   the number of pixels per frame

k   --   the slot/strip number within a set

u   -   distance (from datum) of the initial edge of slot k (in pixels) datum in this case is the beginning of the frame

v   -   width of slot k

s   -   distance of initial edge of strip k

t   -   width of strip k

Ω_1 -   A void (clean) site. [Qualifies for arm accession only during the loading cycle.]

Ω_2 -   A site supporting an intact component set. [To be accessed only during the unloading cycle.]

Ω_3 --   A partially full site, or a site supporting one or more misplaced components.

The scene of interest in this case extends to a slot-set plus a fine peripheral margine around it, as in Fig.6.

Data acquisition in such a scene takes place through seven equispaced frames, which span over the whole site as indicated in Fig.6. The structured light along with the adopting viewing technique (Fig.3) result in the desired contrast irrespective of the component or background reflectance. Recognition is based on a sequential classification scheme. The site under consideration either fails to conform with the required state or qualifies for the subsequent stage of classification. The former leads to immediate inhibition of the arm from accessing a site. A site is inferred to be fit for the arm accession if and only if it satisfies all stages of the classification.

The processing associated with L is identical to that of L . These lines. as evident. lie immediately before and after the slot-set. The aim here is to establish whether there exists a protruding bearing cap (i.e. a misplaced component) or not. The solution is based on convolving the binary thresholded image data due to the acquired frame with the following operator.

$$H = \begin{matrix} 1 \\ 1 \\ 1 \end{matrix}$$

and infering that a protruding component exists whenever the following prevails

$$\left[p_{ij} = \frac{1}{3} \sum_{k=-1}^{1} a_{i-k} h_k \;\middle|\; i = 2, \cdots, N-1 ; \; j = 1, 7 \right] = 1 \qquad (1)$$

In the remaining frames the aim of the processing is to establish the state of each slot. and verify that the adjacent pallet strips are not obscured by any of the components.

There are three states that can be associated with each slot.
ω 1 - empty
ω 2 - full
ω 3 - partially full

These states can hence constitue the pattern of interest in a classical pattern recognition problem.

The number of white pixels within the boundaries of each slot. i.e.

is found to provide an effective feature for assignment of the slots into any of the above catagories.

Classification of various sites proceeds according

$$\{\varphi_{kj}\} \subset \begin{cases} \omega_1 & \forall \quad \varphi_{kj} - \Phi_{kj}^{(e)} \leq \tau_{kj}^{(e)} \\[2ex] \omega_2 & \forall \quad \Phi_{kj}^{(f)} - \varphi_{kj} \geq \tau_{kj}^{(f)} \\[2ex] \omega_3 & \text{if} \quad \{\varphi_{kj}\} \not\subset \omega_1 \cup \omega_2 \\ & \text{i.e.} \quad \Phi_{kj}^{(e)} + \tau_{kj}^{(e)} < \varphi_{kj} < \Phi_{kj}^{(f)} - \tau_{kj} \end{cases} \qquad (3)$$

$$k = 1, \cdots, 5$$
$$j = 2, \cdots, 6$$

$\Phi^{(e)}$ and $\Phi^{(f)}$ denote the prototype (reference) measurement for classes of empty and full respectively; and

$\tau^{(e)}$ and $\tau^{(f)}$ represent the maximum allowable deviation of samples of each class from their respective prototypes.

The inner frames (2 to 6) are further processed to reveal clear pallet strips. Interest is confined to verification of their clearness and hence detection of any existing misplaced component.

Namely, when the following codition prevails the site is dismissed as one accommodating a misplaced component.

$$\sum_{i=\Delta_k}^{\Delta_k+t_k} a_{ij} \geq \tau_k \qquad k=1,\ldots,6; \quad j=2,\ldots,5 \qquad\qquad (M)$$

where $\tilde{\tau}_k$ is a dedicated threshold.

Corollary- Class assignment proceeds as follows

$x \in \Omega_1$ if and only if (iff)

$$\sum_{i=1}^{l+2} a_{ij} < 3, \qquad l=1,\ldots,N-2; \quad j=1,7$$

$$\{\varphi_{kj}\} \subset \omega_1, \qquad k=1,\ldots,5; \quad j=2,\ldots,6$$

$$\sum_{i=\Delta_k}^{\Delta_k+t_k} a_{ij} < \tau_k \quad \forall \quad k=1,\ldots,6; \quad j=2,\ldots,6$$

$x \in \Omega_2$ iff

$$\sum_{i=1}^{l+2} a_{ij} < 3 \qquad l=1,\ldots,N-2; \quad j=1,7$$

$$\{\varphi_{kj}\} \subset \omega_2 \qquad k=1,\ldots,5; \quad j=2,\ldots,6$$

$$\sum_{i=\Delta_k}^{\Delta_k+t_k} a_{ij} < \tau_k \quad \forall \quad k=1,\ldots,6; \quad j=2,\ldots,6$$

$\underline{x \in \Omega_3}$ if any of the following conditions prevails:

$$\sum_{i=1}^{l+2} a_{ij} > 2 \qquad l = 1, \cdots, N-2; \quad j = 1, 7$$

$$\{\varphi_{kj}\} \not\subset \omega_1 \cup \omega_2$$

$$\sum_{i=s_k}^{s_k + t_k} a_{ij} > \tau_k \qquad \forall \quad k = 1, \cdots, 6; \quad j = 2, \cdots, 6$$

Training: The objective here is to determine all unknown parameters in the site classification algorithm, met above. Training proceeds at three successive stages:

(i) Measurement of

$$\{u_k, v_k : k = 1, \cdots, 5\} \text{ and } \{s_k, t_k : k = 1, \cdots, 6\}$$

These parameters relate to the boundaries of successive slots in a site. and are extracted from a highly reflective model of a component set.

(ii) Measurement of prototype features

$$\{\phi_{kj}^{(f)} : k = 1, \cdots, 5; \quad j = 2, \cdots, 6\}$$

by exposing the vision system to the least reflective components that is expected to be encountered, and

$$\{\phi_{kj}^{(e)} : k = 1, \cdots, 5; \quad j = 2, \cdots, 6\}$$

from observation of a clear site. The remaining parameters, i.e. $\tau_k^{(e)}$. $\tau_k^{(f)}$. and τ_k are introduced with a view to width of the dead band met in the pallet location run.

<u>Pallet Location</u> The position of each row in a pallet is signified by a corresponding ribbon like mark, referred to as the pallet marker for that row. as in Fig.2. The correct positions of the pallets are inferred from coincidence of the pallet marks with a virtual reference marker.

The marks are white and have an improved contrast. The virtual reference marker is also of the same width as the pallet markers.

Everytime a pallet is traveresed to present a new live row. the reference marker and the pallet marker corresponding to that row must coincide. The task here is to detect and quantify any misalignments and feed it as positional feedback or command input to the servo system responsible for the pallet movement. if required.

In the course of the training. the distances from the beginning to the leading edge of the reference marker. 'F'. is computed as [Fig.7]

$$F = \min [I]. \text{ with I satisfying}$$

$$\sum_{i=1}^{l+2} a_i = 3 \qquad I = 1 \cdots N$$

Similarly the distance between the trailing edge of the reference marker to the end of the frame, 'R'. is computed from

$$R = \min [I] \text{ with I satisfying}$$

$$\sum_{i=1}^{I+2} a_{N-i} = 3 \qquad I=1,\ldots,N$$

When the hot run starts, similar distances (f and r) are measured with the pallet marker in sight. Thence the required forward or reverse motion pallet motion (M_f or M_r) are deduced.

$$M_f = F - f$$

$$M_r = R - r$$

To verify the authenticity of the pallet marker against those spurious marks that may compete with it, the width of the observed marker is matched against that of the reference marker as follows.

$$[\ N - (\ F + R\ )\ ] - [\ N - (\ f + r\ )\ ]$$

$$= (\ f + r\ ) - (\ F + R\ ) \leq \tau_L$$

PERFORMANCE

The processing associated with each frame takes less than 30 msec (exclusive of the frame acquisition time). The nominal arm speed of 0.26 m/S leaves an inter frame interval of 80 msec. Hence, even with a relatively long line scan times of up to 50 msec. the processing of each frame can terminate before acquisition of the subsequent frame. Each site is attended individually in this installation (i.e. each site is accessed immediately after being viewed, rather than building a total picture of the pallet first). This option has allowed to further extend the frame acquisition time at the expense of deferring the processing of the frames to the end (i.e. after acquisition of the last frame in each site).

The reliability of the system is enhanced by the on-line built-in test provision which checks for proper functioning of the camera, the light source. and the consistency of the command signals from the arm controller. The vision system is also supported by an extensive diagnostic software which facilitates fault analysis.

The vision system has proved to be capable of detecting almost any conceivable misplaced components including overlapped and inverted components, when inspecting a site. Some of the possible variations of sites supporting misplaced components are shown in Fig.8.

CONCLUSION

The vision system described was installed into a Ford engine assembly line at Dagenham, Essex. in August 1983. It was fitted to a transfer gantry loader without the loader design being significantly modified, and without any alterations being made to the components and pallets to improve the difficult appearence of the scene. Despite its uncomprimising situation, and although it uses inexpensive hardware, the system is able to perform its tasks reliably.

REFERENCES

1 Maali.F. and Besant.C.B. "Machine vision in automated handling of engine components". Proc. of 1st Int. Conf. on Automated materals handling, London. pp.161-172, (April 1983).

Fig. 1   The handling machinery

Fig. 2   The slot-set arrangement within a pallet

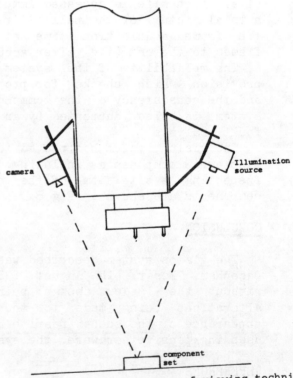

Fig. 3   The geometry of viewing technique

Fig. 5   The front panel

434

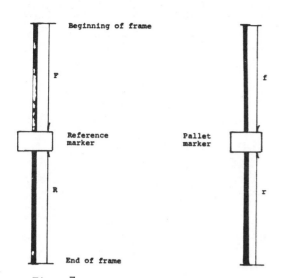

Fig. 7   Parameters in the pallet location run

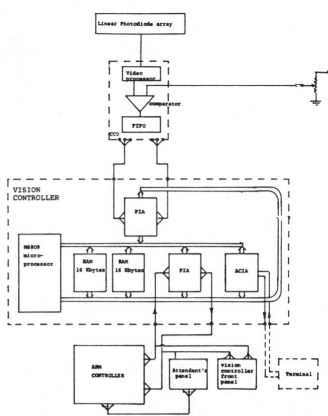

Fig. 4   The overall configuration of the vision system.

ig. 8   A depiction of possible misaligned components

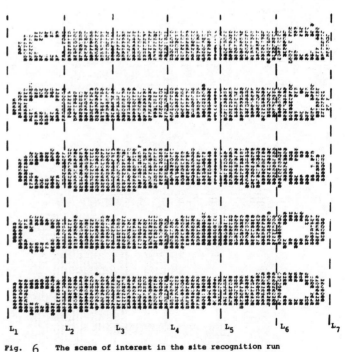

Fig. 6   The scene of interest in the site recognition run

Design of Cost-Effective Systems for the Inspection of Certain Food Products during Manufacture

E. R. Davies
Royal Holloway College, University of London, UK

ABSTRACT

The food processing industry is concerned particularly with small products on rapidly moving conveyors. To be viable, fast inspection hardware must cost less than £10K, yet it must scrutinise foodstuff surfaces closely. This paper describes methods for the rapid inspection of circular products such as biscuits. These include an inherently rapid procedure for the precise location of the centres of circular products, and radial intensity histograms to aid detailed surface scrutiny. These techniques permit individual product items to be inspected by software in 1-5 seconds, so that only relatively modest hardware is necessary for real-time implementation.

1. INTRODUCTION - THE FOODPRODUCT INSPECTION MILIEU

Manufacture of foodproducts has over the past few centuries changed from an activity carried out by the housewife to an enterprise carried out in large factories at relatively few central locations. This process of centralisation has been spurred on by the advent of automation in its various forms, and by the insistent demand of the consumer for high quality. In addition, the consumer has become more demanding in his tastes, and the food industry now has to produce products by quite complicated sequences of processes while maintaining consistent control of quality. Unfortunately, profit margins are not large in the food industry, and it has become difficult to maintain quality with the new multi-process products at the same time as keeping prices competitive.

One answer to this problem is to inspect products automatically, with the particular aim of providing feedback and control in the

manufacturing plant. Even if direct control of manufacturing parameters
by computer is not to take place, the line and factory manager can be
kept in minute-to-minute contact with the situation on the line.
Clearly, another reason for automatic inspection is to reject
unacceptable products, where control of a line has temporarily been
interrupted or where individual items have been found defective. In
this context an item may be either the foodproduct itself (e.g. a
biscuit), or a package containing a number of such items: these two
categories of product tend to have rather different characteristics, and
in this paper packages will be omitted from the discussion.

Foodproducts may be inspected automatically using a variety of
different types of sensor, including optical sensors such as cameras and
photocells, mechanical sensors which guage size or weigh the product, or
temperature sensors. Mechanical and thermometric sensors have been in
use for many years, but provide relatively little information about the
product. A similar situation exists with respect to photocells.
However, technology is at present ripe for a considerable exploitation
of visual sensing systems, since modern vidicon and solid state cameras
are able to provide enormous quantities of information about the
product, and to provide it at enormous speeds. In addition, image
processing and pattern recognition theory is now of age to interpret
visual images efficiently, and technology is beginning to allow this to
take place both reasonably cheaply and at speeds that correspond to real
time processing. Another important factor, in the case of foodstuffs,
is the appearance of the product. A product can be technically up to
specification and its taste unsurpassable, yet if its appearance is
mottled or otherwise unbecoming then it may prove unappealing to the
consumer and it may not be purchased. Thus there is bound to be an
emphasis on the appearance of products, so that visual inspection
systems will be of especial importance.

Another relevant factor is that a very large proportion of
foodproducts are round: biscuits, cakes, jellies, chocolates and so on
tend to be circular - a factor that makes the generation of special
inspection algorithms worthwhile. For this reason we have devoted
considerable effort to the development of algorithms for the detection,
location and scrutiny of foodproducts having a roughly circular shape.
Often these products are also thin, so the depth dimension need not
complicate the inspection problem unduly. However, surface scrutiny is
likely to be essential, and high speed analysis of grey-scale images is
required. This is a problem we have tackled and solved for the various
round products mentioned above. In this paper, we will discuss this
problem for general circular foodproducts. A typical practical
application of the techniques we have developed is that of a simple
biscuit which is to be covered with chocolate: clearly, it will be
useful to measure excess amounts of chocolate around the edges of the
biscuit, and to detect any gaps in the chocolate coating, since these
features affect both the cost of the product and its attractiveness to
the consumer.

2. A METHODOLOGY FOR THE INSPECTION OF FOODPRODUCTS

As remarked in the previous section, it is frequently important to
inspect foodproducts very closely during manufacture, one stage of the
inspection being scrutiny of the surface of the product in a moderately
high resolution image that contains a definite grey-scale. Although on
some occasions it may be useful to inspect coloured images, this tends
to add to analysis problems and may not contribute much vital
information. The result is that the image to be inspected contains
typically 256x256 pixels of 6 to 8 bits, corresponding to 64 to 256 grey

levels. This situation represents a significant processing problem. In general, a single conventional computer of say 0.5 Mips requires of the order of 1 second to locate edges or contours in the image, and perhaps 30 seconds to perform genuine feature recognition and measurement tasks[1]. However, in a factory situation, just one conveyor carrying a single type of product will be viewed by a given inspection system at any one time: this practical constraint eases the problem considerably and may permit an economically viable hardware accelerator to be designed. On the whole there is a tradeoff between the generality of such an accelerator and its cost. In fact, so many round foodproducts exist that an accelerator capable of generalising merely over this set will be guaranteed to be highly useful. For this reason it is reasonable to study effective algorithms for the inspection of this a priori restricted range of object shapes.

We will return to the problem of special purpose hardware below. Meanwhile, it is relevant to discuss briefly the types of camera system that may be appropriate. First, the vidicon type of system is reasonably adapted to the acquisition of stationary images, e.g. when products are visible within their final packing. On the other hand, many foodproducts are manufactured in a continuous process and appear on conveyor systems moving at nearly constant speed. This means that it is natural to employ a line-scan camera for their inspection, with a lateral resolution in the range 64 to 4096 elements. We imagine that a line-scan camera will generally be the appropriate choice. However, this factor will not have a great effect on the algorithms to be discussed below.

The work to be described here falls into the following sequence of stages: (1) a line-scan camera is used to obtain an image; (2) product edges are located, and the position and orientation of the product is computed; (3) some method such as template matching is employed to scrutinise its surface. If the product is circular, at stage 2 the position of its centre is determined, and at stage 3 radial histograms may constitute a convenient means of scrutiny. Stage 1 will not be discussed further in this paper: stages 2 and 3 will be examined in some detail in sections 3 to 5 below.

3. EDGE DETECTION TECHNIQUES

As remarked earlier, the first main stage in the inspection process after image acquisition is that of object location. Traditionally, this is carried out by thresholding the pixel intensity values to produce a binary image, after analysis of the intensity histogram of the original grey-scale image[1,2]. Whereas this approach may be satisfactory for locating silhouetted mechanical components, it is frequently inadequate for the location of a number of types of foodproduct - as for example when they appear on a conveyor which may become smeared with chocolate or crumbs. In any case scrutiny of foodproducts for contours indicating the presence of blemishes is seldom possible by intensity thresholding methods, which tend to remove a considerable proportion of the picture detail. In such cases it is necessary to employ edge detection techniques.

The most well known of the available edge detection operators are the Sobel and Prewitt differential gradient operators, and the Robinson 3 and 5-level template matching operators[2,3]. In general, template matching operators are capable of estimating edge magnitudes with reasonable accuracy, but are rather inaccurate for the estimation of edge orientation[4]. In fact a set of 8 Robinson-type templates, each aimed at locating edges of a different orientation, is only capable of estimating angle to within ~20°. Since the methods to be described below rely on estimates of edge orientation being within 1° of the actual value, template matching operators were rejected for the present application. A number of authors have investigated the accuracy of the Sobel, Prewitt and other differential gradient[5] operators for the estimation of edge orientation (see for example [5]). It has recently been shown that the Sobel operator is the most accurate of these 3x3 operators, the reason being that it is one of a whole class of operators which simulate a circular shape[3]. This theory is useful as it shows how operators of similar accuracy may be designed in larger neighbourhoods than 3x3.

The Sobel operator first computes the local components of intensity gradient in the image by applying the convolution masks

$$S_x = \begin{bmatrix} -1 & 0 & 1 \\ -2 & 0 & 2 \\ -1 & 0 & 1 \end{bmatrix} \quad \text{and} \quad S_y = \begin{bmatrix} 1 & 2 & 1 \\ 0 & 0 & 0 \\ -1 & -2 & -1 \end{bmatrix} \qquad (1)$$

(These are here rotated through 180° from the usual convolution notation, in order to show clearly how they correspond to derivatives in the x and y directions.) It then estimates edge magnitude and orientation by the equations

$$g = [g_x^2 + g_y^2]^{1/2} \qquad (2)$$

$$\theta = \arctan(g_y / g_x) \qquad (3)$$

The first of these equations is usually approximated by taking the sum of the magnitudes of g_x and g_y, or by taking the maximum of the two magnitudes. Computation of θ is clearly a heavy floating point operation (though it may be facilitiated by a lookup table). We have found that the value of θ is not actually required, and that it is sufficient to work with the sine and the cosine of θ (see below), which are easily computed as ratios of g_x, g_y and g:

$$\cos \theta = g_x / g ; \qquad \sin \theta = g_y / g \qquad (4)$$

Thus, although the value of g is inherently less important than the value of θ for our purposes, an accurate value for g is still required: in fact a lookup table will normally replace the computation of equation 2 in real-time foodproduct applications.

Strictly speaking, the Sobel and other operators only enhance edges, and a decision operation is required to detect edges. This may be carried out by thresholding the gradient image - a process generally less prone to error than that of thresholding the original image. However, the result of such an operation is that edges appear thick (and are thereby inaccurately located) in some places, but may peter out in others. Methods exist for thinning and joining edges having these deficiencies: thinning of binary[6] images is a surprisingly complex problem, which has been solved[6] but which involves considerable computational effort. Edge joining, on the other hand, is easily solved on an ad hoc basis[2], but the results may be misleading because noise or image complexity can cause some line ends to be joined erroneously. Relaxation labelling methods[7] may overcome this problem, but at considerable computational penalty. For this reason, it is sometimes best to reduce the gradient threshold value slightly, so that edges are (in a practical situation) guaranteed to be continuous: then edge thinning can cope with the remainder of the problem, while still locating edges to within ~1 pixel.

A recent trend has been to thin edges by direct processing of the original grey-scale image[8]. This approach may ultimately be capable of greater accuracy, though this has so far not been proven. In our work we have avoided the complex computation-intensive processes of joining and thinning by working at higher level: objects such as biscuits are searched for in their entirety, starting with the Sobel edge detection operator in which steps are taken (as noted earlier) to ensure high accuracy in the estimation of edge orientation. As will be seen below, the object location algorithm used in our work does not demand the use of thin edges that are guaranteed to be continuous. Though the main purpose of this whole approach is to increase the inherent speed of the algorithm, its use also appears to result in a definite increase in accuracy.

4. THE DETECTION AND LOCATION OF CIRCULAR PRODUCTS

Once edge points of a circular product have been located and the edge orientations noted, a simple procedure based on the Hough transform technique[9] may be employed for locating the centre of the product. The edge orientation is used to deduce the direction of the centre, and assuming a particular value for the object radius R, a possible location (x_c, y_c) for the centre is recorded in memory. For this purpose only the sine and cosine of the edge orientation angle θ are required, as stated earlier:

$$x_c = x - R.\cos \theta; \qquad y_c = y - R.\sin \theta \qquad (5)$$

(The minus signs in equation 5 arise automatically from the assumption of a dark product on a light background: this causes the edge gradient vector to point outwards from the centre of the product - see Figure 1.) Then it is merely necessary to average the x and y coordinates of the set of possible centre locations in order to obtain an accurate location for the centre of the product.

For an image containing a number of products, each cluster of centre points has to be averaged to estimate the product centre positions accurately. We have found that this procedure may be carried out in a second image space rather than in computer memory, since the problem is essentially two-dimensional: each pixel in the second image space then contains values representing the total number of times its particular x, y coordinates have arisen as a possible centre location. This approach has the advantage that a conventional image processing algorithm capable of locating local maxima in a small neighbourhood may conveniently be used to average the clusters of centre points. This simple method of averaging is not viable in two sets of circumstances: the first is when too much noise is present in the original image; the second is when the products are rather irregular in shape, so that their inverted edge vectors do not point directly to the centre of the product but to some position two or more pixels away. The latter circumstance certainly applies to a considerable number of foodproducts which we have investigated - though others are surprisingly close to a true circular shape, considering the number of mechanisms involved in production that might distort the product.

When greater accuracy in centre location is required, we have found that a two-stage process is invaluable for locating the centres. In the first stage, significant peaks in the cluster space are located. Then the second stage locates the median position in the vicinity of each peak. This procedure is effective since, as is well known, the median of a distribution provides excellent outlier rejection, and outlier points tend to arise from protuberances or other faults on otherwise reasonable product outlines. The result is that centres of circular foodproducts can usually be located to within one pixel. (Special cases like cakes in decorative paper cartons that have serrated edges are clearly more complex to handle.) To achieve this sort of accuracy it is crucial for the edge detection scheme to indicate the orientation of edge pixels within 1°, as stated earlier, since 1° at a radius of 57 pixels represents an error of almost exactly one pixel: this one-pixel error has to be added to a one-pixel error arising from inaccuracy in edge location; under these circumstances averaging is able to reduce the overall error to just one pixel. It should be remarked that the accuracy may actually be limited by what is meaningful considering the precision of the product.

This averaging characteristic makes the approach described above considerably more accurate than other possible methods, such as one based on the bisection of chords cutting the product. In addition it should be noted that the method is particularly robust and does not demand enormous computational effort. Its robustness is indicated by its performance with the broken and overlapping biscuits depicted in Figure 2: see also Figure 3. It has quite rapid speed of operation because it integrates the tasks of thinning, edge joining, object location and centre location into a single powerful process.

5. SCRUTINY OF CIRCULAR FOODPRODUCTS

Surface scrutiny of foodproducts is a task that depends very much on the nature of the particular product being inspected. For this reason just a few main schemes for surface scrutiny will be mentioned.

First, for a product whose surface is supposed to be fairly flat and uniform, testing of the intensity of every pixel within the product to check that it is within prescribed limits may be all that is required. This procedure is of course akin to intensity thresholding, but relies heavily on accurate prior location of the centre of the product, and knowledge of the radius. The latter may be checked by a variety of methods, including the radial intensity histogram technique (see below), but this method itself depends on accurate centre location.

Second, for a product with a textured surface, special techniques will probably need to be used: these are generally outside the scope of this paper, though checks of the maximum and minimum pixel intensity within the product area provide useful preliminary information.

Third, matching of the product luminance to that of a template over the whole product area may be practicable. Normally template matching is difficult when there are many degrees of freedom inherent in the situation, since comparisons with an enormous number of templates may be required. However, when the template is in a standard position relative to the product (as is the case when the centre has been located accurately), and when the product has been orientated correctly (as may sometimes be achieved by the location of small pattern or 'docker' holes on the surface), then template matching methods may constitute a practical solution to the inspection problem. This problem is very data dependent and further discussion will be omitted.

Finally, an approach that is both useful and fairly general for circular products is analysis of the radial intensity histogram. For this purpose the intensity within a product is plotted as a function of radius (or radius squared, to obtain a roughly constant intensity pattern for a uniform product). Not only may defects be detected relatively easily from the histogram, but also it may be used to give a

442

value for the radius of the product. Naturally the method is most appropriate when the product is of a type that is intended to be symmetric under rotation: in that case results can be obtained rapidly and with a minimum of computational effort. Figure 4 shows some examples of this technique. Further details of the method are outside the scope of this paper.

6. CONCLUDING REMARKS

This paper has considered the problems of visual inspection in the food manufacturing industry. Ignoring the rather complex area of packaging, the main problems that arise are (1) the need for moderately high speeds of image analysis to take account of product flow rates in the range 5-30 items per second, and (2) the need to implement relevant algorithms in hardware costing less than about £10,000 - a lower figure than is perhaps normal for inspection equipment. Above this figure, the rather low profit margins characteristic of foodproduct manufacture might be eroded excessively. This situation represents a significant engineering problem. Furthermore, there is a tension between the need to generate general purpose algorithms and hardware systems, and the need for low cost.

In our work we have found that a surprising number of foodproducts have significant features in common. Not only are many products roughly circular in shape, but also they are similar in size and are often fairly flat; in addition, they tend to exhibit a degree of individuality and irregularity that sets them apart from precision-machined parts manufactured in say the automobile industry. The existence of such a large set of similar products makes it profitable to devote considerable effort to the study of suitable inspection algorithms. Moreover, circular products appear in such variety that a solution covering this set automatically has considerable generality. This paper has described algorithms and procedures that solve the problems of inspection for this set of products.

Algorithms for the inspection of circular products must not only be capable of locating product centres quickly and reliably, but they must also be able to scrutinise product surfaces efficiently. We have completely solved the first of these problems, but the second problem is much more dependent on the characteristics of the particular product being inspected, and on the information about it that must be assessed. We have been able to arrive a number of general methods for handling the problem of surface scrutiny, including in particular the radial intensity histogram approach.

Radial histograms are particularly well suited to the scrutiny of symmetrical products that do not exhibit a texture, or for which the texture is not prominent and may validly be averaged out. In order to maximise the information obtainable from the radial histogram, and to ensure its meaningfulness, considerable reliance falls on accurate prior location of the product centre. A method based on the Hough transform, but involving a number of adaptations to cope with the properties of foodstuffs, has been found to be capable of locating the centres of products to within one pixel. This whole approach has been found capable of detecting and identifying correctly a variety of faults and fault conditions, including broken and overlapping biscuits, products having only a partial coating of chocolate or jam, or having other surface blemishes, and items having protuberances on their edges.

In addition to its effectiveness, this approach has the advantage that it embodies several tasks - thinning, edge joining, object location and centre location - in one powerful procedure. This has the effect of making processing inherently rapid, so that each product item can be inspected by software alone in 1-5 seconds. There appears to be no particular problem in implementing the algorithms described above in special purpose hardware that is capable of inspecting foodproducts in real time (e.g. one item per 30 msec), within the price range stated earlier.

Finally, it should be remarked that the centre location algorithm described here itself depends on the accuracy with which edge detection algorithms can estimate edge orientation. For this purpose it is vital that edge orientation be estimated to $\sim 1^\circ$ or less: means by which this may be guaranteed for edge detection operators operating within a neighbourhood of any size have already been found.

Acknowledgements

The author is grateful to the SERC and to United Biscuits and Unilever for financial support during the course of this work. He is also indebted to his colleague Mr A I C Johnstone for valuable discussions on methods of hardware implementation.

REFERENCES

1. Davies, E. R. "Image Processing", pp. 223-244 in "State of the Art Report on Supercomputer Systems Technology". Pergamon Infotech, Maidenhead, 1982

2. Pratt, W. K. "Digital Image Processing". Wiley, New York, 1978

3. Davies, E. R. "The theoretical basis and effectiveness of fast edge detection operators", paper presented at the 2nd BPRA Int Conf, Oxford (19-21 Sep 1983)

4. Davies, E. R. "Estimation of edge orientation by template matching", paper to be presented at the 7th International Conference on Pattern Recognition, Montreal (30 Jul - 2 Aug 1984)

5. Abdou, I. E. and Pratt, W. K. "Quantitative design and evaluation of enhancement/thresholding edge detectors", Proc IEEE, 67, no.5, pp. 753-763 (1979)

6. Davies, E. R. and Plummer, A. P. N. "Thinning algorithms: a critique and a new methodology", Pattern Recognition, 14, nos. 1-6, pp. 53-63 (1981)

7. Zucker, S., Hummel, R. and Rosenfeld, A. "An application of relaxation labelling to line and curve enhancement", IEEE trans Computers, C-26, pp. 394-403 (1977)

8. Paler, K. and Kittler, J. "Greylevel edge thinning: a new method", Pattern Recognition Letters, 1, nos. 5-6, pp. 409-416 (1983)

9. Hough, P. V. C. "Method and means for recognising complex patterns", US Patent 3069654 (1962)

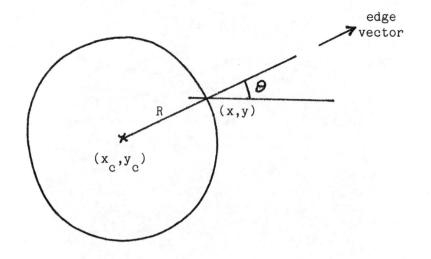

Figure 1 Geometry for computation of centre coordinates

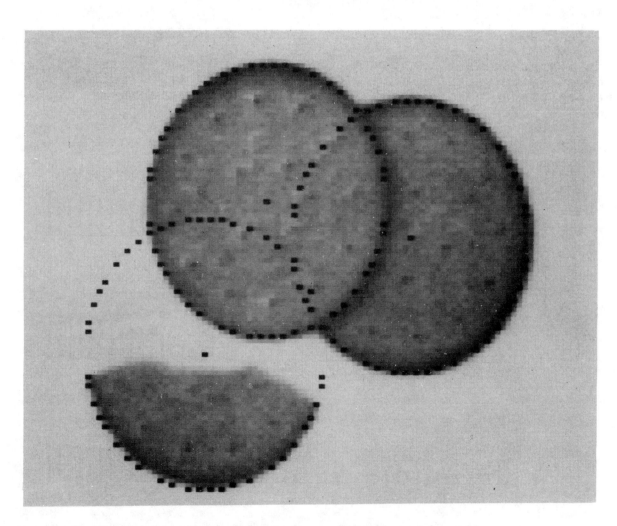

Figure 2 Inspection of broken and overlapping biscuits

This figure shows the robustness of the centre location technique. Accuracy is indicated by the black dots which are each within one half pixel of the radial distance from the centre.

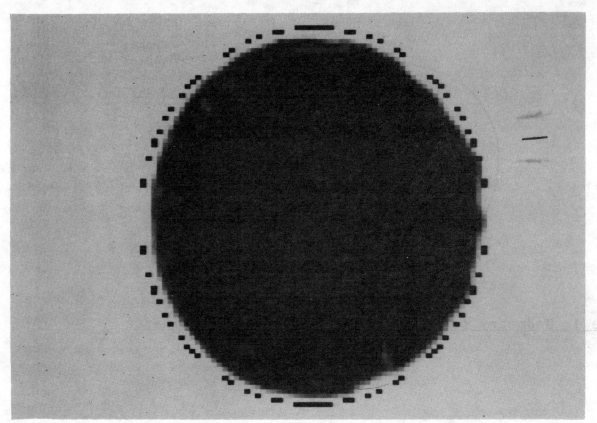

Figure 3 Inspection of biscuits coated with chocolate

This figure shows a chocolate-coated biscuit with excess chocolate on one edge. Note that the computed centre has not been 'pulled' sideways by the protuberances. For clarity, the black dots are marked two pixels outside the normal radial distance.

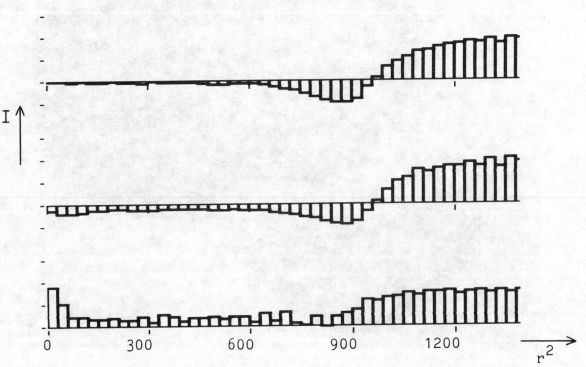

Figure 4 Examples of the radial intensity histogram technique

This figure shows radial intensity histograms for the three biscuits of Figure 2: the order is from top to bottom in both figures. Intensity is here measured relative to that at the centre of an ideal product.

446

Inspecting complex parts and assemblies

B. G. Batchelor

and

S. M. Cotter

Automation and Robotics Centre, University of Wales Institute of Science and Technology, UK

Abstract

The inspection of complex parts and assemblies is discussed in relation to the system requirements, including parts/camera manipulation, lighting, optics, image processing algorithms and computer architecture. The use of source knowledge, in the form of C.A.D. models of the part or assembly, is suggested as a means of simplifying the image processing. The integration of all of these disciplines into a complete inspection system needs careful planning; the purpose of the present paper is to highlight the fact that there are grounds for believing that this would be wothwhile, even with today's limited image processing technology. Preliminary experimental results on a variety of applications are presented.

Introduction

During the last decade, there has been a rapid growth of interest in Automated Visual Inspection and Robot Vision. These two subjects have developed along parallel but separate paths; inspection has largely been concerned with detecting faults in mass-produced goods, whereas robot control has been developed principally for simple positioning of a gripper. There has been very little work on the inspection of large complicated components and assemblies, a task which requires greater flexibility and intelligence than conventional automated visual inspection, and greater sophistication of image processing algorithms than traditional robot vision. There is an outstanding need for systems which can examine such items as those listed below:

(a) simple assemblies, such as electrical plugs, consumer gift packs, in-flight meal trays, hair driers;

(b) medium complexity assemblies, such as automobile carburettors, gear boxes, populated printed-circuit boards;

(c) complicated assemblies, such as automobile and aircraft engines;

(d) complicated components, such as engine blocks, hydraulics systems components.

These and similar tasks require the use of systems which incorporate a wide variety of technologies: illumination, robotic manipulators, optics, image sensors, data processing in analogue and digital hardware, software, algorithms. The proper integration of sub-systems incorporating these disciplines is needed before a reliable working inspection machine can be produced. This is, of course, a situation which has long existed in what we might call "traditional" visual inspection. (1) The present article discusses the design of a system for examining very much more complex parts and assemblies than have been commonly studied to date. In particular, the integration and inter-relationship between the technical disciplines just mentioned is considered in detail.

Illumination

The great importance of lighting for industrial vision systems, whether they be intended for inspection or robot vision, is emphasised elsewhere. (2) One of the major difficulties encountered when trying to design a good lighting arrangement for a very complicated object is that the optimal illumination pattern for one of its features is most unlikely to be satisfactory for all others. For example, a certain illumination direction might be chosen to provide shadow-free viewing of one feature but may cause glinting to occur on a nearby surface of brightly polished metal or glass. There are two obvious approaches to this problem:

(i) to adjust the lighting for each feature to be inspected

(ii) to use diffuse or omni-directional lighting to view all features, whatever their form or surface finish. The first of these is likely to result eventually in the best performance, but to achieve this requires a great deal of "intelligence" on the part of the lighting control unit. On the other hand, the use of diffuse/omni-directional light is both easier and cheaper, as well as being immediately available as a working solution. The value of omni-directional lighting may be judged from the illustrative example given in Figure 1. In our experience, diffuse/omni-directional light provides the best possible compromise, if a simple non-adaptive illumination scheme is to be adopted. The structure of a simple omni-directional light source is sketched in Figure 2. Let us now consider the possibility of arranging for the lighting pattern to be adjusted to optimise the viewing conditions for each particular feature of the object to be examined. The direction of the illumination can be adjusted in one of two ways: by moving the object or light source relative to each other and the camera, or by selecting a suitable lighting pattern using an array of lamps which envelope the object. (Figure 3) While this may seem to be potentially much more powerful than using a fixed illumination pattern, such as that provided by the omni-directional light source, it does suffer from the disadvantage that it requires very careful control of the object-camera-lamp geometry, to obtain even moderate results. How could such an adaptive lighting sub-system be controlled? The most attractive idea seems to be that a program which analyses the data generated by a C.A.D. exercise could be used to choose a good lighting arrangement. In theory, this is perfectly feasible; the C.A.D. model of a component, like an engineering drawing, specifies what that part should look like. A C.A.D. model, or an engineering drawing define the "shape" of the unit, while the surface finish could be derived from qualitative statements about the surface finish, such as "painted matt grey" or "polished brass". This leads us, of course, into an exercise in computer graphics, but clearly it will require a great deal of time and effort to achieve meaningful results on industrial components, which are far from simple in their surface reflectance characteristics. An alternative scheme might be to develop self-adaptive "intelligent" programs which can analyse images obtained from an arbitrary view of an object, decide how the lighting is deficient and then take remedial action e.g. dim or move the lamps. This approach represents far greater speculation; it is not yet clear whether this can be achieved in any realistic sense.

The conclusion we must draw is therefore that fixed diffuse/omni-directional light, though sub-optimal, is a reasonable basis on which to begin our investigation of the other aspects of our more general problem of designing complete vision systems for inspecting complex parts. Clearly, there is still much research to be done in the specific area of developing adaptive lighting configurations.

Optics & Mechanical Scanning

Deciding on a suitable optical arrangement is equally difficult; there is no one configuration which can accommodate all possible situations which can occur in practice. There is, however, a relatively small number of techniques which can cope with a majority of the problems. The principal examples of these are, of course, the use of automatic focus, zoom, pan and tilt. These facilities may be provided by the use of mechanical means of camera/lens translation. There are in existence various mechanisms for pan-and-tilt, amongst which the scanning mirror systems appear to be the most suitable for industrial use. (3) This type of unit provides the ability to move the camera's field of view without moving the whole camera body. (Figure4) The manufacturers quote what appear to be very good response times, fast enough to allow the viewing of different portions of the same scene on successive video field scans. An alternative would be to use a fibre-optic image bundle. (Figure 5) This has already been suggested by several workers,(4,5) but we question whether the bundles available commercially are robust enough for the continual flexing that they would experience in operation on the end of a robotic manipulator. From the discussions that we have had with suppliers of such bundles, there does not seem to be a convincing answer to this question. There are mechanical devices available now for altering the zoom on a video camera and which have a fast enough response for most needs. (3) If necessary, two such devices could be operated in cascade to move cylindrical lenses, thereby achieving different zoom in the x- and y-axes. This would allow the aspect ratio of the camera's field of view to be modified at will. A Dove prism, rotated under stepper-motor control, may be used to alter the orientation of the camera's raster scan, compared to the scene being viewed. (6) Alternatively, there is the option of mounting small objects on an (x,y,theta)-table or holding them in a robot gripper, during inspection. For larger objects, a camera mounted on the end of a robotic arm is an attractive option. (Figure 6) Such a camera might use conventional (fixed or mechanically controlled) lenses, or a fibre-optic image bundle. The latter option is particularly useful if it is required to "see through the palm" of the robot's gripper. The iris aperture also needs to be controlled, since the light intensity reaching the sensor array is highly variable. Auto-iris lenses exist but the response time to temporal light level changes that they offer is often too large for use in a practical situation. The same remark also applies to the a.g.c. control offered by the standard vidicon camera. The only really effective means of control appears to be to alter the illumination level, by adjusting the current supplied to the lamps. The final aspect of the optical sub-system that needs to be considered is the use of coloured filters to adjust the camera's spectral response. Various manufacturers provide switchable filter sets mounted on a turret which can be rotated under computer control. Such an arrangement can only be contemplated for use with a static camera; to cope with colour variations with a sensor mounted on the end of a robotic arm, a miniature colour camera would be useful.

One of the most important sub-problems is that of examining

449

holes and other parts of a complex object in which access is limited. To do this, an endoscope fitted to a camera on the end of a robotic arm or on a linear translation device, such as a stepper-motor driven platform, might be used. There is, however, a danger that the optical device could easily become damaged if it collides with something during the movement of the mechanism. This danger can, to some extent, be alleviated by the use of non-entrant imaging techniques. Deep hole endoscopes and fish-eye lens endoscopes (5) are likely to be particularly helpful here.

A wide angle lens (or its mirror equivalent) provides a number of advantages not enjoyed by standard optical devices:

(a) The depth of focus can be very large. For example, a lens with a focal length of 4-6mm can focus from about 150mm to infinity. This can reduce the system complexity by removing the need for focus control.

(b) The viewing distance (from the object being examined to the camera) is small. For an object such as a car engine, measuring say 500mm across, the viewing distance can be a small as 350-400mm.

(c) Combined with an intelligent frame store, a wide angle lens can provide the means to perform limited but useful pan, tilt and zoom operations electronically, i.e without mechanical scanning.

However, the price paid for using a wide angle lens is the image distortion which occurs, particularly near the periphery of the field of view. This can be corrected by straightforward image warping operations. Furthermore, in many instances, it is of no great consequence, particularly when existential rather than metrological inspection is required.

Image Sensor

When inspecting complex objects, the choice of image sensor depends upon the type of application, just as it does in "traditional" Automated Visual Inspection. The image dissector provides an attractive electronic alternative to the mechanical scanning methods just mentioned, whereby pan-and-tilt, zoom, aspect-angle adjustment, image rotation etc. are effected. The image dissector effectively provides the opportunity of allowing the computer to control the scanning. (Figure 7) Random scanning is also possible. A similar facility may be provided by using a conventional raster-scanned sensor, plus a frame store. (Figure 8) This allows any of the solid-state or C.R.T.-based cameras to be employed, thus freeing the equipment designer to make his selection of sensor on other grounds e.g. spectral response, sensitivity, highlight response characteristics. Electronic zooming may be achieved by altering the time-base wave-forms applied to a vidicon type of camera; the author's laboratory, for example boasts a Newvicon camera which has this facility. (FIGURE 9) We have already pointed out that the intensity levels encountered when inspecting complex objects are likely to vary considerably and, that for this reason, it may be advantageous to employ a camera which has in-built a.g.c.. This is much less important if omni-directional illumination is being used. (Figure 1 illustrates this to good effect.) When a camera is to be mounted on the end of a robotic arm, it is very important to ensure that it is robust enough to withstand the knocks and vibration that it will experience, both in normal use and in fault conditions. A small solid-state camera (e.g. Hitachi KP120) is more suitable in this type of situation. It is also much lighter in weight, than its C.R.T. competitors. In certain applications, it is virtually impossibe to avoid generating some bright highlights in a scene. These can have a very serious effect on the quality of the image as "seen" by certain types of camera. In Figure 10, we illustrate how much the choice of camera can alter the image quality. In this example, a brightly polished valve-stem is present at the bottom of an exhaust port in an automobile-engine cylinder-head. This application is clearly one in which all of the light striking the valve-stem enters the port from the the same direction; omni-directional lighting is quite impossible. The brightly polished metal surface of the valve-stem causes severe glinting in this situation. To accommodate this, a camera is needed which can both tolerate a high optical overload without suffering any (short or long term) damage, and produce reasonable contrast in the darker parts of the image. We have already touched on the topic of protecting the camera against mechanical, as well as optical, damage. This is particularly important if a camera is to be mounted on a robot arm. It may be necessary to fit the camera inside a protective shell for this reason. Thermal damage, due to the overheating of the camera is not significant when inspecting cold objects, but might, of course, be a problem if, say, red hot forgings or castings were to be the subject of the examination.

Algorithms for Inspection and Learning

Inspecting a large complex object, such as an engine block, is essentially equivalent to examining it for a large set of features, each of which is likely to be fairly simple to locate and measure, on an individual basis. The types of task which are required are typified by those listed below:

(a) Is a certain hole drilled?

(b) Is the hole chamferred?

(c) Is the hole tapped?

(d) Is a certain bolt present in its hole?

(e) Is there a washer in place beneath the head of a certain bolt?

(f) Is there any debris inside a given hole?

(g) Is there a certain spring fitted in the correct place?

(h) Is a certain lever bent or straight?

(i) Are there any scratches, blow holes, etc. on the face to which a gasket will fit?

(j) Is there any "flash" present (on a casting)?

The algorithms needed for these and similar sub-problems are simplified very considerably by the observation that we can verify that all is as it should be. In other words, inspection is not performed "blind", because there should always be a model of what we expect to find in a certain place on the complex part. Let us give an example, in order to illustrate the point more convincingly. The simplest one to explain is that of inspecting populated printed circuit boards. There always exists a plan, either on paper or more commonly nowadays in a computer, from which an intelligent person or program can predict what to find in a certain place on the board. Such a component layout will normally specify the type of component to be found at a given (x,y) "address" on the board. At or near such a point, there should be a component whose appearance we can anticipate, simply because we know what type it is supposed to be. Verifying this is almost cerainly very much easier to accomplish than recognising what is there when we assume no prior knowledge. Verification is almost invariably far easier, faster, and cheaper than recognition.

Inspecting a complex object is far too difficult, if we try to do "everything at once". By breaking the problem down into more manageable proportions, we can expect to achieve very much more, even with the limited amount of "intelligence" available today in the form of computer programs. Let us now consider two possible strategies for the inspection of complex parts. These differ in that one of them is reliant upon human teaching, while the other depends upon a C.A.D. model as the basis for the actions which it undertakes.

Human guidance

Imagine a camera mounted on the end of the arm of a point-to-point robot, which is taught to perform a series of movements by a human operator. The camera is connected to a vision system which, like the robot controller, can learn when told to do so. The images captured by the vision system are assumed to be represented parametrically. This allows the possibility of it learning by showing. A "good" example of the complex object to be examined is placed within the working space of the robot. The operator then guides the arm to a number of different locations around the object, so that the camera can in turn view each of its important details, whose inspection is required. Parameters representing each such view of the object are then calculated and their values learned, by a Pattern Recognition sub-routine attached to the vision system. In this way, it is possible, at least in theory, to form the basis for a robot vision system, which can learn by showing to recognise images like those which have been previously seen. There is, however, a serious difficulty with this approach, since there is no provision for the system to learn tolerance parameters. We shall return to this again in a moment, but first let us explain why the above approach was suggested. Consider a simple sub-problem, namely that of inspecting the chamfer at the mouth of a hole in a car engine block. The camera on the end of the robot arm will never "see" exactly the same picture twice, even if it is inspecting the same object again, because the robot is not able to repeat its movements exactly. However, if we assume that the hole is well within the field of view of the camera, a small amount of error in the positioning of the camera will not seriously affect the parameter values obtained. (Figure 11a. The parameters are, of course, assumed to be position invariant.) Now it has been assumed so far that the "good" object used during the learning phase is exemplary; it is a perfect component. However, no such thing exists in practice, even an exemplary model, such as might be made as a prototype, does not conform exactly to the theoretical archetype specified by the design engineer. In other words, it is impossible to make a perfect model. The measurements, which occur as a result of the allowable tolerances cause the parameter vector to wander about (Figure11b) The allowable limits of movement of the parameter vector cannot be calculated precisely, because we do not know how the vector obtained by learning relates to the "ideal" vector. This is a fundamental point and its effects cannot be calculated exactly, unless we can somehow derive the "ideal" vector directly from the design specification. This is the basis of the approach which will now be described.

C.A.D. Guided Inspection (See Figure 3b)

Let us now assume that a complex part has been designed using C.A.D. techniques and that a model of it resides within a computer. Nowadays, it is common practice to generate "solid" images of such parts, using advanced graphics techniques. The images generated thus could be used as the basis of a similar exercise in designing a vision system by learning. To do do this properly, we should need to develop graphical image generation further than hitherto and introduce a facility for modelling different lighting configurations, optical assemblies, and cameras. It would then be possible to anticipate the form of the images which the real camera will see. All this could, of course, be achieved before any metal is cut, plastic moulded etc. Using this approach, it would be possible to circumvent the problems described above, which affect the human-guided learning-by-showing system. Moreover, the tolerances relating to the part could be translated, again within the computer, into allowable perturbations of the parameter vector. All of this is possible in theory, but there is a great deal of work to be done before it becomes a practical and worthwhile technique. There will be enormous problems in achieving a realistic model of the surface reflectance of real industrial parts, and this will almost certainly form one of the chief obstacles in developing this approach to its full potential.

Computer Architecture

The fact that we cannot reasonably expect to be able to examine all features of a complex object simultaneously means that we can take our time in analysing the image data from each one of them. Suppose, for sake of argument, that there are 100 different features of special interest on a part such as an engine block. Furthermore, let us assume that it takes a camera mounted on the end of a robot arm 0.5 seconds to move from one position, where it can see one feature, to the next viewing point. The total time available to examine the complete engine block is therefore 50 seconds. Now, we maintain that there is no real need to obtain the decision about one feature in much less time than this. Suppose then that we employ a set of fairly cheap (slow) image processing devices, each of which is able to analyse the data from one feature. Each such processor has almost 50 seconds available to it to perform whatever image processing it is required to do. The same camera will, of course, be used to capture the original unprocessed image data from each feature, so there must be a multiplexing arrangement to route the data to the appropriate processor. The resulting architecture is called a concurrent processor. (7, Figure 12) Let us assume that the elemental processors are all identical and that they run a software suite which is capable of performing a wide variety of image processing tasks. Such a facility is provided by the interactive image analysis packages exemplified by Susie, Autoview, Grid, Ispec, MV 100, Saidie (8-13) The possibility of incorporating such a package in say a single-board processor is now a distinct reality. These and similar software packages are capable of providing almost all of the image processing facilities required to perform the inspection of such features of complex objects as those detailed above. (The fact that we have prepared all of the illustrations for this paper using Autoview is evidence of this.) Now, let us assume that this is in fact the case; we shall assume that a concurrent processor consisting of a set of identical devices like those just discussed is to be built. At first sight, this appears to be a very extravagant means of solving the problem of obtaining the speed that we need, but it is in fact quite economical and cost effective, as we hope to show. The structure of an element of the concurrent processor is sketched in Figure 13. A concurrent processor like this uses standard software; the cost is therefore the same as for a single processing unit. Moreover, the software is common for many different applications, thereby reducing the cost still further. The hardware cost rises, of course, as the number of units is increased, but even here there is an "unexpected" benefit which results from the reduced unit cost for large quantity purchases. To represent this situation in algebraic form let us define the following notation: (a) Let N be the number of parallel processing elements. (b) Let S be the cost of the software. S is independent of N. (c) Let H be the cost of each of the processing elements. (d) Let V be the additional overhead costs, due to lighting, optics, camera, robot, multiplexor, system design, operator training, documentation, etc.. V is independent of N. The total cost is then given by:

$$S + V + N.H$$

Now, in practice, H is small compared to both S and V, so that even for

quite large values of N, the cost is dominated by these two fixed
variables, (i.e. S and V). The total processing capacity of the system
rises, of course, linearly with N, assuming that it remains processor
bound. To emphasise the point, let us substitute some notional values
into these expressions:

$$S = `10000$$
$$V = `25000$$
$$N = 50$$
$$H = `500$$

Then, the total system cost is `60000 (`35500 if N=1), but the
processing capacity is increased by a factor of 50 over a
single-processor system. There are other, less easily quantified
benefits, arising from the fact that the concurrent processor is an
assembly of identical sub-systems. The chief attraction is that repair
becomes simply a matter of exchanging boards. Spare-parts for the
processor consist of just one type of board. (It may, in some
circumstances, be necessary to change ROMs as well.)
        To summarise, a concurrent processor is ideal in situations,
like this, in which a series of separate tasks is to be performed which
are initiated at different times and do not individually require fast
response times.

Experiments and Results
        So far, we have discussed the inspection of complex parts in a
general non-specific way, without justifying the tacit assumption that
it is feasible to make a useful contribution here with the existing
techniques of image acquisition and image processing. In this paragraph,
we attempt to do this by presenting some results that have been obtained
in the laboratory. These represent a preliminary attack on the general
problem but show, we hope, that our optimism is not without good
foundation. Results are given on the following applications:

(i) Internal thread, existential inspection (Figure 14)

(ii)   Detecting debris in a grooved hole. (Figure 15)

(iii) Verifying that a simple linkage mechanism has been assembled
correctly. (Figure 16)

(iv) Verifying that a cross-head screw is in place (Figure 17)

(v) Verifying that a washer is fitted beneath a screw (Figure 17)

(vii) Checking that an 'O'-ring is in place (Figure 18)

Conclusions
        The general question of inspecting complex parts and assemblies
has been addressed and the major questions identified. There is a strong
interlinking between complex parts inspection, on the one hand, and
robot vision on the other. From the illustrative examples presented, it
is apparent that there is a case for believing that machine vision
systems will soon be able to make a major contribution to the inspection
and measurement of complex piece-parts and assemblies. There is a great
deal of source knowledge available about industrial parts in the form of
C.A.D. models. This information could be invaluable in guiding the
vision system to look for specific features. However, to make use of
this, a great deal of development work remains to be done. It seems
that, with present technology, the inspection of complex parts
presents economic, rather than technical, difficulties.

References

(1) B.G.Batchelor,"UWIST laboratory for industrial inspection research", in "Digital signal processing", N.B.Jones (ed.), Peter Peregrinus, London, 1982, pp 467-484.

(2) B.G.Batchelor, "Illumination & image acquisition techniques for industrial vision systems", proc. workshop organised by IRSIA/IWONL & CETEA, Antwerp, Oct. 1983, in "Industrial uses of image analysis", N.J.Zimmerman & A.Oosterlinck (ed.), DEB Publishers, Pijnacker, Netherlands, 1983, pp 97-118.

(3) K.Pelsue, "Precision, post-objective, two-axis, galvanometer scanning", Proc. S.P.I.E., vol. 390, 1983, pp 70-78.

(4) I.A.Routledge, J.M.Middlemas, D.M.Allen, "Low-cost optical systems for location of objects & robot end effectors", proc. 2nd int. conf. on Robot Vision & Sensory Controls, Stuttgart, F.D.R., Nov. 1982, pub. by I.F.S. Ltd., Bedford, England, 1982, pp 171-176.

(5) D.A.Hill, "Fibre optics", Business Books, London, 1977.

(6) B.G.Batchelor, "Circular pictures, their digitisation and processing", Computers & Digital Techniques, Proc. I.E.E., pt. E, vol. 1, no. 4, 1978, pp 179-189.

(7) B.G.Batchelor, "A laboratory-based approach for designing automated inspection systems", proc. Int. Workshop on Industrial Applications of Machine Vision, N. Carolina, May 1982, proc. pub by I.E.E.E. Computer Society, pp 80-86.

(8) SUSIE, University of Southampton, Southampton, England, also see article by B.G.Batchelor, "Interactive image analysis as a prototyping tool for industrial inspection", Computers & Digital Techniques, proc I.E.E., pt. E, vol. 2, no. 2, 1979, pp 61-69.

(9) Autoview, British Robotic Systems Ltd., London, England, also see article by B.G.Batchelor, D.H.Mott, G.J.Page, D.N.Upcott, "The Autoview Interactive image processing facility", in "Digital signal processing", N.B.Jones (ed), Peter Peregrinus, London, 1982, pp 319-351.

(10) Grid, G.E.C. Hirst Research Centre, Wembley,England, also see article by J.A.Losty & P.R.Watkins, "Computer vision for industrial applications", G.E.C. Journal of Research, vol. 1., no. 1, 1983, pp24-34.

(11) ISPEC, Micro Consultants Group, Kenley, England.

(12) MV100, Vision Dynamics Ltd., London, England.

(13) SAIDIE, Transaction Security Ltd., Guildford, England.

Legends

Figure 1 Omnidirectional lighting using the arrangement shown in Figure 2.
        (a) This piece-part consists of a number of bright metal components.
        (b) Edges calculated from (a).

Figure 2 Omnidirectional lighting arrangement.

Figure 3 Computer controlled lighting.
        (a) Lamp arrangement. The object under examination, shown here
as a black ellipsoid, is held by a robotic device and can be viewed by
any one of 3 cameras (C1 - C3), against different backgrounds. Lamps L1
- L8 are all under computer control.
        (b) System organisation. Notice the presence of the box labelled
"C.A.D. Modeller". This is is an important feature in the present
discussion,as will become evident later.

Figure 4 Scanning a wide field of view using mirrors mounted on a pair
of of galvanometers.

Figure 5 Fibre-optic image bundle and camera. Illumination is supplied
by a fibre-optic ring light. Also see Figure 14.

Figure 6 Camera mounted on the end of a robotic arm. The wrist mechanism
was designed at UWIST and is fitted to a Pendar Placemate robot. The
camera is the Hitachi KP120 solid-state device.

Figue 7 Image dissector operating under the control of an image
processing computer.
        (a) System organisation.
        (b) Possible scanning patterns:
                A - large area, low-resolution scan
                B - small-area, high-resolution scans, may overlap
                C - very high resolution scans, progressive zooming
                D - very high resolution / progressive zoooming
                F- trapezoidal scan, corrects off-normal viewing
                G - rotated scan
                H - parallelogram scan
                Random scanning and curve following are not shown.

Figure 8 Using a frame store and high-resolution raster-scanned camera
to obtain scan patterns like those shown in Figure 7b. The computer
accesses the frame store in the same way as it would address the image
dissector; the camera plus frame store,taken together, may be regarded
as being equivalent to a random-scan image sensor.

Figure 9 Electronic zooming
        (a - c) Zooming using scan-control of a modified Newvicon
camera.
        (d) Zoom by selecting the central area of (c) when it was held
in a frame store.

 Figure 10 Showing the effects of choosing an inappropriate camera.
        (a) Cross-section of the assembly. Glinting cannot be avoided
altogether in such a situation.
        (b) View obtained using a Sivicon camera
        (c) View obtained using a Newvicon camera

Figure 11 Camera position invariance and component tolerance effects.
        (a) Parameter invariance. Idealised view of a feature such as
a stud, hole, or slot in a plain surface. Moving the sampling window
from position A to B does not seriously affect the parameter values that
are obtained. The outer circle represents the edge of a chamferred
threaded hole on a plain background surface.
        (b) Tolerance effects. The vector (P,Q) represents the ideal
parameter vector. (p,q) is the estimate of (P,Q), obtained from an
exemplary model (a prototype) by learning. Box A represents the
allowable tolerance range (centred on (P,Q)), while box B shows the

estimated tolerance range, obtained by showing the vision system a number of "good" or acceptable parts.

Figure 12 Concurrent processing.
    (a) System structure. IP1 - IP4 are four identical but highly flexible image processors, whose functions can be controlled /selected via the line L. Only four processors are shown here, in order to simplify the picture; in practice, there may be many more than this. The outputs B1 - B4 are gated in turn to the decision maker, an intelligent device capable of collating and interpretting the the data about individual features. The outputs A1 - A4 are provided principally for system maintenance and operator training. (See Figure 13.)
    (b) Processor activity chart.

Figure 13 Internal detail of the image processors IP1 - IP4.

Figure 14 Checking that an internal thread is present by illuminating and viewing from the mouth of the hole. A fibre-optic image bundle was used to obtain this image.
    (a) Original image
    (b) Central portion of (a).
    (c) Processed image in which the thread contours are traced It is a simple matter to verify that there is a good thread present from an image like this.

Figure 15 Locating debris in a grooved hole, using an endoscope fitted to the camera.
    (a) Original image
    (b) Processed image showing the swarf highlighted

Figure 16 Verifying that a simple linkage mechanism has been correctly assembled.
    (a) Original image.
    (b) Edges computed from (a).

To verify that the bright lever is both present and well formed, the analysis program might correlate such an edge map with a standard template. Such a procedure would need to be performed "intelligently" in view of the fact that the edge map in (b) contains discontinuous contours.

Figure 17 Locating a cross-head screw and detecting a washer.
    (a) Original image.
    (b) Head outline, screw-driver slot and centroid of the head identified.
    (c) A composite image showing how the analysis might proceed in practice.
        top left: 2 circles whose centres are located at the head centroid. (The analysis program might reasonably sample around these circles.)
        top right; polar-coordinate scan of the head. Abscissa, radius; ordinate, angle. Notice the toothed effect and the dark vertical streak.
        bottom left: a "waveform" obtained by integrating the intensity around a series of concentric circles. The right-most trough is due to the presence of the (dark) washer. Its radial position is indicated by the larger circle.
        bottom right: intensity profile, along the vertical slice indicated by the straight line and showing four peaks. Each is due to one of the lobes of the slot cross.

Figure 18 Checking that an 'O'-ring is in place in a grooved hole.
    (a) Original image.

        (b) 'O'-ring contour.
          If the ring were displaced, the analysis program would yield a
bent/missing contour in (b).

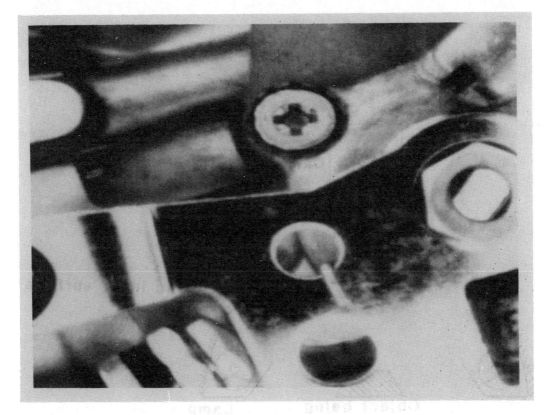

FIG. 1A

FIG. 1B

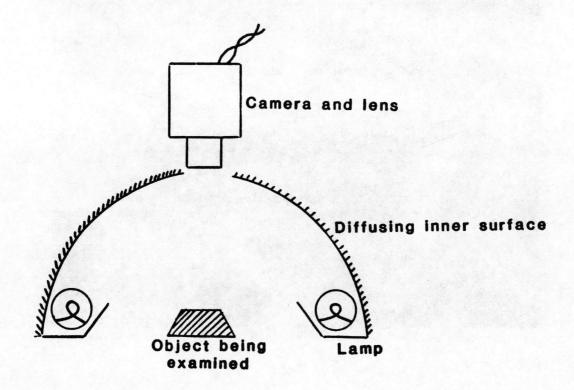

Camera and lens

Diffusing inner surface

Object being examined

Lamp

FIG. 2

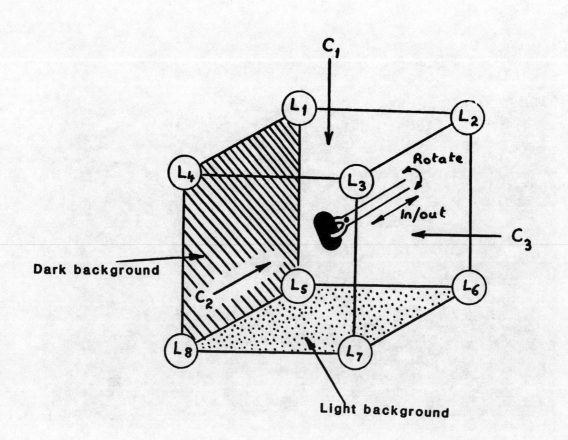

C_1

L_1

L_2

Rotate

L_4

L_3

In/out

C_3

Dark background

C_2

L_5

L_6

L_8

L_7

Light background

FIG. 3A

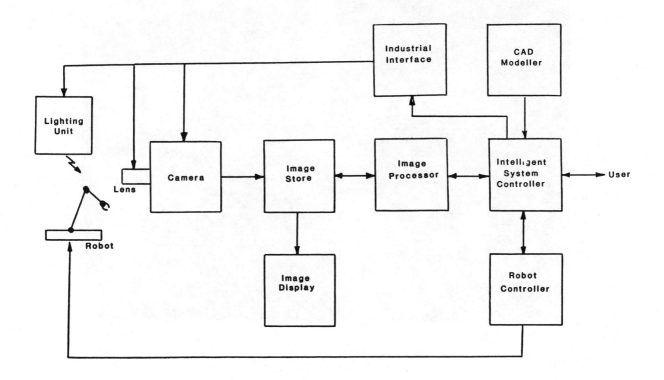

FIG. 3B

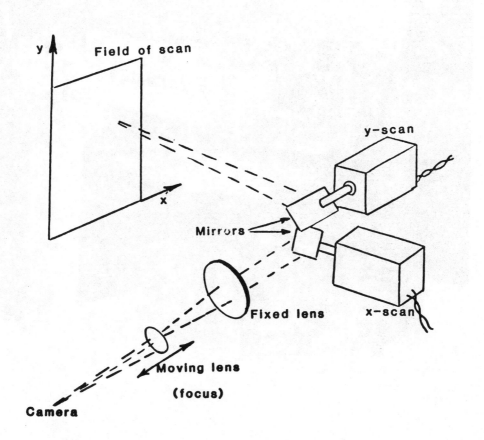

FIG. 4

461

FIG. 5

FIG. 6

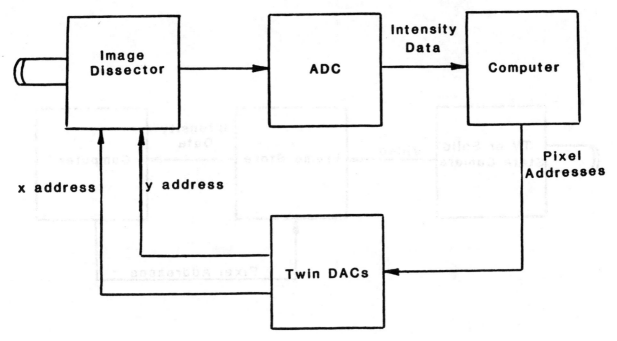

FIG. 7A

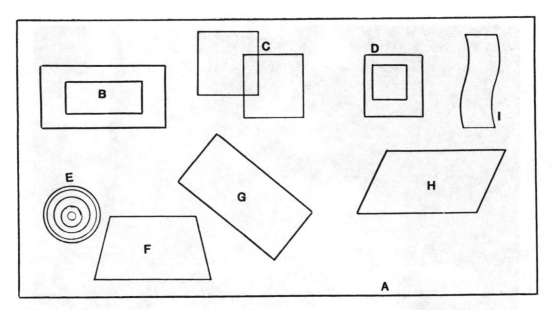

FIG. 7B

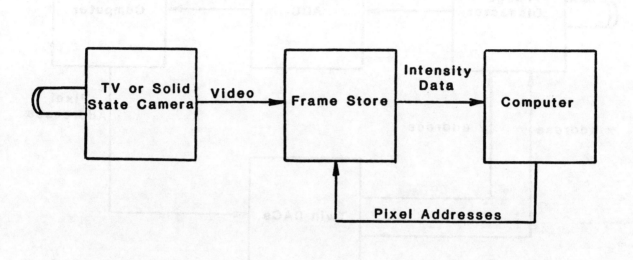

FIG. 8

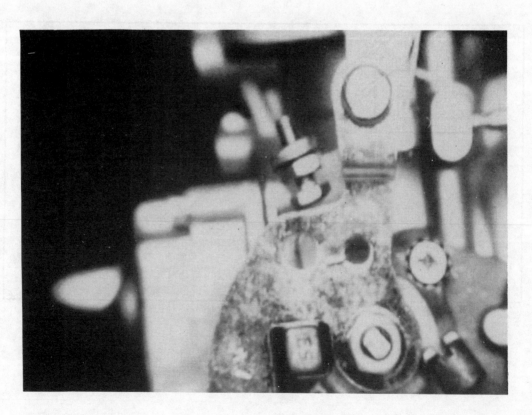

FIG. 9A

464

FIG. 9B

FIG. 9C

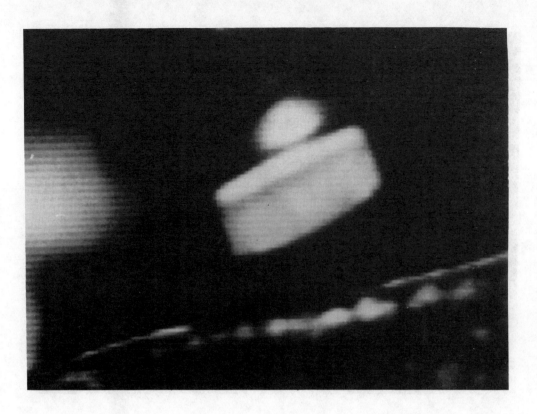

FIG. 9D

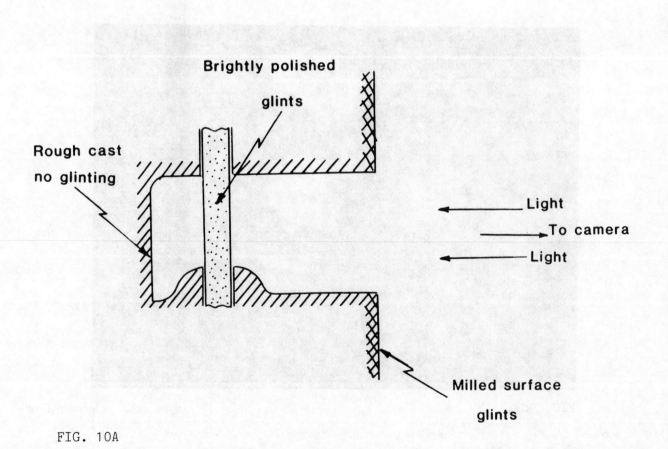

FIG. 10A

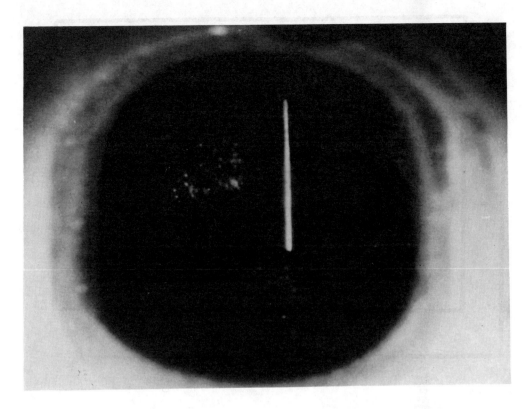

FIG. 10B

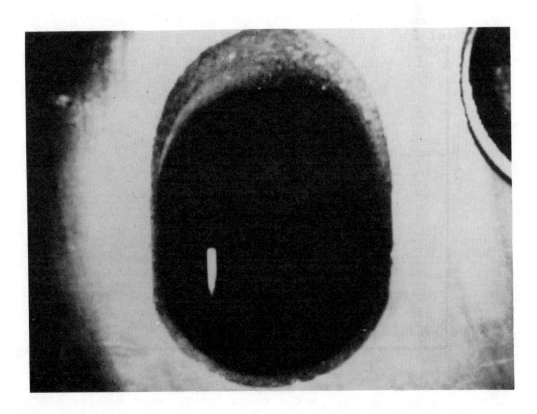

FIG. 10C

467

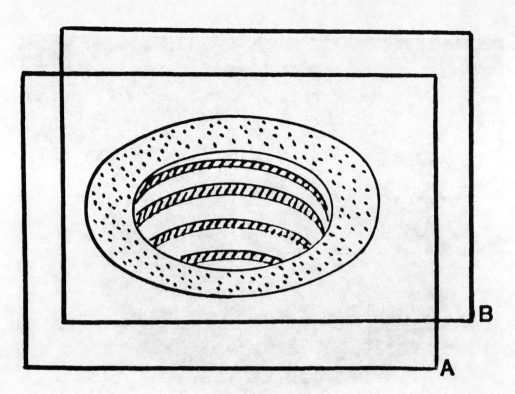

FIG. 11A

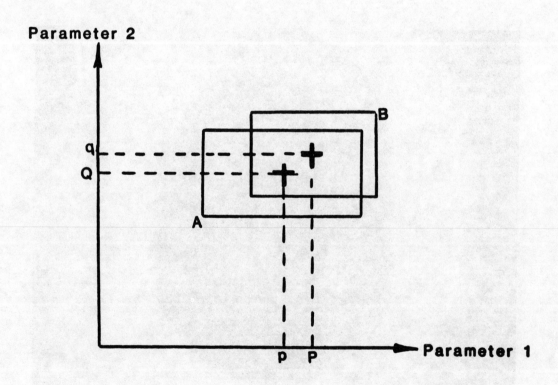

FIG. 11B

468

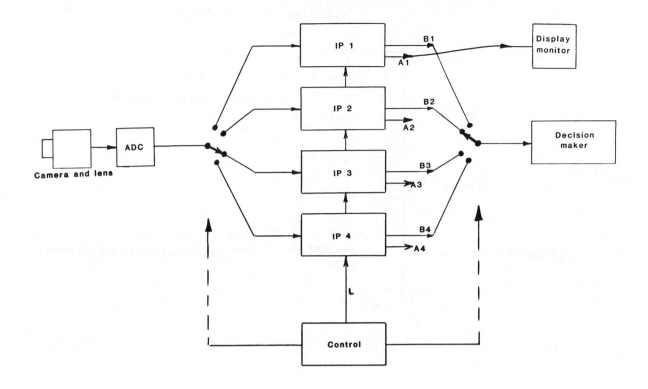

FIG. 12A

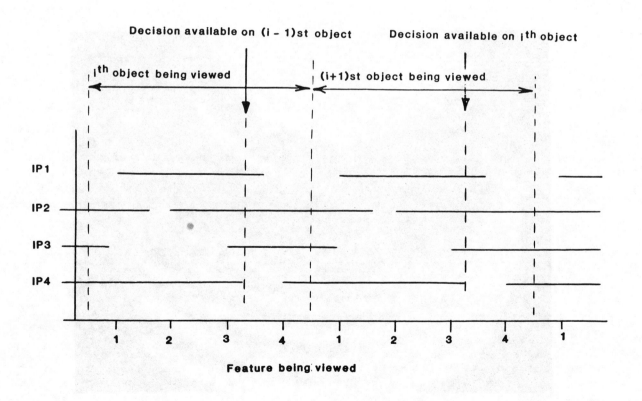

FIG. 12B

469

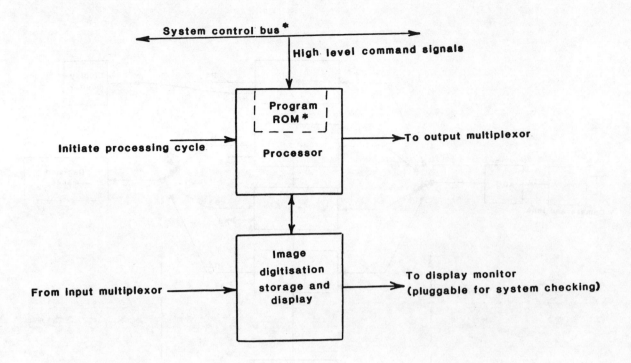

FIG. 13

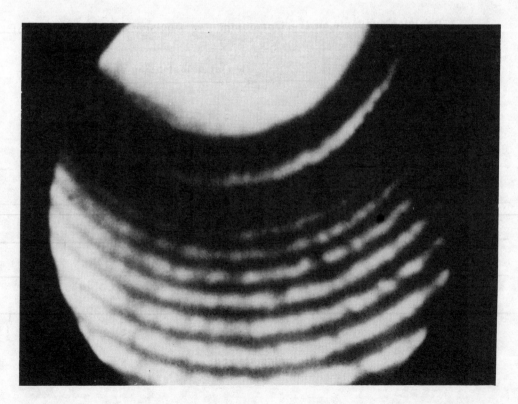

FIG. 14A

470

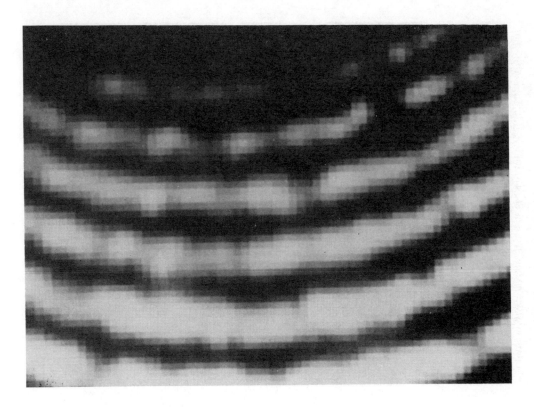

FIG. 14B

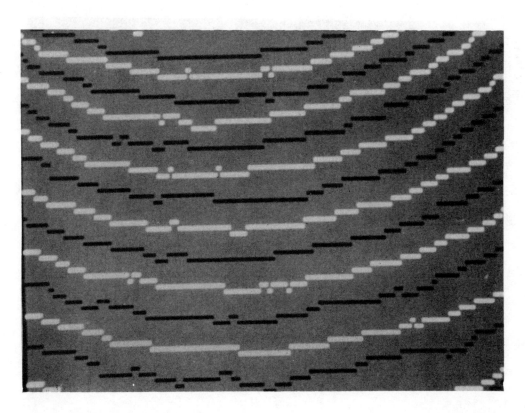

FIG. 14C

FIG. 15A

FIG. 15B

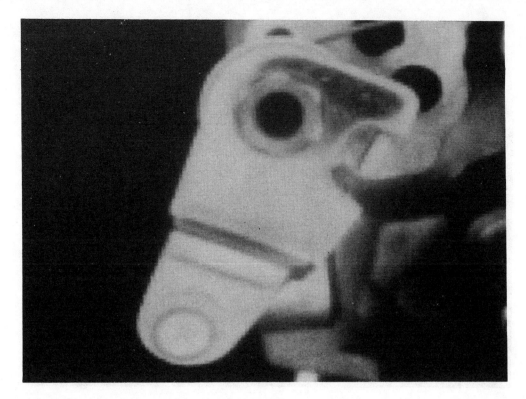

FIG. 16A

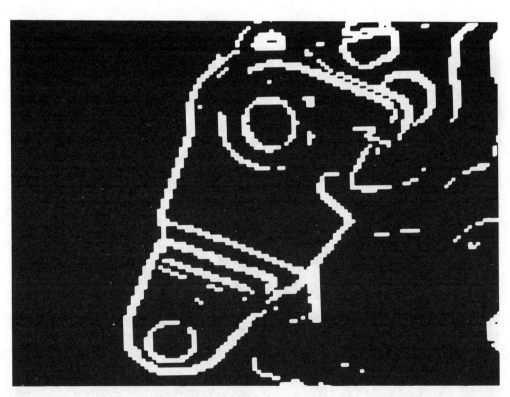

FIG. 16B

473

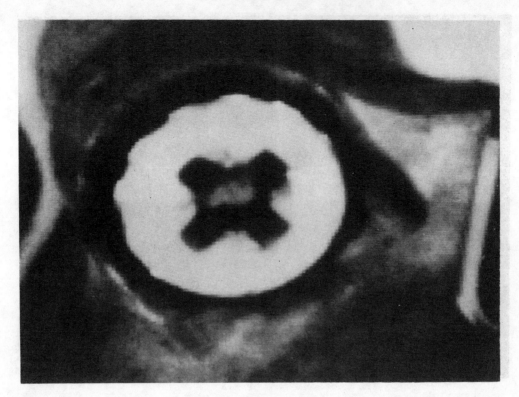

FIG. 17A

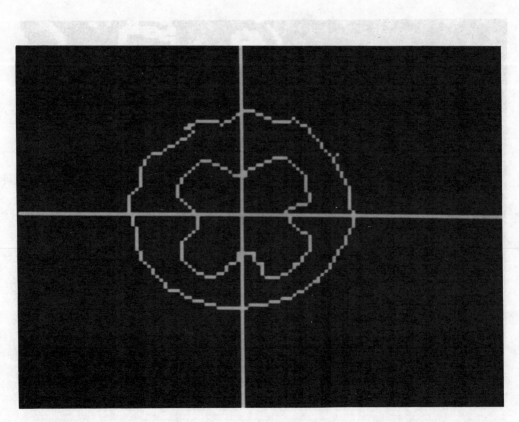

FIG. 17B

474

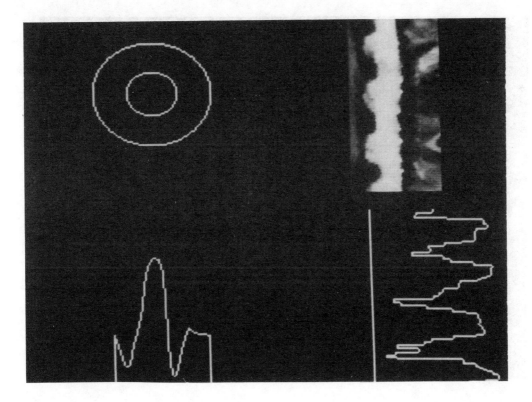

FIG. 17C

FIG. 18A

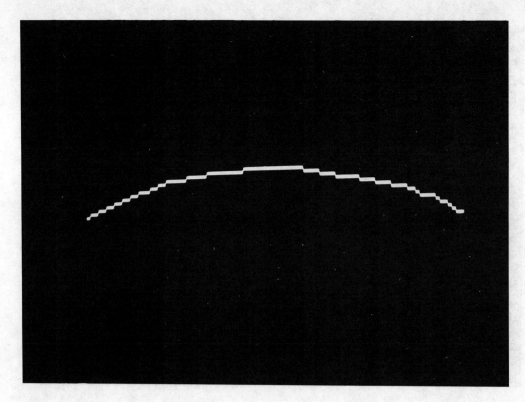

FIG. 18B

LATE PAPER

Project 'Harunobu': toward a locomotive vision robot

H. Mori

Yamanashi University, Japan

Abstract

To study how the robot gets a stabilized image from a sequence of TV images obtained while he is moving, a locomotive vision robot "HARUNOBU - 2" has been developed. He rides on a velocipede and goes through the obstacle scattered passage of a computer room and avoides obstacles and searches for a required target. At first he look around the passage and make two cognitive maps, named body map and room map and localizes himself and the obstacle on the maps. The body map represents the environment egocentrically, whereas the room map represent it objectively. The above process is performed by a visual behavior interaction system.

INTRODUCTION

Assume that you have gone home thinking about something and cannot recall any episode that may have happened on the way home. Why can you have successfully gone home without meeting any accident ? We intend to explain these phenomena by two vision systems, shape analyzing system and visual behavior interaction system. The former is a sequential system that analyses an object by its shape with knowledge data base and send to a motor system an action command unique to the object. The former has a connection to a consciousness system. The latter is a parallel system that takes quickly a fixed action to visual stimulus. In the above case visual behavior interaction system may have succeeded in avoiding an obstacle on the road and in stopping at a cross. The shape analyzing system may have operated efficiently at every significant points on the way home.

Since 1982 we have started a project toward locomotive vision robot under the philosophy of two visual system, shape analyzing system and visual behavior interaction system [1]. It is promoted for three reasons, first for industrial use, second for the study of space perception, third for development of picture interpretation language PILS.

Since 1979 we have been developing a picture interpretation language PILS [2]. At first it was designed for the programming of gastric X-ray film diagnosis, but now it is revised for the programming of robot vision. At first S of PILS stand for silhouette because it was able to process only silhouette image, but now it stands for scene because it is able to process colour scene of the natural world.

VISUAL BEHAVIOR INTERACTION SYSTEM IN ANIMAL

HONEY BEE

Among the animals which may make something like cognitive map around his nest, one of the most interesting animal is the honey bee [3]. Because she is separated from vertebrate branch near the root of dendrogram in phylogenesis and she takes highly shared behavior in her highly organized society. The radius of scope of bee's behavior is about 4 kilometers. In this scope she is able to go back and forth between her nest and a flower to carry honey and pollen. To explain this behavior it is assumed that the bee must have her own cognitive map and must localize her nest and the flower on the map. The bee's life is about a month or so. In the first half of her life, she works in her nest for nest building and infant nursing. In the second half of her life she goes out and works for getting honey and pollen. When she changes her work from indoor to outdoor, she flies out of her nest with her fellows of the same age and floats about half an hour above her nest orientating her head to it as if she try to remember the location of her nest in the environment. More quick orientating behavior is seen when she leaves a flower of the first visit. She gets out of the flower and flies up in the sky drawing two or three circles and then she goes back to her nest. The schemata of how to orientate a nest or a flower is innate one. Any parents and sisters cannot teach her the schemata. Only thing that she can learn and remember is parameter of the schemata. It is empirically confirmed that the bee can see through her compound eyes such geometrical figures as circle and square but cannot distinguish and remember them by their shape.

Through the bee's behavior we can learn that some kinds of space perception and behavior are possible by some innate non-plastic schemata without help of shape analysis.

478

PRIMATE

Recently some psychologists [4,5] and physiologist direct their attention to extra geniculate system projecting from retina into superior colliculus, pulvinar and thence to area 18 and 19. It is called second visual system* in contrast with primary visual system projecting from retina, lateral geniculate nucleus and thence to area 17. Area 17, striate cortex in another word is devoted to edge detection which is supposed to be the first stage of shape analysis.

A monkey whose striate cortex is perfectly removed is not able to perceive stationary object but able to perceive and orientate moving object and succeed in avoiding obstacles. But when superior colliculus of the monkey is moreover removed, the monkey falls into perfect blind. From this experiment an evidence that second visual system is responsible to object orientating and reflective behavior to obstacles is found.

VISUAL BEHAVIOR INTERACTION SYSTEM IN ROBOT "HARUNOBU"

Why we can neglect the role of the visual behavior interaction system of the bee or the ablated monkey ? Is it unreasonable to think that function of lower animal still exists in human being ? In our project a robot will have two visual system in the near future, a visual behavior interaction system and a shape analyzing system however, lets begin the robot with the visual behavior interaction system.

Cognitive MAP

For a locomotive vision robot one of the most difficult problem is how he gets a stabilized image of environment from TV camera in locomotion. To conquer this problem two cognitive map, body map and room map, are applied. The former is a map which is obtained by projection of the environment on the globe that is imagined to be located on the place where TV camera is fixed. The coordinate system of the map is fixed on the robot itself. In the body map the environment are represented egocentrically. The body map is useful for vision motor reflex. The latter is a map which represents the environment just the robot standing on. The coordinate system of the map, which is called framework, is fixed on some physical objective base such as the boundary of floor or the entrance of room. The room map is useful to analyze the environment comparing with knowledge data base.

Vision Robot "HARUNOBU-1"

To confirm the effect of the cognitive map we have developed a vision robot called "HARUNOBU-1" [6], which has a head, a body and a leg. A TV camera is fixed in his head which rotates horizontally and vertically against his body which also rotates horizontally and vertically against his leg which is fixed on a floor. A crane is given for shifting an object.

---

* Primary visual system and second visual system are termed by Bronson [7] through his study in child development. They are named focal vision system and ambient vision system respectively by Trevarthen [8] through the study of split-brain macaques. In our project they are named by their functions shape analyzing system and visual behavior interaction system respectively.

Task of "HARUNOBU-1"

A checkerboard is placed on the floor in front of him as Fig.1. Make a room map of the surroundings with the coordinate system fixed on the checkerboard. Localize an object on the map and pick up it by the crane.

Given Knowledge

Although some knowledge about his body and the environment are necessary to perform the task, It is desirable not to use superficial knowledge but to use fundamental knowledge applicable to any environment.
(1) **Knowledge about his body** Structure of the body and size of the each part are given. Angle between parts are also given at every moment.
(2) **Knowledge about framework** The checkerboard is composed of the same sized squares and make a rectangle as a whole however. The number of the square and the orientation of the rectangle are unknown.
(3) **Knowledge about tool** Structure of the crane are given.
(4) **Knowledge about the object** The object is able to distinguish from the checkerboard by its colour.

Schemata

To perform the given task two schema, cognitive map making schema and object localizing schema are used. To show the process of these schema dramatically, a standard wide angle lens which does not cover the whole of the checkerboard at a glance is used in TV camera.

Cognitive map making schema

(1) Make the robot stand in erect position by initializing pulse motors.
(2) Make the TV camera catch his foot by bowing his head and body.
(3) After fixation of his body, move his head down until the TV camera catch the lower side of the checkerboard at its center.
(4) Rotate his head left looking along the lower side of the checkerboard until the TV camera catch the left lower corner at its center as shown in Fig.2 (a), and localize the corner on a body map by measurement of head angles.
(5) Move his head up counting the number of squares along the left side of the checkerboard until the TV camera catch the left upper corner at its center as shown in Fig.2 (b) and (c), and localize the corner on the map.
(6) After returning his head to the left lower corner, rotate his head right counting the number of squares along the lower side of the checkerf board until the TV camera catch the right lower corner at its center as shown in Fig.2 (d), (e) and (f), and localize the corner on the body map. Fig.3 (a) illustrates the body map obtained by the above routine.
(7) By the function obtained by 3-D geometrical calculation, the body map is projected on a room map as shown in Fig.3 (b).

Crane localization schema

Because the crane is separated from the robot, its localization has to be done at first.
(1) Search for the arm of the crane by moving his head in up to down direction.
(2) After fixation of his head, stretch the arm from A to B by rotating arm motor 6 and then rotate the arm from B to C by rotating arm motor 7. Next shorten the arm from C to D by rotating motor 6 in opposite direction. Plot A, B, C and D on the body map with the angles of crane motor 6 and 7 as shown in Fig 3.(a).
(3) Project A, B, C and D of the body map onto the room map. Obtain on the room map the intersection of the extensions of line BA and that of

line CD as shown in Fig.3 (b). The intersection represents the fulcrum of the arm. Cocentric circles round the fulcrum represent loci of the arm obtained by rotating motor 7 and radial lines through the fulcrum represent locus of the arm obtained by rotating motor 6. Next the circles and lines are projected on the body map as shown in Fig.3 (a). After the projection the map is called body map with motor map.

Object localization schema

(1) Eliminate the checkerboard pattern of TV image by covering the lens with a red filter.
(2) Scanning the checkerboard search for an object and catch it in the center of TV image.
(3) After uncovering the red filter, localize on the body map the arm by the relative position of the arm to the checkerboard.
(4) Move the crane to the object position by the motor map and pick up the object.

Result and Conclusion

(1) The first half of the task that is composed of cognitive map making process is fairly successful however, the second half is not. The robot failed in picking up the object on the checkerboard. It is responsible for the incorrect localization of the object and the arm due to the play of gear of the head.
(2) By the application of the room map, the size constancy and the shape constancy are explained easily, because the size and the shape of an object on a room map is unchangeable by its location.

Locomotive Vision Robot "HARUNOBU-2"

To confirm that the visual behavior system is sufficient for locomotion, we have developed the first locomotive vision robot called "HARUNOBU-2" [9]. He rides on a velocipede which runs at 20 cm/sec. He has a TV camera in his head which is about 85 cm high. The camera has a wide angle lens of 60 degree and 40 in horizontal and vertical visual angle respectively. His head rotates around his neck horizontally at 5.8 degree /sec and vertically 3.3 degree/sec. The TV camera is connected to IKEGAMI image memory IM1182P which is composed of 2 planes, 512(H) X 768(V) pixels/plane , 8 bits/pixel and which has a real time image subtraction mode operated between input image and digitized image. It is connected to MOTOROLA VMC/2 68000 system of 1.8 MB memory.

Task of "HARUNOBU-2"

White boxes are randomly scattered on the passage in a computer room. as shown in Fig.4. one of the boxes is marked by black circle on its top. Go forward avoiding white boxes until reaches the circle marked box, and make a room map of the passage.

Edge detection

To detect an object or discontinuity on the passage, a high speed edge detection technique is desired. The image subtraction mode of the image memory is applied for this purpose. After digitizing a passage image as shown in Fig.5 (a), the TV camera is rotated 1 degree in horizontal and in vertical respectively and then gets a subtracted image obtained by subtraction of an input image by the digitized image as shown in Fig.5 (b). PILS V-1 system reads the subtracted image and gets segmentssomething like run length code as shown in Fig.5 (c), connect them into groups. A group is called unit. An edge of passage is distinguished from obstacles or discontinuity by the location and direction of its unit.

Room map making schema

After making "HARUNOBU-2" stand on a obstacle scattered passage and directing his body forward, the following routine is applied.
(1) Turn his head 30 degree downward from erect position.
(2) Turn his head right and take a subtraction image as shown in Fig.6 (a) and make the image into segments as shown in Fig.6(b). Detect the right side edge of the passage as shown in Fig.6 (c) and plot it on a body map.
(3) Turn his head left and do the same work as in step 2. From the direction of the left edge of the passage, get his direction relative to it, and memorize his direction on the body map.
(4) Make a room map by projecting the body map and localize himself on the room map as shown in Fig.6 (d).
(5) Turn his head forward again and take a subtraction image as shown in Fig.5 (b). After edge detection, distinguish a discontinuity from the passage edge by its location and its direction and plot it on the body map. Classify it into the obstacle category and the target one by the presence of some discontinuity above it.
(6) If it is an obstacle, make an avoidance path from its location on the body map and make him move along the path. Go to step 3.
(7) If it is a target, make an approach path from its location on the body map and make him move along the path. The task is completed.

Result

In the experiment three white boxes of 40 cm X 30 cm and 20 cm in height are scattered at intervals of 100 cm to 150 cm in a passage of 210 cm in width. The farthest one of them is marked with a circle on its top side. "HARUNOBU-2" was successful for the given task at the rate of three to one. He allways succeeded in avoiding the first box, but often failed in avoiding the second box or in approaching the third box. A process time in the case of success is shown in Fig.7.

Conclusion

Through the experiment of HARUNOBU-2, the following designing princip-les are obtained.
1) Something like Vestibulo-Ocular Reflex system in animal is necessary for stabilizing the direction of TV camera against rotation of robot body. Without such system the robot cannot keep the TV camera gazing a obstacle during manipulating the handle of the locomotor to avoide th obstacle.
2) To pass through a narrow space TV camera should be fixed at the position where it is able to look down the front of the robot.

Locomotive Vision Robot "HARUNOBU-3"

In 1984 we have begun to develop the second locomotive vision robot "HARUNOBU-3". The goal of "HARUNOBU-2" is to move around the campus through cluttered obstacles and find some target object and to get a cognitive map of the environment.
"HARUNOBU-3" is designed to have two TV camera on a gyroscope equipped head. One TV camera has a standard lens and is connected a shape analy-zing system which detects features of a object and classifies them with standard pattern into some given categories. The other TV camera has an ultra-wide angle lens and is connected to a visual behavior interaction system which detects an obstacle in front of him and makes a motion to avoid the obstacle. It also guides the TV camera of the former to orien-

tate the object to be analyzed.

Two visual distributed systems have the advantage of reduction of image processing because they are different in image resolution, algorithm, data structure and processing interval.

REFERENCES

(1) Mori, H. "Attention and Behavior of Vision Robot HARUNOBU" (in Japanese), Saikoroji, No.40, 1983, Pp 60-65

(2) Mori,H. and Akabane, H. "PILS: Picture Interpretation Language for Silhouette" (in Japanese), Johoshori-gakkai ronbunshi, Vol.23, No.3, 1982, Pp 243-250.

(3) Kuwahara, M. "KIsooHonno"(in Japanese), NHK Press, 1970

(4) Osaka, R. "Space perception" (in Japanese), in Torii (eds) "Gendai Kiso Shinrigaku 3", Tokyo University Press, 1982, Pp 183-231

(5) Kawachi, J. "UnvisibleVision" (in Japanese), the same book as in (4), Pp 233-255

(6) Mori,H."Cognition of the Environment in Vision Robot HARUNOBU"(in Japanese), Eizojoho Industrial Vol.15, No.18,1983, Pp 25-30

(7) Bronson,G."ThePostnatalgrowth of visualcapacity",Child, Development, 45, Pp 873-890.

(8) Trevarthen, C.B. "TwoMechanismsof VisioninPrimates", Psychologishe Forschung, 31,Pp 299-332

(9) Mori,H., Niki,H., Kakinoki,T., and Watanabe, K. "A Locomotive Vision Robot HARUNOBU-2in a IndoorPassage" (in Japanese), Johoshori gakkai 28 taikai, 1989, Pp 979-980

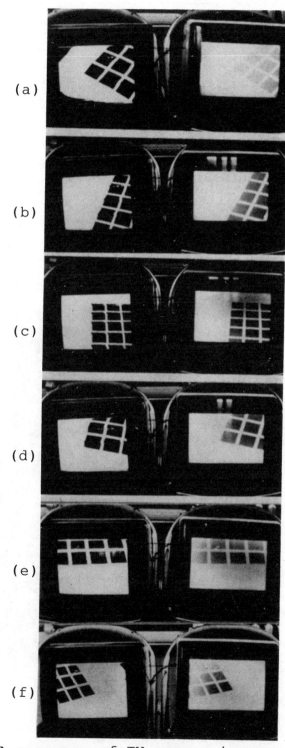

(a)

(b)

(c)

(d)

(e)

(f)

Fig.2 A sequence of TV camera images obtained by head motion during a cognitive map making

Fig.1 **HARUNOBU-1** and his task

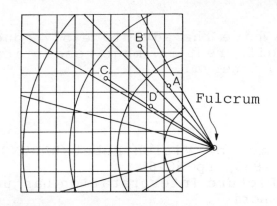

Fulcrum

Robot's
foot
(b) Room map

Fig.3 Cognitive maps

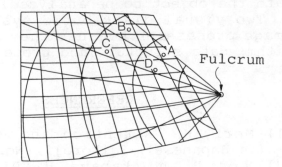

Fulcrum

Robot's
foot
(a) Body map with motor map

(a) Digitized image of a passage

Fig.4 **HARUNOBU-2** and the passage

(b) Subtraction image after slight
head motion

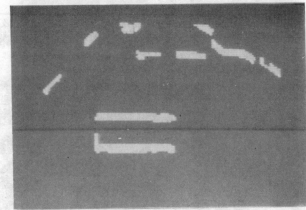

(c) Segments obtained from the
subtraction image

Fig.5 Edge detection by subtraction

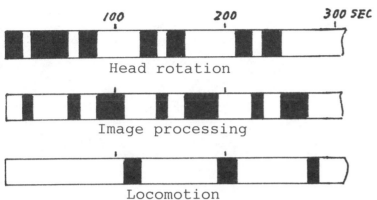

Head rotation

Image processing

Locomotion

Process	Time (sec)	%
Head rotation	124	44
Image processing	120	42
Locomotion	41	14
Total	285	100

Fig.7 Process time for 5 meter run

(a) Subtraction image of the right side of the passage

(b) Segments of the right side

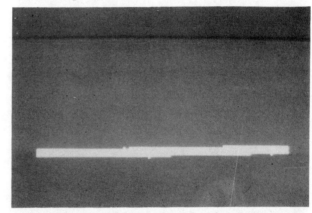

(c) Obtained edge of the passage

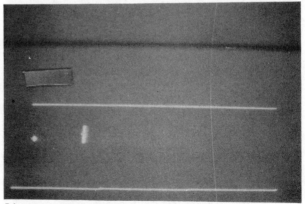

(d) A room map
Two parallel bars represnt both side of the passage. A pole and a dot shows an obstacle and HARUNOBU himself.

Fig.6 Room map making process

486

SUPPLEMENTARY PAPER

A Method of Image Analysis for Object Detection and Measurement of Their Geometric Parameters

A. N. Anuashvili

and

V. D. Zotov

Institute of Control Sciences, Moscow

and

Y. N. Tsveraidze

Tbilisi State Medical Institute,USSR

Abstract.

The present paper offers a method for localizing objects on images and measuring their geometrical parameters (linear dimensions, area, perimeter, maximal diameter etc.). The optical centre of object serves as the starting point to describe the object by radial-circular integral functions. Every radius of the description terminates on the last element of the object image. The processing time is reduced several times as compared with available techniques. A method of detecting low contrast movable objects is described. The method is based on the holographic principle permitting to detect movable objects even on the case where they do not differ at all from the background as far as their external properties are concerned.

Introduction

A principle of image analysis design has been developed according to which certain operations are carried out at the preprocessing stage. The final goal of image analysis is, as a rule, recognition of objects (images) i.e. description and classification of their quantitative or gualitative characteristics obtained as a result of some transformations of the input data by appropriate hardware and algorithmic means. An image analysis procedure includes several sequential stages of input data processing carried out by a hierarchical system of dedicated subsystems each of which is responsible for data compression in accord with a certain algorithm. The output information from a previous (lower level) subsystem is the input information for subsequent (higher level) subsystems the number of subsystems being a function of the number of compression steps required to transform input information for the final recognition purpose.

To obtain an efficient description and, consequently, to provide high quality recognition and classification of images the first step normally requires preprocessing of information aimed at reducing its redundancy and noise effects.

Isolation and selection of indicators, their classification and structural analysis have been widely treated in literature [1] therefore this paper is reduced to consideration of some problems of data preprocessing whose hardware realization largely depends on the sensor used to convert the visual information on the object to be detected. Special attention is paid to the detection of objects on images and their localization, the use of special image displays permitting convenient measurement of their geometric parameters and to the detection of low contrast movable objects.

Informative fragments of the object's image are usually placed chootically and masked by various obstructions (artefacts). Therefore a complete analysis of the image requires mutual displacement of the object and the videosensor with respect to each other and detailed analysis to solve the recognition problem should be carried out only after the object is detected and localized.

An appropriate combination of different sweeping techniques allows search and detection of objects and measurement of their geometric parameters.

Localization of the image informative fragments may be considered as a time-based process which includes the following operations:

- presentation of the considered object's image fragment in the field of view of the videosensor;
- search and detection of the object on the image;
- localization of the object and special sweeping;
- issuing of the signal requesting information processing by higher level subsystems.

The method

The essence of the suggested method is in the following. The image points are scanned with a step not exceeding the apriori known minimal diameter of the microobjects to be detected. The brightness of each point is then measured indicating whether the point belongs to the object i.e. whether it stays within its contour. After an intracontour point is found the line scanning is suspended and the image is type-displayed within a circle with the center in the point belonging to the object. The results of the radial-circular (type- P) display are used to obtain the geometric parameters of this microobject ofter which the line scanning continues with the initial step until and intracontour point of the next microobject is detected and the whole procedure is repeated. The detection of an intracontour point decreases the line scan step down to a minimally permissible value. The line scanning with the reduced step is carried out for a segment of the image inside a square whose side is no less than the double size of the apriori known maximal diameter of the detected objects and whose center is in the found intracontour point. The object's optical center is then found by comparing brightnesses of points inside this square. This is followed by suspension of line scanning and type- P display inside the circle with the center in the point of the given microobject's optimal center and the radius no less than one half of the apriori known maximal diameter of the detected microobjects. Line scanning is then continued from the point of the object's optical center.

An example considered to illustrate the advantages of this technique is the procedure of detection of a lengthy (oval-shaped) microobject.

Assume the long axis of the microobject is vertical and the found intracontour point is in the object's lower part. The available techniques imply in this case that the type-P display should be carried out with a radius no less than the maximal diameter of the object while the suggested method makes it possible to use a radius which equals approximately a half of this diameter (since the display center is shifted toward the optical center of the image). The advantage here is in reducing the information redundancy approximately by a factor of 4 and in the corresponding improvement of performance speeding-up the process of microobject detection.

If the long axis is located horizontally and an intracontour point found is at the end of the microobject's minimal diameter (that is, on the contour) then the available techniques do not provide, for example, accurate determination of the object's maximal diameter as its geometric parameter. The suggested method, in this case, provides a more accurate determination of this geometric parameter thanks to shifting the display center into the optical center. The accuracy is increased by about 30 percent.

Technical realization of the method

The urge toward the effective use of modern videoinformation processing means with high throughput and the need to solve multiple routine problems which are characterized by large amounts of data requires automation of the process of localizing the informative fragments of the considered object's image via designing special means of search and detection.

Such specialized means include semiconductor image convertors (SIC) suggested earlier by the authors [3,4] .

A SIC is a semiconductor plate with central point and circular metal electrodes deposited upon it. To these electrodes a DC source is applied producing a radial drag electric field. Between the central point and circular electrodes a circular display $p-n-p$ structure is located on the plate. The semiconductor plate also has a number of position sensing electrodes to find the optical center of the image.

The object's image is projected onto the sensitive surface of the SIC. Current carriers appeared as a result of plate illumination drift along the radii under the effect of the radial electric field and reach the sweeping structure. In this structure the cells are circularly requested and the videosignal is thus generated on a differentiating device.

SIC is responsible for the so-called integral radial-circular (type-P) image display. This is a display in the polar coordinates in which illumination intensity is summed up for all the points of a given radius rather than found for each particular point.

The signal from SIC comes into the analog input device and further into the computational unit. The system is comprised of a light source, lens,

SIC, analog signal input device, processor, signal output device and drive.

Providing image preprocessing, the SIC reduces redundancy of information. The signal then goes through the analog input device into the processor of the PS-300 microcomputer where quantitative characteristics are computed and the density of distribution of the image informative indicators is obtained. The output device provides drive control.

The image is scanned by a resolution spot whose size corresponds to that of the object reduced via the lens to the size of the sensitive surface of the SIC with position sensing electrodes. When the object appears in the field of view the semiconductor converter SIC is shifted until the image is fully centered.After this the signal of the position sensing electrodes leads the videosignal into the processor, and its probabilistic characteristics are computed. Then the object is shifted and the entire procedure is repeated.

The above device permits isolation of microobjects on images under noises (noninformative segments of the image) and processing of the most informative parts of images thus reducing information redundancy and eliminating repeated processing of image segments with those objects which were already analyzed. Besides the device improves the accuracy of measuring geometric parameters of microobjects (area, perimeter, maximal diameter) compared with that attained by the available technology since the radial-circular (type- P) scanning of the microobject is performed from the point of its optical center. The device is small and economical.

It should be noted that the reduction of object dimensions to the sensitive area of the videosensor should rest upon the use of apriori information on the classified objects (image fragments) obtained at the initial stage of processing.

Detection of low contrast movable objects

Consider the problem of detection low contrast movable objects. The method suggested for the purpose employs the holographic property of the absense of light diffraction on fuzzy sections of a hologram. Fuzzing of the diffraction picture is due to the presence of movable objects in the field of view.

The essence of the method is in the following. The restoring rodiation light flox is modulated by the object hologram. Then in the modulated light flux the areas are revealed with light intensity exceeding the threshold value found from the expression

$$I = I_0 e^{-\alpha d} \tag{1}$$

where I_0 is the intensity of the light radiation source; d is the thichness of the absorbing layer and α is the absorption factor depending on the wave lenght.

The result is used to detect movable objects and find their spatial coordinates.

Realization of this method is exemplified on the Figure showing laser

(1), semitransparent mirror (2), spheric mirror (3), subject beam (4), reference beam (5), objects and their environment (6), registration medium (7), photoreceiver (8), image analysis unit (9) and control unit (10). With the help of the semitransparent mirror (3) the laser radiation is divided onto the subject (4) and reference (5) beams which illuminate the objects and the environment (6) with the coherent light and interfere in the area of registration medium (7) generating, in case of absence of movable objects, a stationary interference field.

The radiation reconstructing the hologram diffracts on the interference pattern (is modulated by this picture); it is partially absorbed by the registrating medium and falls onto unit 8, after which it is analysed by unit 9. The control of the system is carried out by unit 10. The image analysis unit may incorporate a photoreceiver and a computing device. Unit 8 should feature sufficient sensitivity in order to provide reliable isolation of those sections whose illumination intensity exceeds the threshold level (minimal diffraction effectivity).

When a movable object appears in the environment i.e. when objects are mixed up, for instance, along the Y-axis with the velocity V_0 under the given exposure (Fig. 1) the distribution of the complex amplitude $U(x,y)$ will go along the Y-axis with the same velocity V_0 generating a travelling interference pattern. In this case the expression for the movable interference pattern light intensity is of the form

$$I(x,y-V_0 t) = U^2(x,y-V_0 t) + z_0^2 + 2 z_0 U(x,y-V_0 t) \cos\left[\varphi(x,y-V_0 t) - 2\pi\nu x\right] \qquad (2)$$

Where t is the exposure, and $U(x,y)$ is the complex amplitude.

The interference pattern under exposure is travelling in the same direction along the Y-axis but the light-sensitive centers of the registration medium have certain sensitivity threshold and will not be able to fix the interference pattern (it will thus be fuzzy). As a result during the reconstruction of such holograms the fuzzy sections of the interference pattern (which correspond to moving objects) will feature minimal light diffraction while the intensity of light radiation which comes through these sections will be maximal. Consequently the analyzed image will contain sections whose brightness exceeds a certain threshold. Detection of these sections and finding their spatial coordinates, shape and dimensions permit one to judge on the presence of movable objects and their location. The search for the brightest sections of the image should be performed with regard for a certain illumination threshold depending on the intensity of the light source in accord with expression (1).

The above method of detecting objects on images provides sensitivity to displacements along the three coordinates up to several fractions of a μm. Indeed, a travelling interference pattern a travelling interference pattern will appear any time the subject or reference waves shifts to the magnitude of the wave length of the recording radiation (for visual range this amounts to 0.3 - 0.7 μm and for the infrared range, 1.2 μm).

The efficiency of the suggested method steps from the fact that it provides a principally new way of detecting movable objects which may be absolutely identical to their environments (background) in terms of external indicators: color, relief, brightness, shape. Moreover, the detection of such objects is possible with a high accuracy when they shift along three coordinates on distances up to several fractions of a micrometer.

References

1. Rosenfeld, A. "Picture processing: 1979". Computer graphics and Image Processing, No.13, pp.46-79, 1980.
2. Anuashvili, A.N. and Zotov, V.D. "Simiconductor image form convertors". Automation and Remote Control, Vol.36, No.11, pp.88-92 (November 1975). Translated from Russian.
3. Anuashvili, A.N. and Bykhovsky, V.K. "A robot with holographic control memory". Proc. 3rd CIRT and 6th ISIR, pp.D1-1 - D1-8. IFS Ltd, Bedford, England (March, 1976).
4. Anuashvili, A.N. and Zotov, V.D. "A simple real-time visual system for an industrial robot". Proc. 7th ISIR, pp.507-514. JIRA, Tokyo, Japan (October, 1977).

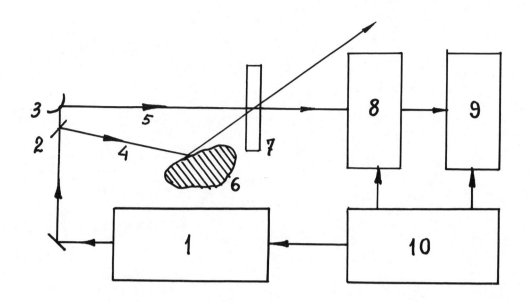

Fig. 1

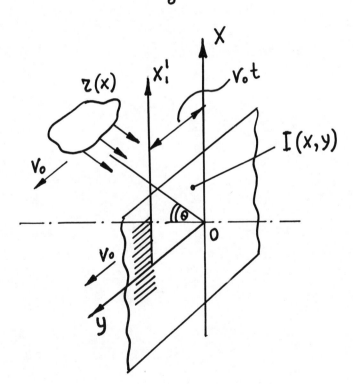

Fig. 2

494

array, resulting in a polar scan of the image. This technique has been first applied for the visual inpection of sintered carbide inserts [1] where besides the dimensional control, the system had to monitor the sharpness of the cutting edges and to detect on these latter the presence of chippage. Due to its performing results, this approach has been further developed and became a powerful tool for the high precision dimensional control of small flat mechanical parts.

HARDWARE DESCRIPTION

The hardware structure of the presented high resolution vision system, can be split in the following main functional blocks:

Image acquisition

A solid-state CCD linear array of 2048 elements is used for image sensing. Each photosensitive element is 13 x 13 um wide; thus, by using high precision macro optics with a slight magnification ratio, the resolution of the imaging system can reach 10 um. Such a high resolution system requires an extremely precise and rigid mechanical assembly of its various components. The line scan rate of the linear camera is synchronized with a stepper motor, which rotates the object in front of the camera with an angular resolution of 1000 steps/revolution. One video line, representing a radial section of the inspected object, is acquired at each angular step; the complete object is scanned with 1000 angular steps, which takes 1 sec at a line scan rate of 1 kHz.

Video signal preprocessing

The analog video signal is transformed into a binary signal at the pixel rate of appoximately 2 MHz. Each video scan is composed by 2'048 samples, and accordingly, a complete polar scan of the image consists of more than two million pixels. On each video line however, only a few pixels, corresponding to the transitions between the object and the background, are significant. More precisely, all significant information is included in the position of these transitions on the video line. Thus, whenever the video preprocessing circuit detects such a black-to-white (or vice-versa) transition, it stores the respective address into a First-In-First-Out (FIFO) memory. This operation reduces the average data rate from 2 MHz to some kHz, depending on the complexity of the object's shape (the number of transitions detected on each cross-section). The sequence of the stored addresses of a complete image scan form a compact polar representation (radius length as a function of the angle) of the object's shape (figure 1).

Digital signal processing

For each scan line, the contents of the FIFO memory (stored addresses) are transfered on a Digital Equipment LSI-11/23 microcomputer for further processing. The microcomputer configuration includes a floating point arithmetics processor, a parallel I/O interface, a graphics display and a dual cassette unit for program loading and for the recording of the inspection statistics. The software is written in Pascal and Macro-11 and runs under the MicroPower-Pascal operating system.

Vision system for high precision dimensional control

P. Kammenos
KUKAM SA, Switzerland

ABSTRACT

This communication describes a self-teaching vision system f
high precision dimensional control of flat mechanical parts.
one-dimensional polar signature of the object's shape (radius length
a function of the angle) is obtained by rotating the part in front of
high resolution linear image sensor. An appropriate processing of th
signature allows to compensate for the errors due to the lack c
accuracy in the centering of the object on the rotating device. Th
corrected signature is matched and compared with a reference, store
during a preliminary interactive teaching session. The measurment
accuracy can be further increased by using as reference an analytic
geometrical description of the object.

INTRODUCTION

The high precision non-contact dimensional control of mechanical
parts is an important application field of vision systems in the
precision mechanics industry. The overall dimensions of the parts to
be examined do not exeed in general some centimeters, and the desired
measuring accuracy is often in the range of 10 - 100 microns. This
requires high resolution imaging systems with a resolving power of some
thousands square picture elements. No matrix solid-state image sensors
with such resolution are yet available, so it is necessary to use
linear image sensors (available in arrays with resolutions up to 4096
pixels) combined with some mechanical scanning for the acquisition of
two-dimensional images. The most common technique is to move the
linear sensor across the image field, which leads to a line by line
image acquisition. This communication presents a different approach,
where the object to be measured rotates in front of a linear imaging

PROCESSING METHODOLOGY

The goal of this development was to design a high resolution visual inspection system that could easily be programmed for the dimensional control of various small mechanical parts. One convenient way of programming the system for the inspection of a new object is the self-teaching method, where the system automatically stores, during the teaching phase, the shape of a reference object; this reference is then compared with the inspected object, which is passed only if it lies inside some predetermined dimensional tolerances. One straitforward method for such a teaching philosophy is to store as reference the complete two-dimensional representation of the object and to perform a direct comparison of this stored image with the one of the object under inspection. This approach however has two major disadvantages: first, it requires an enormous data storage capacity (a complete image is represented by 2 million pixels); and second, it requires either an exact alignement in position and orientation of the reference object with the inspected one (with an accuracy that cannot be reached with common industrial handling devices), or the use of a time consuming two dimensional correlation procedure, which searches for the best match between the two images before proceeding to the shape comparison.

When the compressed polar coded information (as described in the preceeding chapter) is used as shape descriptor, a similar problem arise. In that case, the shape of the discrete one-dimensional polar representation of the object $r_i(\alpha_i)$ is very sensitive on variations of the relative position of the object in respect to the axis of rotation in front of the camera (figure 2a). An enough accurate positionning of the examined objects on the rotating device is not practically feasible. In order to cope with this problem, a processing method has been developed that transforms the rough polar signature data into a position invariant polar signature of the object.

This new polar representation is generated by using as origin the center of gravity of the object's boundary (which is not dependent on the relative position of the object in respect to the axis of rotation). First, it is necessary to transform for each angular step α_i, the incoming polar coordinates (r_i, α_i) into cartesian coordinate pairs (x_i, y_i) by using the relations:

$$x_i = r_i \cdot \sin(\alpha_i) \tag{1a}$$

$$y_i = r_i \cdot \cos(\alpha_i) \tag{1b}$$

The coordinates (x_{CG}, y_{CG}) of the center of gravity of the object's boundary can be computed as the average value of the N coordinate pairs (x_i, y_i) :

$$x_{CG} = \sum_{i=1}^{N} x_i / N \tag{2a}$$

$$y_{CG} = \sum_{i=1}^{N} y_i / N \tag{2b}$$

By using the center of gravity as a new origin, the data can be transformed back into polar form by computing for each coordinate pair (x_i, y_i) the corresponding radius r'_i and angle α'_i :

$$r'_i = \sqrt{(x_i - x_{CG})^2 + (y_i - y_{CG})^2} \qquad\qquad (3)$$

$$\alpha'_i = \arctan [(y_i - y_{CG}) / (x_i - x_{CG})] \qquad\qquad (4)$$

With an appropriate interpolation and resampling of these new polar coordinates it is finally possible to create a position invariant polar signature of the object's shape (figure 2b).

This computation is first performed during the teaching phase, and the position invariant polar signature $R_{REF}(i)$ is displayed on the graphics monitor. At this stage the operator can interactively feed the processing system with the dimensional tolerances that are acceptable at various locations of the object's boundary. These values form the upper and lower tolerance functions ($T_U(i)$ and $T_L(i)$), which can be viewed as the limits of a variable width window along the object's boundary. These functions are stored, together with the reference signature, in the computer memory.

During inspection, the position invariant polar signature $R_T(i)$ is first determined. The origin of this signature is by definition the same as the one of the reference (center of gravity of the object). A rotation angle between the inspected object and the reference one is however not to exclude; this is reflected as a circular shift (figure 1a-1b) between the two corresponding polar signatures. In order to align the two polar shape descriptors, it is necessary to compute the values of a circular comparison function (e.g. correlation or integral of the absolute differences). This corresponds to a circular shift of one signature in respect to the other, in order to find the angular step (k) corresponding to the best match [2],[3].

After the optimum alignement of the two signatures, a point-by-point difference function of the radius lengths ($D(i)$) is computed.

$$D(i) = R_{REF}(i) - R_T(i+k) \qquad \text{(k: best match angular shift)} \qquad (5)$$

Finally, this difference function is compared with the upper and lower tolerance functions; the object is accepted only if the measured difference at each angular step lies inside the tolerance limits.

$$T_L(i) < D(i) < T_U(i) \qquad\qquad (6)$$

EXPERIMENTAL RESULTS AND CONCLUDING REMARKS

Experimental results have proven that the presented system is a reliable tool for the high precision dimensional control. In its present configuration it can detect dimensional errors down to 10 microns. The processing speed for a full dimensional control is shorter than 4 seconds. The system does not need an accurate positionning of the parts and can be adapted to common, high speed handling devices. The teaching of a new object shape is simple and does not require programming knowledge.

In case of objects where the shape can be described (fully or partly) by means of some analytic geometrical model, the measuring accuracy can be further increased. This is possible by mathematical association of several local measures. The square object of figure 1 can be considered as a typical exemple; its shape can be described as an assembly of four straight line segments, connected with four circular arcs (corners). In that case, several coordinate pairs of the polar signature can be associated for the estimation of the exact position of the four straight line segments (computation of straight line equations by linear regression). Preliminary experimental results have shown an important upgrade of the system performances.

REFERENCES

[1] Panayotis KAMMENOS, Boris LOZAR: Electro-optical Computerized System for Visual Inspection of Sintered Carbide Inserts, Proceedings of SPIE, volume 397, April 19-22, 1983, Geneva, Switzerland
[2] F. DE COULON, P. KAMMENOS: Polar Coding of Planar Objects in Industrial Robot Vision, Neue Technik, 1977, no. 10, pp.663-672
[3] P. KAMMENOS: Performances of Polar Coding for Visual Localisation of Planar Objects, 8th International Symposium on Industrial Robots, Stuttgart, West Germany, 30.5-1.6.78

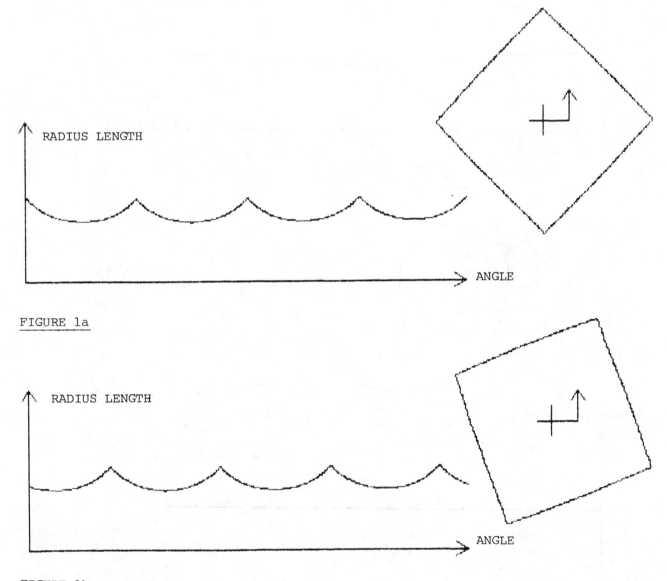

FIGURE 1a

FIGURE 1b

Figure 1: Two views of a square object)(carbide insert) and their corresponding polar signatures

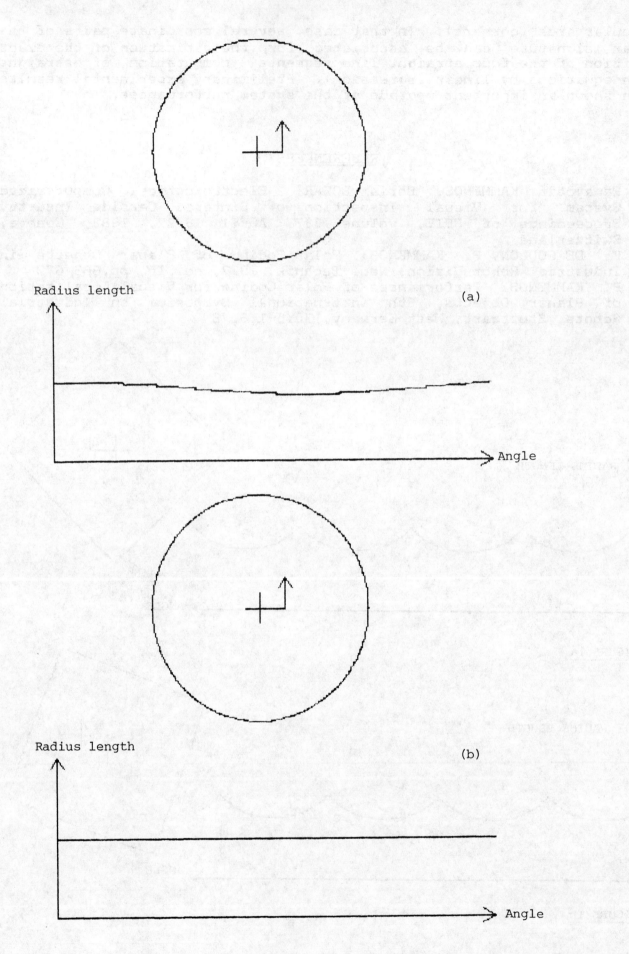

Figure 2: Round object and its polar representation before processing (a)
and after center of gravity estimation